CONSTITUTIONS OF MATTER

CONSTITUTIONS OF MATTER

MATHMATICALLY MODELING THE MOST EVERYDAY OF PHYSICAL PHENOMENA

MARTIN H. KRIEGER

THE UNIVERSITY OF CHICAGO PRESS
Chicago & London

MARTIN H. KRIEGER is professor of planning at the University of Southern California. He is the author of *Advice and Planning* (1981), *Marginalism and Discontinuity: Tools for the Crafts of Knowledge and Decision* (1989), and *Doing Physics: How Physicists Take Hold of the World* (1992).

The University of Chicago Press, Chicago 60637
The University of Chicago Press, Ltd., London
© 1996 by The University of Chicago
All rights reserved. Published 1996
Printed in the United States of America
05 04 03 02 01 00 99 98 97 96   5 4 3 2 1

ISBN (cloth): 0-226-45304-9

Library of Congress Cataloging-in-Publication Data

Krieger, Martin H.
    Constitutions of matter : mathematically modeling the most
everyday of physical phenomena / Martin H. Krieger.
        p.   cm.
    Includes bibliographical references and index.
    ISBN 0-226-45304-9 (cloth : alk. paper)
    1. Matter—Constitution—Mathematical models. 2. Mathematical
physics. I. Title.
QC173.K857   1996   96-16986
530—dc20   CIP

For Jennifer, Joseph, Judy, and Marc

CONTENTS

FIGURES

PREFACE

> It is sometimes felt that mathematical physics deals
> with epsilontics irrelevant to physics. Quite the contrary,
> here one sees the dominant features of real matter
> emerging from deeper mathematical analysis.
>
> W. Thirring, introduction to Lieb's *Stability of Matter*

As a mode of philosophical analysis, mathematical physics shows ever more precisely which are the essential features of our material universe, or at least of our models of our universe, and why they are essential. Moreover, the mathematical formalism and techniques we employ to work out physical problems, and to push through derivations and to prove theorems, are ontological, in that they say how the physical world (or, again, our models of it) is constituted. And the mathematical manipulations and approximative schemes we employ along the way are fully meaningful, in that what appear as devices and tricks and techniques are filled with physics. So I shall argue in the following chapters. I should note that, as generic claims, much of this is a commonplace. It is an article of faith of practitioners:

> When a successful mathematical model is created for a physical phenomenon, that is, a model which can be used for accurate computations and predictions, the mathematical structure of the model itself provides a new way of thinking about the phenomenon. Put slightly differently, when a model is successful it is natural to think of the physical quantities in terms of the mathematical objects which represent them and to interpret similar or secondary phenomena in terms of the same model. Because of this, an investigation of the internal mathematical structure of the model can alter and enlarge our understanding of the physical phenomenon. Of course, the outstanding example of this is Newtonian mechanics.[1]

One of my purposes here is to present several detailed examples of such mathematical modeling as philosophical.

The distinction between theoretical and mathematical physics is subtle. While much of theoretical physics employs sophisticated mathematics, as well as physical arguments and models, mathematical physics is characterized by the greater mathematical rigor of its arguments (which I shall argue is of physical interest), and by its level of abstraction and generality or its level of computational complexity in solving particular problems.[2]

THERMODYNAMICS AND STATISTICAL MECHANICS

Thermodynamics studies the interactions of heat and work, as in a steam engine, and the interactions of species of particles, as in a chemical reaction. In general, systems come to equilibrium defined by a small number of macroscopic variables such as temperature and pressure; energy is conserved; entropy increases in time; and the entropy is small when the temperature is very, very low.[3] For a system that is at constant temperature, a measure of its energy that is minimized at equilibrium is the Helmholtz free energy, $F(V, T)$, where V is the volume and T is the temperature.

Statistical mechanics is the study of the macroscopic properties of systems composed of large numbers of similar components such as molecules. As in classical mathematics, in which a partition function is used to enumerate the number of ways of adding up an integer from smaller integers, the statistical mechanical partition function (for a system at constant temperature, the "canonical ensemble"), Q, does more or less the same thing. (In German, it is the *Zustandsumme*, the sum over states.)[4] To connect the thermodynamic with the statistical mechanical, we find that $\exp - F/k_B T = Q$, where k_B is Boltzmann's constant, 1.4×10^{-23} joule/kelvin, the scale of thermal phenomena, $k_B T$ being a measure of the kinetic energy possessed by a molecule at a temperature T. (Other constants that will concern us are Avogadro's number, roughly the number of molecules in an everyday small bit of matter, 6×10^{23} [note that $k_B \times$ Avogadro ≈ 10 joule/kelvin, an everyday amount of energy]; and $\hbar$, Planck's constant/2π, the scale of quantum phenomena, 1×10^{-34} joule-seconds.) And the changes in the free energy that occur with changes in some property connect it with measured quantities. So the magnetization, M, of a piece of iron is given by how the free energy changes when a magnetic field is applied.

Since the partitionings here are often rather more complex than in the usual mathematical context, the computation of the partition function, Q, is a very great physical and mathematical challenge. For example, a characteristic feature of bulk matter is its capacity to change state, say from gas to liquid. For our purposes, this change of state may be taken to be a change in orderliness or symmetry. (There are phase transitions that are neither abrupt nor symmetry-breaking.) Phase-transition points, such as freezing, will here be called critical points (although that term usually has a more restricted usage). At such points, the partition function develops singular behavior, and it is a central task of statistical mechanics to show how that occurs, both physically and mathematically.

THE ARGUMENT

In mathematical physics, what some might take as syntactical and formal devices of mathematics carry enormous semantical content about the physics embedded in the models. Moreover, that mathematics itself is in fact not so formal, but it too is a model of the world. And that formalism draws out the underlying physical assumptions in a model.

Formal definitions of physical phenomena are technical, defined in terms of the realm of calculation; philosophic, saying what is necessary, crucial, and significant about a phenomenon; experimental, making possible the conversion of actual observations to an ideal form that they imperfectly imitate; and physical, dependent upon particular features of how the physical world actually works. Crucially, such formal definitions are never merely a matter of greater mathematical rigor, for it seems they lead to new physical insight as they are employed.

In the first chapter, I set up the problem of "adding up the world" out of its components, statistically, as in the ideal gas or as a sum of random variables.[5] The device for doing so is the infinite volume limit, $1/N$—when, in a model, the volume is allowed to go to infinity but the density remains constant and so the number of particles also goes to infinity. Analogously, the central limit theorem concerning the sums of random variables prescribes a norming constant, $1/\sqrt{N}$. (Both of these are asymptotic limits; namely, they converge to a particular form or shape. I should note that it is perhaps more appropriate to refer to the bulk limit, rather than to the infinite volume limit.)[6]

By the way, I will mention but not focus on the contemporary, "more rigorous," approach that studies infinite systems directly rather than taking the infinite volume limit for finite systems. The basic tools there are the notion that subsystems of energy E sit in a temperature bath ($\beta = 1/k_B T$) of the infinite system (a conditional probability of $\exp - \beta E$), the uniqueness of states, and the breakdown of that unique-ness with transitions of phase.[7]

The constitution of bulk matter may be seen to involve taking infinite volume limits. But how are they to be put to work? In chapter 2, I describe a canonical analysis of the stability of Coulombic matter, composed of protons, electrons, and neutrons, which says just how one is to prove that stability. Mathematical technology and physical models are shown, in chapter 3, to fulfill that canonical analysis, literally filling in the blanks. In particular, I shall be concerned with why bulk matter, matter composed of a very large number of molecules, has extensive properties, such as the volume, that scale with N, the number of molecules. And with why such matter is stable, so that when it is compressed, it becomes more dense (rather than less dense).

Along the way, a variety of models of bulk matter are proposed. A model is a simplified version of the physical world, one that we hope will allow us to understand the physics of a situation. Models are diagrammatic, mathematical, and physical: a notation that represents what we believe is going on, a mathematical formulation, and a simplified actual object.

I am particularly interested in showing how physical notions are built into the mathematical technology. Chapter 4 continues in this vein, developing further formalism (or models) that is effective for describing how matter may make a phase transition, such as an iron bar gaining or losing its magnetization when it is below or above 1043 K, respectively. In particular, the two-dimensional Ising model of ferromagnetism and of its transition to becoming a permanent magnet, will be the focus. It can be shown that one might consider the lattice of atomic magnetic spins or dipoles that make up an Ising ferromagnet to be formally equivalent to a lattice of spatial sites where atoms may be present or not: a lattice gas, which then exhibits condensation into a liquid if the temperature is low enough. In each case, the formalism may be seen as a device both for taking the infinite volume limit and for saying how matter is constituted out of its components ("adding up the world"). In Chapter 5, I return to the task of analysis, saying just what it is—canonically—we have been trying to define in modeling phase transitions, almost always a sharp transition in temperature and a change in orderliness. How do technical definitions do physical and philosophical work? Chapters 6 and 7 display how we discovered the right degrees of freedom of Ising matter, and how they are built into the formalism. The last chapter brings the discussion more or less up to date, in particular saying how the right degrees of freedom are nowadays built into the formalism and the models.

I should note that the problems I shall be discussing are for the most part aspects of classical physics, even if their modern formulations depend on some quantum mechanics and lots of analogies to quantum mechanics and quantum field theory. The stability of matter surely requires the Heisenberg uncertainty principle of quantum mechanics and the fact that electrons obey the Pauli exclusion principle, but the forces involved are classical electromagnetic ones (Coulomb forces between charged particles), and much of the task is a matter of classical statistical mechanics. As for the Ising model, this is true a fortiori. However, what is striking here is how quantum field theoretic notions of fermions and of Wick products provide the powerful technology for conceptualizing and solving these problems, without their becoming quantum mechanical in any sense (no $\hbar$)—although the actual ferromagnetic forces between spins are quantum mechanical in origin.

THE PHYSICS

In the end, what I am most interested in is how "the physics" emerges from all the various modes of solution of a problem. How do we discover and make evident what is really going on in a physical situation —the mechanisms and the basic principles and the symmetries and conserved quantities—at least through the work of mathematical physics? How do we make what is true evident?

The physics is a sequence of analyses, each analysis illuminating its predecessors and opening up new avenues. The physics is not an endpoint, where we finally learn just what is really going on. In subsequent analyses, we will learn even more of what is really going on.

But it is also true that, if a solution shows that the problem at hand is nicely analogous to a well-known conventional problem, we feel much closer to the physics, more likely to take it as obvious. That set of conventional problems—the ideal gas, the harmonic oscillator, the hydrogen atom, the vibrating crystalline solid (actually it is a quite finite list)—is bedrock. It is part of the physicist's toolkit, mastered in school, and then remastered through years of experience.

For example, much of my discussion here is readily rephrased in terms of modern quantum field theory, and there are a variety of texts from the last fifteen years whose titles are a combination of the words *statistical, critical phenomena, quantum field theory, renormalization*. The interchange of ideas between statistical mechanics and quantum field theory has revealed a great deal of the physics involved in their respective models.

Still, I do believe the physics is about the world and not only about the models. Like most persons who are trained as scientists, I am on my working days a rather unphilosophical naive realist, taking my models to be a description of the natural world. The theory and the natural world are very close indeed.

But I also believe the best one can do in a discussion such as this is to make a comparative study of the various modes of modeling and solving problems, looking for shared features and illuminating contrasts. It is surely the case that some of the time the physics that would seem to be essential in one solution is rather different than the physics that would seem to be essential in another solution—although we might hope that eventually we might show they are the same physics.

It is perhaps safest to say that we must be comparativists. But we hope that the various aspects of the models of nature that we discern may eventually be seen as aspects of a whole, as aspects of nature. Put differently, the physics is realized in rather different models and different mathematical forms in different modes of solution, and it is perhaps presumptuous to believe that we can strip away those models and that mathematical formalism and have the physics, as such.[8] (But, to be honest, I do so presume.)

With respect to the physics and the mathematics, this book is for the most part perhaps fifteen years behind, with only an epitome of current thinking provided in the last chapter. To those who are au courant in this field, the star-triangle relation is now embedded in the Yang-Baxter equation and the factorizability of the S-matrix; and much of my teased-out implications in chapters 4–7 are now commonplaces.[9] So here is one recent epitome (see chapter 8 for others); my point for the lay reader is the tone rather than the details of what is being said.

> The fundamental role played by the Yang-Baxter equations (or their higher dimensional generalizations) comes from their ability to generate global results from local properties. The deep geometrical meaning of the Yang-Baxter equations had actually been underlined by R. J. Baxter as soon as 1978 with the concept of the so-called "Z-invariance" [invariance of the partition function to star-triangle-type transformations under very general lattice structures],[10] a long time before Witten's theory wrapped everything up into a single package (Jones polynomials, braid groups, spin models, Yang-Baxter equations, Yang-Mills equations, Donaldson theory, everything...). These local relations are the "deus ex machina" in a number of domains of mathematics and physics (S-matrix factorizability, quantum inverse scattering, quantum groups, knot theory, Bethe Ansatz in Solid State Physics,...)....a long time before being rediscovered, revisited or restyled in a quantum-group-knot-theory-topological-field-theory language, the concept of Yang-Baxter equations first emerged and developed in lattice statistical mechanics [the star-triangle relation for the Ising model connecting hexagonal and triangular lattices].[11]

"To generate global results from local properties" is to be able to derive thermodynamic properties of bulk matter, such as the specific heat, from atomic properties—namely, to provide a constitution of matter. And that "deus ex machina" is in the end deeply physical, while "everything" is only apparent after fifty years of collective scientific effort.

As I indicate in chapter 1, I have chosen not to be so Whiggish as to reduce a 1944 paper to 1994 terms, at least for the most part, although that is a perfectly acceptable thing for a professional physicist to do. But I have let subsequent developments influence my reading of earlier papers, in part because in the case of Onsager subsequent developments seem to be naturally derived from his work. I should note how difficult it is to discern the physics in these papers, although it helps to be aware of all the background and concurrent literature.

A number of implications might be drawn for the teaching and practice and writing about natural science. Alternative proofs and derivations, all the work subsequent to an original discovery, are often major advances, for they allow us to discover the physics hiding in complex successful formalism. In pedagogy, motivation that says what is going on in the formalism, imprecise though it may be, teaches students how physics is actually done—although that motivation may be mathematical and formal as well as physical. And in scholarly papers, it would be helpful for introductions to papers to give away their plot and story early on—at least as much as the author may possess at the time. This means that subsequent review papers by others may be major contributions to physics, just because they reveal motivation and physics not available originally.

TECHNICALLY

Much of the material here is highly technical, depending on detailed knowledge of the physical arguments I am discussing. I have tried to encapsulate much of that (the flag is "technically"), so that lay readers could follow most of the argument, perhaps only schematically appreciating the technical sections—although I believe the argument becomes cogent only when the technical detail is available. I do not believe there is a nontechnical remedy for this (much as classical philology depends on detailed knowledge of particular languages and texts).

Technically, we would seem to discover from such analysis, metaphysics, and interpretation that some of what is fundamental and crucial about the material world as physicists conceive of it is that it is described by antisymmetric functions of many variables. (A function, $f(x, y)$, is antisymmetric if $f(x, y) = -f(y, x)$; note that then $f(x, x) = -f(x, x) = 0$.) So we shall find that matter is stable because electrons obey the Pauli exclusion principle, which says that no two electrons can be in the same state: $f(x, y) = f(\phi, \phi) = 0$, where x and y refer to the states of the two different electrons, and here the state is ϕ. Electrons are "fermions," described by an antisymmetric wave function. (A symmetric, or "bosonic," world is not stable to the Coulomb force; and even the fermionic nature of the electron does not save us from gravitational collapse.) Moreover, descriptions of phase transitions—such as that of an iron bar becoming a permanent magnet—would seem to require (at least for the most tractable realistic model of two-dimensional matter) that we employ antisymmetric functions in our description, reflecting the yes-no character of the bonds, $\uparrow\downarrow$ or $\uparrow\uparrow$, between atoms in this model. Along the way, we shall see how quite technical matters of uniformity of convergence of approximations and of the interchangeability of limiting processes are crucial indicators of very deep physics being present in the middle of apparently bland formalism.

As befits the thermodynamic systems we are studying—bulk matter at a given temperature and pressure and composition—we shall see again and again how these technical devices are connected with the possibility of a temperature throughout the system, namely, that these systems are characterized by an average energy or noise level per particle. Yet these are also mechanical systems, in that they are a consequence of the interactions of their atomic components, much as a gas is a consequence of the collisions of its atoms with each other. Much of the mathematical physics is an attempt to find the right sort of atomic particles—ones with well-defined energies, more or less independent of each other, and interacting much as the molecules do in an ordinary gas.

A Note to the Scholars

This book is not a history of the episodes it draws from.[12] A potted eponymous history of the stability-of-matter problem would feature Onsager, van Hove, Yang and Lee, Dyson and Lenard, Fisher and Ruelle, and Lebowitz, Lieb, and Thirring. For phase transitions and the Ising model, there is of course Lenz and Ising, and then van der Waerden; Kramers and Wannier; Montroll; Onsager; Kac and Ward; Lee and Yang; Yang; Schultz, Mattis, and Lieb; McCoy and Wu; and Baxter—and many, many others significantly contributed to this development.[13] Making no claim as to historical priority or significance, I focus on just a few papers, notably those by Onsager, Lieb, and Baxter and their respective collaborators, to do the work I need to do here.[14] An article by Lieb and Lebowitz provides the main title for this book.

I have in the appendix reprinted Lars Onsager's two seminal papers, one on the stability of matter (1939), and the other on the Ising model in two dimensions (1944). Although they were published in the major journals of their fields, they are not so readily available, the 1944 paper being known more for its difficulty than through direct acquaintance. I want to express my gratitude to the American Institute of Physics and to the American Chemical Society for permission to reprint.

I have restricted my consideration to the actual published papers and books, what would have been more generally available to the community of scientists, what they would have had to rely on for the most part. I regret having only fragmentary access to the oral culture in which motives and themes are more directly expressed—in published lectures, conference proceedings, and review articles that are meant to expose more generally "ideas that have circulated privately but have never been published."[15] I have less concern for the archival material that would concern the historian, although those materials might have been informative about ideas that were in circulation but not otherwise published. In any case, I have not tried to do anything like a laboratory study of mathematical physics; I have been for the most part concerned with ideas and models and the interrelationships among them.

My hope is to place studies of science most directly within an arena that actual scientists will recognize as their own. That requires that historical and philosophical studies will have to be technical and detailed, in part because actual scientists know that in those details lies their actual everyday work.[16]

A NOTE TO READERS

I have envisioned a variety of readers for this book. Readers curious about what scientists actually do should find it reasonably straightforward to read chapters 1, 2, 4, and 8, and by judicious scanning—looking for common themes and skipping over mathematics and physics that are unfamiliar—I believe and I am told by lay readers that the rest of the book is accessible. I hope that physics students and physicists will find the technical details of chapters 3, 5, 6, and 7 illuminating about what is going on in the scientific papers and the textbooks. Of course, it will seem to some that the analyses I provide are after-the-fact hand waving, only meaningful once we have the technical solutions. I tend to agree, but I have over the years learned a great deal from such interpretations.

In many ways, this book is a continuation of the discussion of modeling in physical science and the recurrent appearance of a small number of methods and models in a "physicist's toolkit," in *Doing*

Physics: How Physicists Take Hold of the World (1992), and for both lay and technical readers that book may be usefully read in conjunction with this one. And that book is a continuation of my *Marginalism and Discontinuity: Tools for the Crafts of Knowledge and Decision* (1989).

One last point. There is somewhat more repetition here than might be needed if readers read from beginning to end, or at least felt obligated to do so (as for a textbook). I wanted to make chapters and sections more self-contained, with less need to go back to an earlier chapter. In any case, technical terms are in the index, and the first page reference should lead to a definition. Similarly, notation is in the index, and the first page reference should lead to a definition.

ACKNOWLEDGMENTS

This work owes a very great deal to a series of formative conversations with Gian-Carlo Rota over the last dozen years. My colleagues at the University of Southern California have educated me. And I am indebted to Eric Livingston, Robert Tragesser, and Harold Garfinkel, whose work, and observations about my own, has been of help and encouragement. Karen Barad and Hubert Saleur provided useful readings of the manuscript. Sam Schweber has been an abiding support. The usual disclaimers apply, a fortiori.

My work was supported by grants from the Exxon Education Foundation and the Lilly Endowment, and by the Zell-Lurie Visiting Professorship in Entrepreneurship at the School of Business at the University of Michigan, Ann Arbor. I am grateful for a sabbatical leave from the University of Southern California.

My son, David, is of course the constitution that in the end matters the most.

1

Modeling the Constitutions of Matter

*Adding up the world: the kinetic theory of matter, infinite volume limits,
mathematical modeling and interpretive cultural practice • Technical
matters are the physics: mathematical manipulations are meaningful,
interpretations of mathematical physics*

People make models of the world, to some extent formal and mathematical, ones that show initial promise of being good representations of the world as we come to know it. Rather than experimenting on the world, we experiment on the model to see if some generic phenomena appear when certain features are added or not, and so discover if the features are fundamental to the model and so presumably to the natural world.[1] Moreover, there are a variety of "technical" moves made within the model, in its mathematical formulation or in computation, so as to solve the model and to derive its implicit consequences. We might well believe that the mechanism of the model and its technical manipulation or "working," mathematical and formal though they be, should point to the physics of the model (and presumably to nature). For example, if you need quantities of a certain kind to solve a model, say antisymmetric matrices, then those kinds or symmetries may be intrinsic to the physics of the situation and may not be merely disposable apparatus. Modeling here is both a mode of philosophical analysis (varying features to find out what matters) and substantive.

I shall describe some of the ways physicists model the constitution of ordinary, everyday, bulk matter out of its purported constituents. How might we add up a large number of more-or-less independent objects to achieve a coherent whole—a problem also available to democratic theory, to the molecular kinetic theory of gaseous matter, to general equilibrium accounts of markets composed of rational actors, and to statistics.[2]

For example, mathematical economists show that if one constitutes a market out of individuals who can make bargains among each other, presumably trying to make the best bargains they can (what is called the "core"), then such a market converges to a competitive equilibrium as one adds more individuals.[3] And, in the central limit theorem, statisticians give an account of how large numbers of similar but independent random variables add up, achieving a Gaussian shape, the variance of 1 the Gaussian (the sum) being equal to the sum of the individual variances.

We shall pay close attention to highly technical and mathematical features.[4] The usual philosophic move of employing an abstracted toy example does not provide the richness that makes for a convincing account or analytic description, one that might actually contribute to the philosophic argument.

As will become apparent, I am agnostic about the "remarkable connection of mathematics and physics," which I take as two modes of production that have closely influenced each other, their terms of trade changing because of relative technical advances. I am rather more concerned with describing in detail some of those connections. And I take mathematics to be as much a description of the world as is physical science.[5]

ADDING UP THE WORLD

The calculus tells us how to add up velocities, $v(t)$, to get a distance traveled: Distance $= \int v(t)\, dt$. And as I have indicated already, the central limit theorem tells us how to add up random variables, adding up their variances and noting that a suitably normed sum of (more or less independent, identically distributed, finite-variance, mean-zero) random variables asymptotically converges to a bell-shaped curve or Gaussian (and more generally to the Lévy stable distributions), which is often of a rather different shape than the summands. But what if the variables were systematically dependent, say, their covariance going as $\langle X_i X_j \rangle = s(|i-j|)$, perhaps $= \exp - |i-j|/\xi$, as are the spins of magnetic atoms in a crystal. How do we add things up then?

The constitution of bulk matter out of its mutually interacting molecular constituents is for physicists a matter of statistical mechanics —treating each of the systems of molecules as being represented by random variables (or each molecule as being so represented). Here I shall be describing a variety of the ways physicists do such constitution, attending to three distinct kinds of matter. The first is the kinetic theory of gases, what is closest to the central limit theorem, both in idea and in historical time. The second is an account of how ordinary everyday matter (matter dominated by Coulomb forces among electrons and nuclei) is constituted and so is stable, neither collapsing nor exploding; and how there can be a thermodynamics, namely, an account of bulk matter that separates out intensive variables such as pressure and temperature from extensive variables such as volume and energy.

A third way of constituting matter is provided by the Ising model of a ferromagnet, that is, a lattice of atoms. The atoms interact with each other as if they are simple magnets or spins, σ_i; they interact locally and linearly as would magnetic dipoles (energy or Hamiltonian $= \Sigma - J\sigma_i\sigma_{i+1}$, where $\sigma_i = \pm 1$). And the constitution of matter is an account of the behavior of the specific heat (the amount of energy needed to raise the temperature by one degree) and of the development of a permanent magnetization with changes in temperature. The holy grail is an account of sharp phase transitions and of the spontaneous emergence of long-range order in bulk matter, namely, a permanent magnetization if the temperature is below the critical temperature (fig. 1.1).

In sum, I am concerned with the possibility of thermodynamics and phase transitions in everyday matter, none of which is obvious theoretically despite the manifest experience of everyday life. In the matter of course, I shall survey the plurality of constitutions that seem to be

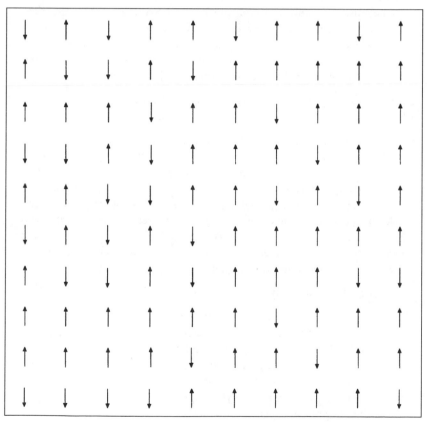

FIGURE 1.1 The Ising lattice in two dimensions, at a high temperature. See also fig. 4.1.

available and are needed to account for bulk matter. I shall be relying on an account of the stability of matter, by Elliott Lieb and collaborators (especially Joel Lebowitz and Walter Thirring), and their predecessors, and on a seminal paper by Lars Onsager on phase transitions in the two-dimensional Ising model (which is reprinted in the appendix) and on fifty years of subsequent work (especially by Lieb and R. J. Baxter and their collaborators).

Here are three relevant passages from an account, by McCoy and Wu, of solutions to the two-dimensional Ising model.[6] The first is about thermodynamics being only contingently possible, depending upon the nature of the interactions between molecules ("locality").

> If bulk properties of any statistical mechanical model in the thermodynamic limit do depend on the boundary conditions, then the distinction between bulk and surface can no longer be sharply made. While we will not be concerned with such problems in this book, it should be remarked that such difficulties can arise if the forces between particles approach zero sufficiently slowly as the separation between spins becomes infinite. In such cases a discussion of macroscopic properties in terms of a free energy. . . [defined as the free energy per spin] is inadequate.

Second, about phase transitions and the difficulty of showing what has to be true actually is true.

> Our final remark is the obvious and at the same time profound observation that T_c [the critical temperature], which has been here shown to be the temperature at which M [the magnetization] vanishes as T is increased from zero, is the same T_c as that found. . . [for which] the specific heat becomes logarithmically infinite.[7] It would seem inconceivable that these two temperatures could be different. Therefore, it would be most useful if a proof could be constructed which would establish the identity of these two temperatures without first having to compute explicitly [the specific heat and the magnetization].

And, third, about plurality.

> It must be recognized that, while the entire development in this book rests on the use of Pfaffians [a kind of determinant], the two-dimensional Ising model does not *have* to be solved this way. Indeed, the original approach of Onsager does not involve Pfaffians at all. By the phrase "nothing more to be done" [on an aspect of this problem] we mean that there is no quantity whose analytic expression is unknown. Without a doubt the known results can be obtained by many different methods, but we feel that such an investigation, which merely reproduces known

results, is pointless. To be of value, such a reproduction must either represent true simplification or result in new physical insights.

I shall be taking a somewhat different perspective on a plurality of modes of solution, reflecting the fact that subsequent alternative derivations of the stability of matter and of the two-dimensional Ising model have provided new insights and true simplifications.[8] Different forms of modeling reveal different physics. And so, in just this sense, this is a book about modeling. I have two main tasks. The first is to demonstrate how matter is constituted in various ways, a display of the modes of modeling. I want to appreciate the richness implied by a commitment to reductionism, how occasional and unideological are the strategies for implementing it. The second task is to say something about the relationship of modes of modeling to the physics itself, here how mathematical physics philosophically analyzes the constitution of matter and so shows what is essential about each of those constitutions, and how the technical features of that mathematics are physically meaningful, not merely formal apparatus needed to solve a problem.[9]

The Kinetic Theory of Matter
How is a gas to be understood as being constituted of molecular components? Kinetic theory says that the properties of those molecules, such as their velocities, are random variables. In the mid−19th century, Maxwell realized that, if the molecules (mass m) were distributed in a spatially homogeneous way and they had no net momentum (and hence their average velocity were zero, in all three directions); and if the variance of those velocity random variables were finite (namely, that the average energy per particle, $\approx v^2$, where v is the velocity, were finite); then, given that space is isotropic and that we expect the distributions of velocities in the x, y, and z directions to be independent of each other, we deduce a Gaussian distribution for their velocities: $\approx \exp - mv^2/2k_BT$. So matter is to be constituted out of statistically independent molecules. Their collective effect—and the relevant macroscopic variable, the pressure, namely, the momentum transferable across a plane—is a sum of the effects of the molecular collisions. And the temperature is identified with the mean energy (per degree of freedom) of those molecules, that is, the variance of the velocity distribution. There is, of course, much more to the kinetic theory of gases, in particular, accounts of how equilibrium is achieved, how those random variables (velocities) with their initially not-so-Gaussian distributions actually become steady state and Gaussian. But the crucial point about the kinetic theory's account of the constitution of matter is that it is

about sums of more or less independent, identical random variables. Now, such an account that centers on independence won't do, at least at first, if we expect long-range interactions among the molecules and their electrons and their nuclei (such as the Coulomb forces), or if there is systematic correlation among the molecules, as in the Ising model. Yet in the end, once we find the right variables for describing the system, the variables that make evident what we know to be true, once we get the degrees of freedom right, the ideal gas of kinetic theory shows its face again and again.

Infinite Volume Limits

All of these various modes of modeling the constitution of bulk matter depend on taking what is called an infinite volume limit. One wants to go to an infinite number of molecules, N, whose spatial density remains constant, and so the volume, V, goes to infinity: $\lim_{V \to \infty, N \to \infty N/V \text{ constant}}$. In this way, one hopes that bulk properties ($\approx V$) rather than surface ones ($\approx V^{2/3}$) will dominate in one's account. For, as the quote above indicates, bulk properties dominate most of the time with actual bulky matter: the shape of the vessel does not determine the enclosed matter's temperature or pressure, for example, and so we might measure the pressure either at the surface or within the medium itself. We believe with good reason that in ordinary everyday life the number of molecules in a cubic centimeter of matter (say about 10^{23}, where Avogadro's number is about 6×10^{23}) is large enough so that we are at that infinite volume limit most of the time.

H. A. Kramers called thermodynamics a "limit science," since many of the empirical properties of bulk matter, such as phase transitions and the irrelevance of surface effects in determining bulk properties, are apparent theoretically only in the limits of infinite volumes and finite densities.[10] Recall the example of economic markets, where there is also an infinite volume limit, namely, going to an infinite number of agents or individuals, such that no single individual's buying and selling can affect prices. Each agent is modeled as a point in an interval on the real line.[11]

Put differently, infinite volume limits are the device mathematical physics uses to model ordinary matter. Now, while we might imagine this fact to be casual and so uninteresting, merely taking a finite model and letting $N \to \infty$ at constant density, in actual fact physically and mathematically that limit is contingent. Matter in which long-range forces are not screened does not possess such a limit, and (as in gravity) it can implode due to attraction and then explode in order to release energy. Models of phase transitions depend on the problematic character of such limits. Physically, at phase transitions the infinite volume limit of the density may not exist: for first-order phase transitions, such as freezing, the limit is undetermined, for there is an indefinite ratio in the mixture of solid and liquid phases. Mathematically, apparently reasonable exchanges of limits no longer work at phase transitions (since, for example, uniform convergence is not guaranteed). Mathematical-physically, we have to construct a model of matter's bulk constitution that possesses an infinite volume limit of the right sort, only when there should be a limit—no mean task, as we shall see. To constitute matter, to model matter, so that it can be bulk matter is an achievement.

More generally, recall that it is not automatic that a normalized sum of random variables possesses a limit. Only for certain kinds of variables will such a limit exist, and rarely do we normalize the sum by N ($\approx V$), Σ/N; rather, more likely and infamously we normalize by $\sqrt{N}$, $\Sigma/\sqrt{N}$, or other powers of N. Given what we know from the central limit theorem and that Coulomb forces are infinite in extent, $\approx 1/r^2$, so that there are N^2 interactions, it is a remarkable physical fact that a physical infinite volume limit exists, a limit of a sum of identical random variables, atomic properties, proportional to or normalized by N.

Mathematical Modeling and Interpretive Cultural Practice

In mathematical modeling, there are two sorts of approximations: the model surely abstracts from the actual physical world; and, in solving the model, one makes various degrees of mathematical and numerical approximation. I shall be focusing on "exactly solved models" (to use the term of art),[12] when one does not make mathematical approximations. But, of course, one is guided in one's abstraction of the physical world by what might be solvable, and perhaps exactly so. Moreover, I shall suggest that the modes of solution are also models of the natural world (the mathematics being semantic, let us say).

As indicated in the preface, my mode of reading the papers and books and textbooks that expound these various constitutions of matter is mildly Whiggish, especially if the works are early and germinal. For it turns out that I understand the richest implications of what is said in a text only retrospectively, from results that are developed in subsequent work, often by others than the author. Namely, I am interested in the deeper features (presumably of nature) embedded in these models, and after a large number of re-provings or re-solvings, one often gets a better hint at what is crucial, now seen in the early papers as passed-off remarks or as technical moves. So, in rereading Onsager's 1944 paper on the two-dimensional Ising model, I keep discovering that, not only did Onsager seem to know it all, but the text indicates he seems to have been quite aware of what he knew. Yet I find there is always more to it (an even deeper story), some of which Onsager did not know in its modern form.[13] In any case, that I take papers and books as texts, exactly meaning the words they say and the symbols they use, is the occasion for then becoming as Whiggish as is necessary.

The reading is incompletely Whiggish, however, for I have not at all insisted on a much deeper reading, for example, seeing all of the results on the Ising model in terms of quantum field theory on a lattice, or in terms of more complex lattices that illuminate simpler ones.[14] I have been restrained in invoking post-1980 results.

Finally, I take my descriptions as descriptions of a culture's practices, my interpretations as attempts to provide meaning to those practices, meaning that might well be recognizable to the culture's members, even if they had not thought of that meaning before. I have followed this strategy in a previous book, *Doing Physics*.[15]

Philosophically, I am touching on two sorts of issues, albeit not in philosophical form. The first concerns the soritical paradoxes: for example, if a heap has N grains of sand, and you remove one at a time, when is it no longer a heap? Technically, in physics, this is just the problem of mesoscale phenomena, when in condensed matter orderly behavior is beginning to set in. The infinite volume limit, in all of its manifestations, provides not so much a solution to the soritical paradoxes as an example of how they might appear in modern guise. The second issue is the relationships of models and mathematics to each other and to the physical world, where my purpose is to describe such a relationship in some detail.[16]

TECHNICAL MATTERS ARE THE PHYSICS

Ordinarily in statistical mechanics, the kind of mathematical questions we wish to examine *are completely ignored*. But in the Ising model, which is exactly soluble in certain respects, it is rewarding to be more careful, and to see precisely what one must assume even here. (Schultz, Mattis, and Lieb)[17]

One of the fascinating features of statistical thermodynamics is that quantities and functions, introduced primarily as mathematical devices, almost invariably acquire a fundamental physical meaning. We had examples in the Lagrangian parameter μ [$= \beta$, in doing the Stirling's approximation deduction of the canonical ensemble], the maximum z [$= \exp \beta$, from the Darwin-Fowler method], and the sum-over-states or partition function [exponential of the free energy]. (E. Schrödinger)[18]

The reduction to elliptic integrals is usually a chore; on this occasion the elegance of the various relations greatly relieved the tedium. (L. Onsager)[19]

[Hermann] Weyl, [Chen Ning] Yang, and [Elliott] Lieb have. . . in common a style and a tradition. Each of them is master of a formidable mathematical technique. Each of them uses hard mathematical analysis to reach an understanding of physical laws. Each of them enriches both physics and mathematics by finding new mathematical depths in the description of familiar physical processes. . . In all three cases, a rather simple physical idea provided the starting-point for building a grand and beautiful mathematical structure. Weyl, Yang and Lieb were not content with merely solving a problem. Each of them was concerned with understanding the deep mathematical roots out of which physical phenomena grow. (F. Dyson)[20]

The renormalization group formalism can be set up in a fairly logical manner to reduce a problem to one involving a finite number of degrees of freedom at each iteration. Unfortunately for practical calculations success or failure can depend on whether the finite number is 1 or 10 or 100, and inevitably one must resort to tricks to achieve 1 or 10 in place of 10 or 100. If one finds this prospect discouraging, one should remember that the successful tricks of one generation become the more formal and more easily learned mathematical methods of the next generation. (K. G. Wilson)[21]

Along the way in our survey of the constitutions of matter, I will provide evidence for two claims mentioned earlier: the technical details in the mathematical apparatus, and the mathematical moves employed in doing theoretical physics, are in fact physically meaningful, and not merely devices employed to do some auxiliary work. And the task of mathematical physics is one of making clear just what are the desiderata for and consequences of some of our everyday notions of the natural world—namely, that mathematical physics is philosophical.[22] Here is Freeman Dyson (discussing "The Algebra of Local Observables" by Haag and Kastler):

These axioms, taken together with the axioms defining a C^* algebra, are a distillation into abstract mathematical language of all the general truths that we have learned about the physics of microscopic systems during the last 50 years. They describe a mathematical structure of great elegance whose properties correspond in many respects to the facts of experimental physics. In some sense, the axioms represent the most serious attempt that has yet been made to define precisely what physicists mean by the words "observability, causality, locality, relativistic invariance," which they are constantly using or abusing in their everyday speech.[23]

The physics is embodied in mathematical practices, in mathematical formulations and mathematical manipulations. Again, this is not to say anything about the "surprising" congruence between mathematics and physics, nor about whether the mathematics leads to the physics, about which there is a need for historical work. There is another issue, whether and how technical devices, and advances in using them, lead to progress in physics, about which historical work is called for. But I suspect that it is much the same whether those technical devices come from mathematics, or from electrical or mechanical engineering, or from tinkering and invention.[24]

More generally, I am concerned here with scientific accounts and so justifications for the most everyday of phenomena, such as the stability of bulk matter. Such justifications often involve quite delicate technical points, whether the justification be about the existence of the universe (through the big bang inflationary scenario), or about the origin of life, or about stable bulk matter.[25] The initial conditions have to be of the right sort, the phase transition of the right sort, the particles of the right sort—if such justifications are to work. I do not mean to suggest that these everyday phenomena are delicate phenomena, subject to some parameter being just right for the world to have turned out as it has, although sometimes that might well be the case.[26] Rather, in analyzing these everyday phenomena, we revise our unthought assumptions about the technical and mathematical devices we use to model them and the physics embodied in those devices.

So if you cannot do something theoretically since it seems mathematically intractable, when you find out how to do it, you not only learn some mathematics, you learn some deep physics, and you learn why—what you were ignoring—you could not do it before. This is a recurrent theme among theoretical and mathematical physicists, whether because they see mathematics as a powerful tool for doing better at the physics, or because in their practiced theorizing they have discovered how significant is the relationship of mathematical tools to physical explanation, even when those tools seem at first only to be devices for doing a complex calculation.[27]

Again, subtle technical questions are exactly the physics. Mathematical subtleties and matters of rigor, in being resolved, reveal the essential physics—the mechanisms and the crucial features of the phenomena as we have modeled them—just what we did not fully enough appreciate until then. The technical questions, as they are resolved, show how what we take as obvious and necessary just might not have happened. So, in answering these technical questions, we shall learn why bulk matter is stable.

Another example: When we are computing the spontaneous magnetization of a crystal, its capacity to be a permanent magnet, we have to take two limits—one to infinite size ($\lim_{N \to \infty}$) and one to the externally applied magnetic field being zero ($\lim_{B \to 0+}$). The infinite-size limit is meant to model bulk matter, and it is known that one can get phase transition only in that limit. The zero-field limit is just what we mean by a permanent magnet, namely, a magnetization that is internally maintained, without an external field.

Now, if we take the zero-field limit first, then the spontaneous magnetization turns out to be identically zero; and the infinite volume limit will then just preserve this identity. Here is not only a mathematical issue, but some physics as well. Namely, we first need to have some condensed or bulk matter (the infinite volume limit) that could possibly maintain such a magnetization, before we find out whether it actually does so. Finite sizes are insufficiently effective at pinning down the atomic spins, to hold such magnetization in the first place. In this sense, the order in which we take limits is physical.

Mathematical Manipulations Are Meaningful
Mathematics in the service of natural science is neither formalism nor technical apparatus nor makeshift tools that we might discard once we have used them—although it would often seem to be just such formalism, apparatus, and tools.[28] But these are meaningful formalism, appropriate apparatus, and custom-fashioned tools. While the physicist might experience the mathematical moves used in solving a problem as devices, tricks, and miracles, there will be at least in retrospect an account of the structure of the physical world that explains why the physicist might have found these tricks and devices effective, why these curious substitutions and schemes are productive and felicitous and so taken as natural.

The mathematics is as factual as the physics. Tools and instruments tell us about the physical world, a not unusual notion for a paleontologist. When someone, here C. N. Yang, tells the following sort of story, we hear about the terrain of the physical world as it is initially explored before more straightforward paths have been found, at least as much as about the author's modes of persistence and creative invention.

>I was thus led to a long calculation, the longest in my career. Full of local, tactical tricks, the calculation proceeded by twists and turns. There were many obstructions. But always, after a few days, a new trick was somehow found that pointed to a new path. The trouble was that I soon felt I was in a maze and was not sure whether in fact, after so many turns, I was anywhere nearer the goal than when I began. This kind of strategic overview was very depressing, and several times I almost gave up. But each time something new drew me back, usually a new tactical trick that brightened the scene, even though only locally.
>
>Finally, after about six months of work off and on, all the pieces suddenly fitted together, producing miraculous cancellations, and I was staring at the amazingly simple final result, equation (96). . . . That was in June, 1951, about a week before my first child, Franklin, was born.[29]

What is local, tactical, tricky, and miraculous will most often turn out in retrospect to have a physically meaningful strategic overview, tricks being natural devices to handle crevasses, miracles being surprises because you did not know just what you were achieving by your devices and desires (not coincidences between events, natural and natal). Here ontogeny recapitulates research, or so we are assured by Kenneth Wilson in one of the quotes that began this section.

A demonstration for such a claim might be the detailed annotation of a paper such as Yang's, where each reported calculational and mathematical trick and move is shown to have a physical meaning of its own, something we might be able to do only after further solutions suggest to us the meanings of the moves in Yang's paper. Now, in Yang's computation of the spontaneous magnetization, one can safely say that Yang is computing the split in the degeneracy of the two largest eigenvalues of a matrix, for temperatures less than the critical temperature, much as one computes the Zeeman splitting for the hydrogen atom in a magnetic field in quantum mechanics. In both cases, one is employing perturbation theory. The formalism of the statistical mechanical problem parallels the formalism for the analogous quantum mechanical problem. But it is much more difficult to appreciate the subsequent flow of Yang's solution.[30]

Onsager's original 1944 paper deriving a solution for the two-dimensional Ising model appeared to many to employ a panoply of seemingly arbitrary, esoteric devices—for example, Clifford algebras and elliptic functions.[31] Newell and Montroll, in a 1953 review article, explicitly provide as an annotation an overview of the plan and the step-by-step motivations for that paper, and much of the subsequent work in this field is meant to interpret these various moves and so make them natural.[32] (That those devices seem to be restricted to two dimensions, with zero magnetic field, is much more problematic. Eventually, those restrictions are also understood physically, poignantly, since there is no way to simply generalize them to higher dimensions or nonzero fields. But other methods may well prove to be more generalizable.)[33]

What is also often noted is how the particular mathematical resources possessed by a physicist not only readily influences the physicist's capacity to apply a certain technique but, more importantly, shapes the kinds of explanations a scientist can provide (and, of course, some of the time the resources are totally unsuited to a problem).[34] So, for example, Bruria Kaufman—as a consequence of her training in using spinors—could apply this particular technique, in her studentship and collaboration with Lars Onsager, to a deeper understanding of the two-dimensional Ising model.[35]

Of course, the power of mathematical devices in working on a particular problem is not always something that is accidentally discovered. Physicists will often set up simplified mathematically and computationally tractable versions of their models, with possibly crucial features left out for the moment, in order to get a feel for the consequences of a model. Such toy models allow for rather more straightforward solutions than does the full model. Those solutions may reveal the more sophisticated mathematical devices or phenomena likely to be present in the full model.

For example, Lars Onsager worked with some toy models of the two-dimensional Ising lattice (strips $2 \times \infty, 3 \times \infty, \ldots, 8 \times \infty$), and from some common patterns in each of the solutions, he suspected what the mathematical structure of a solution had to be (those Clifford algebras).[36] Lee and Yang discovered the unit-circle theorem from a sequence of such toy models.[37] Kramers and Wannier, and Montroll discovered from toy models the utility of transfer matrices in the solution of the Ising model.[38]

One last example: Baxter describes how a sequence of toy models of the "hard-hexagon model" led to an extrapolation to an infinite-width two-dimensional lattice, which allows for various estimates of the coefficients of a series expansion of the partition function. The crucial numbers, slightly rounded, would seem to be the roots of a quadratic equation with integer coefficients. That led to analogies to solutions of other models, and a guess that the exact solution is expressed in elliptic theta functions. And then Baxter says, "Having guessed the right answer, the next step was to look for a way of deriving it."[39] By the way, the solution involved the so-called Rogers-Ramanujan identities, and Andrews gives a lovely account of how Baxter met that literature.[40]

In sum, from such toy models we might hope to distinguish crucial features of the full model—ones necessarily present (that necessity perhaps even proved mathematically) if the model is to work appropriately at all—and so eventually to understand how we might solve the model. Just as we might say that some features of a physical problem are central and others are ornamental, so we might hope that some technical devices might be highlighted as central in the toy model, while others (that we may or may not know how to handle already) will be seen as both meaningful and not very productive and so are taken as peripheral.

Interpretations of Mathematical Physics
In an elegant study, Jed Buchwald has shown that the mathematical physics of Maxwell's equations was interpreted very differently by the English Maxwellians (and by Maxwell himself) than we do nowadays in the ordinary training of physicists. Further work by Daniel Siegel has shown how the fabled displacement current does not come out of some need for symmetry in those equations, as is taught to students in modern textbooks; rather, the current was suggested by a physical model of vortices in an ether than informed Maxwell's original formulation of the equations.[41] Hydrodynamic and fluidic notions, characteristic of the thinking of the Maxwellians, rather than particulate ones, characteristic of modern pictures, allowed for rather different explanations of how the fields actually worked, albeit the mathematics appears the same. Actually, small differences in formulation (such as on which side a term was placed in an equation, the sign suitably adjusted) indicated enormous differences in interpretation, but the formalism was opaque to such differences. Another such radical difference in interpretation is provided in a gauge field theory other than electromagnetism, gravitation. J. A. Wheeler's geometrodynamics is as substantial as Maxwell's fluid mechanics, the mathematics of differential geometry at all points being a servant to a physical picture of the world that, too, is about flows and sinks.[42]

My claim here, stripped of all its technical examples, is perhaps unexceptional. What some physicists may view as technical devices are in fact laden with physics—tricks and devices being the names for procedures whose physical content we have forgotten or do not yet appreciate. In time, we see those tricks and devices as reflecting the facts of the physical world, and then of course such technical mathematical devices, as Wilson notes, are merely more of what it means to do physics.

2

Analysis

THE STABILITY OF BULK MATTER

The stability of matter • The stability of atoms • The stability of bulk matter • Recapitulation and prospect

The subfield of physics called statistical mechanics accounts for the macroscopic properties of matter, the properties of matter in bulk, in terms of the atomic or molecular properties of its large numbers of similar components interacting with each other. Statistical mechanics is "mechanical," since it was originally concerned with the collisions and interactions of particles, as was Newton. And it is "statistical," since the properties and phenomena of everyday matter in equilibrium are shown to be the consequence of an average over the properties of all possible systems, subject to the same constraints, whether those systems are in equilibrium or not (namely, an average of all arrangements of those components)—much as a mean is an average.

Infinite volume limits, that is, going to an infinite number of particles at constant density and therefore to infinite volume, $\lim_{N, V \to \infty, N/V (= \text{density}) \text{constant}}$, are the technical devices of statistical mechanics for saying what it would mean to be "bulk" matter, matter that also might be "stable" and even undergo "phase transitions"—notions that themselves demand technical definitions, and the quotation marks are meant to indicate this. Here I show how physicists conceive of bulk matter and its properties, attending to some of the technical features in its analysis.

Again, the claim is that what would appear to be technical devices and methods are encrusted with physical meaning. What appear as tricks and turns in a derivation are in fact about the nature of the physical world, "a trick" or "a device" being what we call some technical procedure whose physical content is not yet appreciated. To appreciate that physical content is to understand the physics of a model or a situation. And in multiple subsequent rederivations (as for Onsager's solution of the two-dimensional Ising model), which often build upon each other and which may make the original derivation perspicuous and 15 motivated (or supply apparently different alternative solutions), we do come to such an understanding.[1]

Moreover, by varying our assumptions in those derivations, we may find out which features of our models are crucial if they are to model stable bulk matter, and which features are accidental. In this sense, such mathematical physics is a mode of philosophic analysis, saying what it means to speak of bulk matter.

THE STABILITY OF MATTER

In a review article on the stability of matter, the physicist Elliott Lieb says:

> The content of this paper can be summarized as follows:
> (i) Atoms are stable because of an uncertainty principle;
> (ii) Bulk matter is stable because of a stronger uncertainty principle that holds only for fermions;
> (iii) Thermodynamics exists because of screening.
> My hope is that the necessary mathematics, which is presented as briefly as possible, will not obscure these simple physical ideas.[2]

Now, the mathematics in the various scientific papers is not so brief, and it is reasonably advanced and technical—although it is much briefer and more perspicuous than the original work, which was described by one of its authors, Freeman Dyson, as "hacking through a forest of inequalities."[3] But after all is said and done, the mathematical analysis shows that what is crucial for the stability of everyday matter is that (1) quantum mechanics describes atoms; (2) electrons cannot be bunched up in space (for they are particles described by antisymmetric functions, so that if two electrons in the same spin and momentum state are also very close to each other, that antisymmetric wave function is close to zero—namely, they are "fermions"); and (3) matter is electrically neutral (having roughly equal numbers of electrons and protons). Mathematical analysis shows that matter is not stable if we use only classical physics, if electrons could bunch up (as do bosons, such as photons or pions, described by symmetric functions), or if there were substantial nonneutrality of electric charge.[4]

More precisely, what mathematical analysis shows, by counterexample (which explicitly takes into account the range of conceivable alternatives, and so performs the philosophical analysis) and by careful proof (which implicitly takes into account a host of examples and counterexamples), is just how crucial are these facts of nature for the existence of what we ordinarily encounter as bulk matter. What analysis suggests is how we are to think of ordinary matter: just how it is quantum mechanical, just how its fermionic nature plays a role, and just how electrical neutrality (in terms of the fact that positive and negative charges can cancel each others' effects at a distance, namely, "screening") works to allow one to treat the material world as made up of separable bulky pieces. At the end of proving that matter is stable and bulky, Lieb says: "In any event, the essential point has been made that Fermi statistics is essential for the stability of matter. The uncertainty principle for one particle, even in the strong [Sobolev] form, together with intuitive notions that the electrostatic energy ought not to be very great, are insufficient for stability. The additional physical fact that is needed is that the kinetic energy increases as the 5/3 power of the fermion density."[5] That is, "the essential point has been made" through the fairly technical theorems that are proved in the paper. (Of course, new counterexamples, inconceivable now, might show that the point has not been made, the proofs needing reworking.)

THE STABILITY OF ATOMS

Let us begin with the stability of atoms. A hydrogen atom is stable if it does not collapse, the electron and proton falling into each other, attracted by their Coulomb or electrical force: an atom must be of finite nonzero size, r, its electrical energy ($-e^2/r$) is finite and bounded from below, and its kinetic energy is finite as well. As Lieb tells it: "It is difficult to find a reliable textbook answer even to the question: How does quantum mechanics prevent the collapse of an atom? One possibility is to say that the Schrödinger equation for the hydrogen atom can be solved and the answer seen explicitly. This is hardly satisfactory for the many-electron atom or for the molecule. Another possible answer is the Heisenberg uncertainty principle. This, unfortunately, is a false argument."[6]

Note that Lieb spoke earlier of an uncertainty principle. But it is not the Heisenberg uncertainty principle, as he shows by a counterexample.

> [A]n electron cannot have both a sharply defined position and momentum [the momentum and so the kinetic energy would be infinite if the "size" were zero; this is often taken to say that a hydrogen atom is stable: the atom is of finite nonzero size and does not collapse]. If one is willing to place the electron in two widely separated packets, however, say here and on the moon, then the Heisenberg uncertainty principle alone does not preclude each packet from having a sharp position and momentum.

Thus, while [the Heisenberg uncertainty principle]. . . is correct, it is a pale reflection of the power of the operator $-\nabla^2$ [the kinetic energy in the Schrödinger equation, which could go to infinity] to prevent collapse. A better uncertainty principle (i.e., a lower bound for the kinetic energy in terms of some integral of ψ [the wave function] which does not involve derivatives) is needed, one which reflects more accurately the fact that if one tries to compress a wave function *anywhere* then the kinetic energy will increase. This principle was provided by Sobolev (1938).[7]

Sobolev's inequality can more effectively settle the question of the stability of a hydrogenic atom, taking into account a much wider range of cases than does the Heisenberg principle. It represents a lower bound of the kinetic energy as an integral of the electron density (or a power of it), $|\psi|^2$, and so is sensitive to density peaks.[8] Again, the point is that quantum mechanics (the Schrödinger equation) is the basis for the stability of the hydrogen atom, but the Heisenberg uncertainty principle is not the expression of quantum mechanics that will do the required work. But counterexamples, such as the earth-moon case, and careful mathematical analysis, here employing the Sobolev inequality, do indicate an uncertainty principle that will do the work.[9]

Just what work has been done? Having provided a technical account, Lieb also provides a colloquial summary: "An electron [in a hydrogenic atom] is like a rubber ball, or a fluid, with an energy density proportional to $\rho^{5/3}$ [where ρ is the density]. It costs [lots of] energy to squeeze it and this accounts for the stability of atoms."[10] But, for the well-trained physicist, the technical account alters the meaning of this colloquial summary, the physicist who now recalls those counterexamples when the Heisenberg principle does not do the work. What the physicist always thought was well-understood textbook material—that the hydrogen atom is stable—is true, but true in a very different way. In other words, we might well continue in our belief that quantum mechanics is the source of the stability of atoms, but we have new respect for the meaning of quantum mechanics and, as we shall see, for the meaning of being an antisymmetric particle or fermion.

To preview the rest of the argument, as Lieb indicates, an uncertainty principle will not do all the work. In a multielectron atom (helium, uranium) or in a solid with many fermions (here, in particular, electrons), N, it turns out that the uncertainty principle alone (namely, ignoring the electron's being a fermion) would suggest that the kinetic energy grows as N (for the above limit for the kinetic energy is for a single-electron atom)—which it turns out is not enough to prevent collapse. But, in fact, from considerations that take into account that with more than one similar fermion present there are antisymmetry effects, we find that the kinetic energy grows as $N^{5/3}$.[11] And the extra $2/3$ in the exponent is just what is needed to prevent collapse.[12] (Recall that kinetic energy goes as, is asymptotically (for large N) proportional to, the square of the momentum; the fermionic character of the electron pushes up the mean momentum, since only two particles can occupy any given momentum state, and with more fermions present, the occupied states increase in momentum.)

> This distinction stems from the Pauli principle, i.e., the antisymmetric nature of the N-particle wave function. As we shall see, this $N^{5/3}$ growth is essential for the stability of matter because without it the ground state energy of N particles with Coulomb forces would grow at least as fast as $-N^{7/5}$ [for bosons] instead of $-N$ [namely, if N is increased, the energy/particle gets even lower, and so there is accretion and collapse].
>
> The Fermi pressure [the extra $2/3$ in the exponent] is needed to prevent collapse.[13]

I should note that in this chapter I have deliberately and regretfully left out much of the detailed technical and physical reasoning in these papers (and I shall continue to do so to some extent), for which one ought to consult the original papers and reviews. Here I want to emphasize how phenomenological measures, such as exponents on N, or the spatial dimensionality, mediate between the everyday phenomena that are characterized by such measures and the highly technical arguments that are used to compute them. The phenomena exemplify just what we might mean by stability, the technical arguments being the means of philosophical analysis. Later chapters are concerned with similar issues, with a greater focus on technical matters.

THE STABILITY OF BULK MATTER

We might recall a bit of everyday phenomenology. It is a known feature of everyday life that if we put two rocks together nothing much happens. "We take a piece of metal. Or a stone. When we think about it, we are astonished that this quantity of matter should occupy so large a volume. Admittedly, the molecules are packed tightly together, and likewise the atoms within each molecule. But why are the atoms themselves so big?. . . Answer: only the Pauli principle, 'No two electrons in the same state.' That is why atoms are so unnecessarily big, and why metal and stone are so bulky."[14] Without the Pauli principle, "[w]e show that not only individual atoms but matter in bulk would collapse into a condensed high-density phase. The assembly of any two macroscopic objects would release energy comparable to that of an atomic bomb."[15] When the everyday rocks are brought together, the energy released is at most proportional to the number of particles—actually it is proportional to the surface area, $N^{2/3}$, the energy of adhesion.[16] We say, matter is energetically stable. (Were the energy released proportional to $N^{1+\epsilon}$, $\epsilon > 0$, then putting the rocks together would have explosive consequences, the world in effect accreting onto itself, the released energy then causing outgoing shock waves, an explosion.)

By the stability of matter, we have come to mean a number of features beyond the stability of a hydrogen atom. Initially, we want to assure ourselves that a multielectron atom or molecule is energetically stable. It has a lower bound of its energy, roughly proportional to the number of electrons. It won't collapse; it won't accrete even more electrons and collapse even faster. The electrons and nuclei do not just crash into each other. (For the multielectron atom, what is required is a nontrivial extension of using an uncertainty principle to show that the hydrogenic atom is stable.) Moreover, a large atom is bulky: "A large atom is like a galaxy. It has a small, high density, energetic core in which most of the electrons are to be found and whose size decreases with z [the nuclear charge] like $z^{-1/3}$. The electron-electron electrostatic repulsion always manages to push a few electrons out to a distance of order one which is roughly independent of z; this is the radius one observes if one tries to 'touch' an atom."[17]

Second, we want to be sure that, as we add more atoms and molecules to make up bulk matter, matter does not collapse as more is added on. There is, again, a lower bound to its energy, proportional to the number of electrons or particles.[18] (Were the electrons not fermions but bosons, that proportionality factor would be an "unpleasant" $-N^{7/5}$, unpleasant in that there would be collapse.)[19] And such matter ought to be bulky, having a radius of about $N^{1/3}$, as we might expect.[20] (If the density is constant, then the linear size of a system goes as $N^{1/3}$, the surface area as $N^{2/3}$, and the volume as N.)

Third, when the number of particles gets large, the energy per particle of the system is roughly constant (a linear law), rather than being proportional to $-N^{\epsilon}$, $\epsilon > 0$, say, due to interactions that are "long range." This is called the thermodynamic limit. Not only is there stability, a lower bound proportional to N or a power of N, but there is a linear relationship in N. (A bound, even a linear one, is not the same as a linear asymptotic limit. *Asymptotic* means convergence to a form, here linear in N, and not merely convergence to a number.) There are well-defined extensive (namely, proportional to N) values for energy, and so forth, in what is called the infinite volume limit or the thermodynamic limit. As Lieb puts it: "The linear law $[\lim_{N \to \infty} N^{-1}E(N, k = N/z, Z) = -A'(z)]$ is not just a lower bound it is, indeed, the correct *asymptotic law*. . . . If this $N \to \infty$ limit did not exist. . . matter would not behave the way we expect it to behave—*even if stability of the second kind* [a lower bound for the energy linear in N] *is assumed to hold*."[21] Again, "In summary, bulk matter is stable, and has a volume proportional to the number of particles, because of the Pauli exclusion principle for fermions (i.e., the electrons). Effectively the electrons [in bulk matter, and not just an atom,] behave like a [mildly incompressible] fluid with energy density [proportional to] $\rho^{5/3}$, and this limits the compression caused by the attractive electrostatic forces."[22]

Through mathematical analysis, these features of stable bulk matter are given precise roles in a model of nature. It is not at all obvious how (1) the stability of the hydrogen atom, presumably shown in elementary quantum mechanics courses, then should lead to the stability of multi-electron atoms, or that (2) that stability should lead to stable bulk matter, or that (3) that stable bulk matter, in effect not collapsing, possesses thermodynamic limits, and so properties such as temperature and entropy, for example. As importantly, these distinctions—implied by this sequence of provings—are given meaning, counterexamples showing that the sequence would not go through if electrons were bosons (described by symmetric functions), for example. Moreover, (4) it is emphasized that (a) stability in terms of a lower bound in the energy, and so stability against collapse, is very different than (b) stability in terms of a lincar asymptotic limit in the energy that suggests stability against explosion, and so thermodynamics applies; and that is different than (c) having stable matter that does not expand when pressure is placed upon it ("thermodynamic stability").[23] There is physics, different physics, behind each of these distinctions, and those differences are reflected within the mathematical proofs of theorems and lemmas that are used to make a physical argument.

I should like to reiterate that perhaps the most crucial feature of the above discussion is just that program of provings (1–4) and distinctions (a–c) that leads to an account of the stability of matter, where the stability of matter is now understood as the hierarchy of stabilities sketched by that program. (This program is due to work by Onsager, van Hove, Lee and Yang, Fisher and Ruelle, Dyson and Lenard, and Lebowitz, Lieb, and Thirring, an achievement of two generations of scientists.)[24] What is an obvious everyday fact is now given a canonical technical definition, one that is both provable and physically deep. The mathematical agenda and machinery and the physical understanding parallel and influence each other.

At each stage in the hierarchy, well-known physical principles are shown to do the appropriate work. But it is both in the staging of the hierarchy and in just how those principles are employed (in effect, their particular mathematical formulation) that the physics comes to be understood.[25] Quantum mechanics might be said to raise the kinetic energy of an electron in a hydrogenic atom enough to prevent its collapsing into the nucleus, but the lower bound on that kinetic energy shows just how quantum mechanics does that work. Quantum mechanics and the Pauli exclusion principle (or, effectively, the antisymmetric nature of fermions) shows why large atoms are stable. But only if the exclusion principle is used appropriately, for many electrons and in the right technical form, will it do the work quantitatively: "the imposition of the Pauli exclusion principle raises E [the energy computed from wave functions that are not restricted to being antisymmetric]. The miracle is that it raises E enough so that stability of the second kind holds. . . . [i]t is not easy to quantify the effect of antisymmetry. Even the experts have difficulty, for it is not easy to think of an antisymmetric function of a large number of variables."[26] Finally, to understand that "miracle" (of stable bulk matter), one must prove and understand how a collection of atoms or molecules, so to speak, allows each atom to have its own space—namely, one must give a technical account of atoms among each other in close proximity. In effect, there is comparatively little binding among the molecular constituents, and so matter is not only stable but bulky.[27]

RECAPITULATION AND PROSPECT

It may be useful to recall our task and sketch where we have gone so far. I am trying to show how the everyday notion of ordinary matter comes to be canonically described by physicists in such a way that they can show that the basic principles of physics at least allow for what we know exists. (It would be even nicer to show that what we know exists is in fact necessary, and to some extent that is also done.) More generically, I am showing how the most everyday of phenomena may demand the most subtle of technical definitions and accounts. And more specifically, I have so far suggested how an account of the stability of bulk matter, namely, that it does not collapse into itself, is a peculiar consequence of quantum mechanics and the fermionic nature of electrons—each of which in effect pushes electrons and so atoms sufficiently far apart from each other so that matter is bulky.

All of these achievements are just what is needed to keep the $1/r$ Coulomb potential from gobbling up all of matter; they are concerned with small distance ("small-r") effects. There is, however, another problem (3 above). The Coulomb force is long-range, and so we might expect that the energy of interaction of many atoms would rise nonlinearly with the number of atoms (since more of them would be involved with each other). Were that the case, then the usual division of the properties of bulk matter into intensive and extensive properties would not work, since such a division implies that the properties of the bulk either are size-independent or are proportional to size. Moreover, we would have some rather interesting consequences of placing two rocks —even rocks that are not radioactive—next to each other, as I described above. Our account of the stability of matter must also show that bulky matter has some nice average properties, some that are size-independent (temperature, pressure) and others that simply add (volume)—that a thermodynamics exists for such a system.[28] Put differently, it should not matter whether we measure those size-independent properties by isolating the sample or by measuring a property in situ.[29]

We know from our experience of everyday matter that there are such intensive and extensive properties or variables, and such possibilities of measurement, the intensive variables quite robust to how we measure them, the extensive variables nicely adding up arithmetically. Somehow the long-range Coulomb forces are restricted in the work they can do. That somehow is called "screening"; the effect of the positive nuclear charge and the negative electronic charges in a roughly neutral piece of matter cancel each other at not-too-large distances (Newton-Gauss law). Put differently, matter can be treated as being composed of additive lumps, or "bulkettes" (as I shall call them), there being a smallest size of bulkette that is still quite small, yet bulky and stable and asymptotic enough. While all of this is obvious physically, it is not at all obvious if we are trying to prove anything, and in so proving we again come to revise what we mean by bulk, stable, extensive matter. We learn, again, what is crucial about it. And, of course, our technical proving and machinery is driven both by our physical insight and by our mathematical technology.

3

Mathematics

INFINITE VOLUME LIMITS AND THERMODYNAMICS

*Temperature and equilibrium—the Darwin-Fowler derivation: the physics,
the possibility of a temperature, recapitulation • The thermodynamic limit:
proving a thermodynamic limit—the free energy, screening •
Thermodynamic stability and mathematical physics as empirical:
mathematical and physical argument, mathematical physics as
philosophical, limits are physical*

Mathematical physics is a form of philosophical analysis, in effect
showing more precisely the meaning and implications of our basic
physical notions, the technical features of the mathematics building in
the physics. In this chapter, I shall examine just how temperature may
be seen as an artifact of formal, technical, mathematical device (to be
sure, introduced well after an empirical and theoretical notion of
temperature was established); and how further devices are employed to
account for the possibility of a thermodynamics in our world, namely, a
proof of the existence of the thermodynamic limit and of thermodynamic stability. Yet those formal, technical, mathematical devices are
laden with physics. The more general claim is that mathematical physics
is empirical and contingent in that its detailed workings, its formulations and their effectiveness as explanations, are intimately connected
to specific contingent facts about the natural world. In this sense, as we
shall see, mathematical limits are physical—the seventeenth-century
lesson in the differential and integral calculus, where tangents as limits
are velocities and where rectifications as limits (for example, covering a
curve with small tangent lines) are distances.

As for bibliography, I will discuss papers chosen to elucidate certain
points, rather than providing anything like a history of the problems I
shall be discussing, even something like a potted history. Perhaps the
germinal paper is by Lars Onsager. Onsager, in 1939, understood the
fundamental facts about the stability of matter in terms of hard cores
and finite sizes of the component atoms. (See the appendix for a reprint
of this paper.) More precisely, the energy of interaction of the charges
would appear to go as N^2. Onsager showed that that interaction energy
is bounded from below by the sum of the self-energies of each of the

nonpoint charges, so it is proportional to N, and therefore there is a linear lower bound and so stability.[1]

Van Hove, in 1949, employed these same ideas plus the assumption of finite-range forces among the component atoms, to prove the thermodynamic limit for the partition function and thermodynamic stability as well (namely, that the compressibility of matter is nonnegative or, equivalently, the pressure measured in the canonical ensemble, externally, is the same as that measured in the grand canonical ensemble, in situ). Yang and Lee, in 1952, used techniques similar to those of van Hove—a geometric series of increasing-size cubes—to prove the existence of the thermodynamic limit for the grand partition function.[2] And, more recently, Dyson and Lenard, Fisher and Ruelle, Lieb, Thirring, and Lebowitz and their collaborators, and many others, have brought the story to a sense of conclusion.

TEMPERATURE AND EQUILIBRIUM: THE DARWIN-FOWLER DERIVATION

In 1922, Darwin and Fowler presented a discussion of "the way in which energy is partitioned among an assembly of a large number of systems. . . with some attempt at rigour. . . . Most of the results are not new; it is the point of view and the method which, we think, differ from previous treatments."[3] It is a beautifully motivated paper, with a series of successively more complex examples leading to the crucial derivation, with references to the work of Planck and Bohr and the like being central. (Much of this is left out of the presentation in Fowler's 1936 textbook, it would seem the source of most of the subsequent expositions I have seen.) Various moves, which in those subsequent presentations seem to be artifices, here have a natural origin. Still, as we might expect, modern readers usually do not go back to this wonderfully didactic paper.

The Darwin-Fowler method is nowadays presented as the more rigorous derivation of the probability of a thermodynamic system's having a particular energy, among systems for which we know a priori only their average energy (or, it turns out, their temperature). We are within the canonical ensemble in statistical mechanics, where the canonical ensemble refers to the statistical mechanics for a very large number of similar systems (an ensemble) possessing a definite temperature rather than a definite energy—namely, an average energy or noise level—and where statistical mechanics refers to the physics of systems with very very many similar components. (Recall that, in the microcanonical ensemble, the energy is fixed, while in the grand canonical ensemble, only the average energy and average number of particles are known.) For the canonical ensemble, we find that the probability of a system possessing an energy E, given a temperature T, is proportional to $\exp - E/k_B T$, where k_B is Boltzmann's constant ($\approx 1.4 \times 10^{-23}$ joule/kelvin). The less rigorous derivations usually involve either a Stirling's approximation for the factorial ($\ln n! \approx n \ln n$), employed in computing the number of ways of distributing energies among a set of systems, looking for the most probable configuration, or the use of an entropy function within a thermodynamic argument.[4] Whatever the increase in rigor, the Darwin-Fowler derivation also allows one to estimate the errors involved and so justifies the approximations employed, since it computes "the average state of the assembly instead of its most probable state."[5] However, this is not my main concern here. Rather, I want to say something about how the derivation allows us to think about bulk matter.

I should note that the limits in this section are limits to an infinite number of systems, N_S, $N_S \to \infty$, an "ergodic" limit. For most of our discussion elsewhere, we focus on an infinite volume limit, $N \to \infty$, the number of particles going to infinity. One might consider each system as a particle, and the ergodic limit then becomes something like an infinite volume limit if the system ("particle") energies are independent of the number of systems (the "volume"), in effect equivalent to keeping the density constant.[6]

In modern presentations, the Darwin-Fowler derivation goes roughly as follows:[7] Consider an ensemble or set of thermodynamics systems, of average energy U, each system characterized by its energy, E_i (i indexing the possible energies), the total number of systems (N_S) and the total energy of all the systems ($N_S U$) fixed. Assign a probability distribution to the ensemble, one that tells how likely a system is to have any particular energy. Now we ask, Given such a probability distribution and such an ensemble, what is the number of distinct ways of sharing the total energy among the systems? Is any one such probability distribution overwhelmingly likely, compared to other such distributions, and so likely to be the one we observe? Are its number of distinct ways much larger than that for all other distributions? One computes the average probability distribution and its variance (which turns out to be very small, and so the average is equivalent to the overwhelmingly likely).

To answer these questions, one computes the combinatorics for each proposed probability distribution, keeping in mind the fixed number of systems and fixed total energy—and then goes to the $N_S \to \infty$ limit. For example, if the systems were otherwise distinguishable, and if each of the systems had the same energy, sharing the total equally, there would be just one way of sharing. If half the systems had half the average energy, and the other half had 1.5 times the average energy (so satisfying the average energy and total energy constraints), there would be $N_S!/\{(N_S/2)!(N_S/2)!\}$ ways, a rather large number (if $N_S = 100$, then the number of ways is $\approx 10^{29}$).

Technically, in order to find the average distribution, one forms a more general function, a generating function,[8] formally: $G(N_S, z) = \sum_{U'} z^{N_S U'} Q(N_S, U')$, where Q, the partition function, is the number of ways of partitioning the energy among the systems for any particular probability distribution, and the average energy, U', is now a variable. (Ideally, an actual generating function could be shown to have a nice closed form, $G = g^{N_S}$, $g(z)$ being nicely computable—for which see condition 4 below.) That generating function allows one to automatically incorporate the total energy assumption if one can pick out the $z^{N_S U}$ term, just the one that corresponds to the constraint that concerns us. Namely, one employs a Cauchy integral (see figure 3.1 for the path of integration) in z to pick out the coefficient, $Q(N_S, U)$, of just this term:

$$Q(N_S, U) = 1/2\pi i \oint G(N_S, z)/z^{N_S U+1}\, dz = 1/2\pi i \oint g^{N_S}/z^{N_S U+1}\, dz.$$

One then shows that the integral may be evaluated by using the method of steepest descents, especially as $N_S \to \infty$. If the saddle point of the integral, where the overwhelming contribution to the integral lies, which saddle point is shown to be on the positive real line, is expressed as $\exp - \beta$ (β turns out to be $1/k_B T$), then the probability distribution we are concerned with is $\exp - \beta E$. The average of the distributions of energies among the systems is in fact just that given by $\exp - \beta E$, and it is the vastly most probable one, and the average energy per degree of freedom is $\approx 1/\beta$.

There are a number of curious features of the derivation as it is nowadays presented, ones that show that apparently formal, technical features—namely, devices that are invoked, it would seem, as devices per se, rather than as physically interesting—are here in fact the physics: a formal use of a complex variable, a use that suggests a physical motive; a formal requirement needed to make the method work, one that goes along with the phenomenology of a temperature as an intensive variable, a variable that characterizes the system as a whole; a formal requirement for a technical argument to go through, one that then turns out to be physically quite reasonable; and specific formal devices, apparently machinery useful for producing a result, that

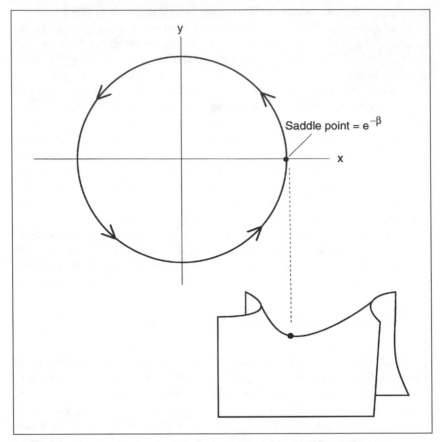

FIGURE 3.1   Cauchy integral contour employed in the Darwin-Fowler derivation. Adapted from Huang, *Statistical Mechanics*, fig. 10.2.

turn out to have physical meaning. Here formalism is physics. (I follow Huang's presentation, as sketched above, supplemented by Schrödinger's deep physical insights. While there will be analogous features in other derivations, I mostly stay with one particular derivation, for only then are these features actually features within a derivation.)

Technically, the following conditions apply to the derivation:

1. The Cauchy integral's variable z (which turns out to be $\exp - \beta$ on the real axis, where the saddle point lies) is of course allowed to be complex. It would seem that allowing for a complex z is just for the purpose of employing the Cauchy integral in the method of steepest descents. But we also know that $1/\beta$ possesses all the qualities of a

temperature, and we might ask, what would a complex temperature mean?

We know that in quantum mechanics, where we end up analogously with terms like $\exp - iHt/\hbar$, rather than $\exp - \beta E$, where $\hbar = h/2\pi$ (10^{-34} joule-seconds) and h is Planck's constant, t is the time, and H is the Hamiltonian function for the energy, a complex energy or eigenvalue of H (that is, its real and imaginary parts are in general each nonzero) would represent unstable and so decaying states—which suggests that a complex β represents nonequilibrium states of the system. For thermodynamic systems, we know that such states are in fact almost always dominated by equilibrium states, and in the end they should not and they do not much figure in the partition function, Q—just what we find from the specific contour integral that succeeds in this formulation (its saddle point is at real β).[9]

2. Again, $\exp - \beta$, β real and positive, is the saddle point in the Cauchy integral. The sharpness of that saddle point depends on N_S, the number of systems considered. Moreover, in the derivation, $\exp - \beta$ turns out to depend only on the average energy, U, and not directly on N_S, at least in the $N_S \to \infty$ limit.[10] Put differently, temperature, which is proportional to the average energy per degree of freedom of a system, here turns out to be a good $N_S \to \infty$ limit object; and if it did not so turn out, we would have neither a good model nor a workable derivation. Here a technical feature that allows one to employ the method of steepest descent, namely, energies, E_i, independent of N_S is also an important physical fact of the bulk matter we are considering: it possesses a temperature as an intensive variable. (Actually, the other crucial technical feature is the existence of a single minimum of the integrand on the positive real line [at $x = \exp - \beta$], that is, the existence of a well-defined single temperature for the medium. This minimum condition would seem to depend on condition 3 below.)

3. In the derivations I am considering, the energies the systems might possess, E_i, are for the technical purposes of the derivations assumed to be integers and have no common divisor (they are rationally incommensurable), and the ground-state energy is assumed to be zero, and hence the energies are nonnegative.[11] This is usually justified physically by saying that the energy unit is arbitrary, the zero is also arbitrary, and in the end that energy unit does not appear in the final result. Schrödinger is rather more careful physically, indicating that a continuous spectrum would imply an unbound state, one unlikely to be thermodynamic system.[12] He also points out that the existence of no common divisor could depend on just one relatively incommensurable state, a reasonable expectation (the alternative being hard to counte-

nance). Recall that the Planck spectrum does not avoid the energy unit $\hbar\omega$, and that commensurability of frequencies or energies was a crucial issue in multiply-periodic Bohr atoms. So these apparently mathematical devices as presented in the Darwin-Fowler derivation are in fact part of the physical world of that time.

Now, formally and mathematically, that incommensurability of the energies allows for the mutual cancellation of their contributions to the contour integral off the real line;[13] the zero of energy allows one to readily show there is but the single minimum in the integrand.[14] So, again, technical features that are employed to make the mathematical derivations work are, it would seem, physically reasonable facts, as we might hope.

In our derivation of the canonical ensemble, the identification of β with the inverse temperature ($\beta = 1/k_B T$) is typically done by considering the probability distribution of two ensembles of systems that are then connected thermally, that connection being defined by the fact that one considers their total energy and total numbers of systems as the crucial variables, and one then finds that the systems have a shared β (whatever different βs they had beforehand).[15] Along the way one may invoke all the above "technical" assumptions about energy levels—the derivation is of the same sort. Moreover, β is independent of the number of systems when that number is large. β^{-1} has the right behavior for a temperature.

4. Finally, and most formally, the Darwin-Fowler method would appear to be merely a way of solving a combinatorial problem, a partition problem, counting the number of ways we might distribute balls into boxes. (The Onsager solution to the Ising model also solves a combinatorial problem; see chapter 4.)[16] The generating function above, G, can be rewritten so as to automatically take into account the combinatorics: $G = g^{N_S}$, where $g = \Sigma g_i z^{E_i}$ (and so g itself is something like another generating function, for the $\{g_i\}$), and G is then a multinomial expansion.

Each factor of g (in G) represents choosing a box for one ball, an energy for a system. Note that, formally, g turns out to be a partition function for just one independent system (since $z = \exp - \beta$), and hence the partition function for N_S independent systems should just be the product of their individual partition functions, which it is. The weights, g_i, are, it would seem (as they are introduced into the derivation), formal devices, deus ex machina, to be set equal to one at the end. This is not an unusual mathematical move; Euler's ζ, which we shall encounter in a moment, was another such weight to be set equal to one at the end. But, as Schrödinger points out, these weights are in effect degeneracies, and he uses this insight to estimate their relationship to the mean occupation numbers of the relevant states.[17]

In their original 1922 paper, Darwin and Fowler have the weights play an integral natural and physical role as degeneracies, from their initial example onward. These are not to be seen as mathematical devices at all. More generally, Darwin and Fowler show through their examples why one might end up with multinomial expansions, and why one might want to pick out a coefficient in such an expansion. And they show, through an interesting derivation of the density of states from the partition function, another way one might encounter a complex temperature.[18] What we observe in modern accounts was to some extent built into the original.

By now, I have presented a number of my main technical points, ones that will recur often. (1) Partition functions are generating functions for the number of states for a particular number of particles, n—the coefficient of, say, z^n. (2) More generally, such functions appear in combinatorial analysis, say in counting the number of ways of partitioning an integer into the sum of smaller integers. (3) Following Bernoulli, Euler (1748) discovered the form of such a partition function, $\Pi_{n=1,\ldots,\infty} 1/(1 - q^n \zeta)$, the $q^{v} \zeta^{w}$ terms in the polynomial expansion enumerating a particular partitioning. He treated q as a parameter, studied the partition function as a function of ζ, and then set $\zeta = 1$ in the end. (Note, also, the similarity of this form to the partition function for an ensemble of Planck oscillators, $\Pi_j 1/(1 - \exp(-\beta \hbar j \omega))$, where $q \approx \exp - \beta \hbar \omega$.) (4) Jacobi (1829) showed how to transform the partition function into what are now called theta functions (a kind of elliptic function), functions which in part arose in solving the heat equation by Fourier analysis in ζ for the temperature, $\theta (\approx \Sigma \zeta^n q^{n^2})$, within an infinite slab.[19] Hence the connection of elliptic functions and number theory. (Technically, Jacobi's scheme was to let $q = \exp i\pi\tau$ and $\zeta = \exp 2iz$, so that the partition function becomes $\Pi_{n=1,\ldots,\infty} 1/(1 - \exp i\pi\tau n + 2iz)$. And then, for example,

$$\theta(z, \tau) = 1/\sqrt{(-i\tau)} \exp z^2/i\pi\tau \, \theta(z/\tau, -1/\tau),$$

the first multiplicative factor being the distinctive character ["modular"] of the transformation of these functions. In the Ising model, a phase transition takes place when $q = 1$ or $\tau = 0$.) (5) Given the generating function, now being taken as a partition function, we might wonder if its coefficients are unique. So Poincaré (1912) showed that for the blackbody or Planck spectrum of electromagnetic radiation (in effect, the partition function), the density of states (or coefficients of the partition function) corresponding to energy jumps, Planck's hypothesis of quanta, is unique or necessary. Actually, Fowler (1921) corrected Poincaré, and his skills in complex variable theory demonstrated there served him well in the Darwin-Fowler derivation a year later.[20]

The Physics
The devices and conditions necessary for this version of the Darwin-Fowler derivation—complex β, E_i unaffected by N_S, rationally incommensurable energies, and the weights g_i—are in fact physically meaningful, and their meanings are reasonable ones. Still, it is not unusual to hear a remark like

> A thorough and rigorous development of statistical mechanics in a form general enough to apply to all systems of physical interest is a difficult job which is outside the scope of this book. Fortunately, we avoid some of the mathematical difficulties by restricting our attention to systems with discrete energies only. Furthermore, we will make the additional technical restriction that the discrete energy levels are commensurable. . . . It is inconceivable that any quantity of physical interest can depend on the existence of ΔE [the energy unit] and, indeed, ΔE will not appear in the final form of statistical mechanics which we shall derive.
>
> Our derivation of statistical mechanics is not going to be logically impeccable.[21]

What might we mean by rigor? In this context, rigor is perhaps about fewer assumptions, more generic and careful mathematics, or new physics. Still, in fact, the devices as they are employed in any particular derivation need not be treated simply as artifactual devices, to be thrown away once one has solved the problem or possesses a more rigorous derivation. As we shall note again and again, they can be seen to say something about the physical world, and their physical implications may well have influenced the technical mathematical strategy employed. A new, more rigorous derivation that does not depend on these particular devices and conditions would also demand some other conditions. And I would expect that its devices and conditions would also point to the physics of the system (perhaps pointing to different features).

It should be possible to show from an even more sophisticated point of view, the one implicitly pointed to in the above quote, that the Darwin-Fowler derivation is a curiosity, the concerns about discreteness and commensurability being artifacts of methods and device. But, in fact, this even more sophisticated account is likely to have its own devices, devices that also can be shown to have physical content.[22] So I believe that what one obtains from greater rigor (say, a more precise mathematical formulation) is another appreciation of what is physical about all the formalism. Put differently, I am quite willing to believe that the points I make above might be consumed by a more rigorous account (say, one that avoids the delicate issues of infinite volume limits altogether).[23] But there will be other points, equally physical, that can then be made.[24] (What is interesting is to see how the physics implicit in the desiderata in one kind of proof is accounted for by the explicit physics in another kind.)[25] Making a derivation more obvious by using more sophisticated mathematical tools does not avoid the physics.

Moreover, subsequent rederivations, even if more rigorous, will in the end still contend with the fundamental physical symmetries and structures of a system that warranted or allowed for the original technical devices—perhaps more and more directly, and this is their real strength. Here I am thinking of Onsager's 1939 paper on the stability of Coulomb matter and his 1944 paper on the Ising model (reprinted in the appendix). No subsequent derivation avoids the devices Onsager needed (either for the stability of matter or for the two-dimensional Ising model). Or, better put, just about all subsequent papers employ devices motivated by the same physics that motivated Onsager—or so I shall argue.

I have more than once referred to "the physics." To be well trained as a physicist is to know what the physics of a situation is, and it is a sign of being well trained that one "thinks like a physicist" and can point to the physics. As a first approximation, and putting it abstractly, the physics of a situation refers to the actual mechanism or interaction process of the particles, fields, and matter, as well as their energies, potentials, and charges, rather than to any abstract formalism, such as a variational principle, in which those physical objects are not seen to interact directly, powerful though that formalism be for working out a problem. (Of course, to some physicists variational principles are the physics. But they too will find some kinds of explanation more mathematical than physical.) And the small, shared, and conventional and repetitively taught and used set of those mechanisms make up part of what I call the physicist's toolkit.

It is one task of the physicist to interpret such a formalism so that its physics is manifest—for the physicist as well as for the community of scientists. This may demand smaller, more tractable, toy models, or electrical, mechanical, or field-theoretic analogies—such as the electrostatic analogy employed by Yang and Lee in their explanation of the consequences of the unit-circle theorem (see chapter 5), or Feynman's models of path integrals leading to classical paths—that draw upon the common experiences of physicists as they worked out a problem or in their original professional training. Of course, such an interpretation may then allow physicists to see the physics in rather more abstract formalisms.

The Possibility of a Temperature

In general, what appears as a technical feature may have enormous physical significance. For example, within the Darwin-Fowler derivation, it would seem that uniqueness of the saddle point in the integral and its sharpness make possible an intensive variable that characterizes the system, a temperature. Technically, the in-effect statistical independence of the addends in that Cauchy integral off the real line (that is, their contributions cancel out, just as the length of a sum of randomly oriented vectors goes as $\sqrt{N}$ rather than as N, as in the central limit theorem)—means that the only contribution to the integral comes from points near the real line, and that fact allows one to pull out a crucial factor and so identify the temperature variable, $\exp - \beta$.

Let us consider some somewhat hypothetical alternative technical features. (Before going further, I want to note that the argument that follows is quite speculative, and I am not sure it could be made to hold up internally—but I think it is suggestive.) Imagine that there were two minima along the real line that could be employed, separately, in the saddle-point integration in the Darwin-Fowler derivation (at least for finite N_S). There would be, in effect, two "temperatures" we might assign to a system, $x = \exp - \beta_1$ and $\exp - \beta_2$—for it would seem that the saddle-point integration would go through in each case. It would look as if, depending on physically arbitrary initial conditions, one might assign either temperature to the system.[26]

Now, one can show that the integrand has only one minimum along the positive real line by examining it formally and algebraically, as I indicated earlier.[27] Physically, it would seem that what is crucial is the nonnegativity of the variance of the energy, what we might take as a sign that the energy is a real number and so we are dealing with equilibrium states.[28]

In other words, that the integral is nicely defined in the sense of there being one very sharp $N_S \to \infty$ minimum of the integrand along the real line is a formal requirement that represents the fact that through measurement we do assign a well-defined temperature to bulk matter. More generally, in other models, convergence to a limit is a sign that we might attribute other intensive properties (pressure, compressibility) to bulk matter.

Moreover, it would seem that the steepest descent derivation works because of $1/\sqrt{N_S}$ falloff of second-order terms.[29] By "works," I mean two different features. The first is that, as a consequence of the falloff, one might identify separate extensive and intensive variables, ones that we now call free energy and temperature, and that we might speak of the free energy per system. Namely, in the Darwin-Fowler derivation, because of this falloff there is a crucial moment when one can pull out or separate out a N_S factor (so we can divide it out), N_S times a function of $\exp - \beta$, the number of systems times a function of the temperature.[30] Of course, this is also the move we need for there to be a temperature.

Second, the falloff by $1/\sqrt{N_S}$ might well remind us of the central limit theorem, in which a sum of random variables is sharpened as $\approx 1/\sqrt{N_S}$. Let me suggest why, albeit speculatively: Recall that the sharpness of the saddle point in the $N_S \to \infty$ limit in the Darwin-Fowler derivation is a matter of the incommensurability of the energies, and so the phase contributions of z^{E_i} cancel out except at the point where the contour hits the positive real line, at $x = \exp - \beta$. (Again, this is reminiscent of Feynman's argument showing how the classical limit appears in quantum mechanics, as a sum of randomly canceling amplitudes, except on the classical path.)[31] In that derivation, the incommensurability of the energies is effectively equivalent to the averaging that occurs in the central limit theorem among the added-up random variables.

Recall that for there to be extensive properties there must exist asymptotic limits of the sort $\lim_{N \to \infty, N/V \text{constant}} N^{-1}E(N)$. As we shall see, physically, this would seem to require that there be microscopically large but still comparatively small macroscopic bits of bulk matter (those bulkettes) weakly coupled to each other, so that they interact through their surfaces alone (the energy of interaction is roughly a surface energy)—there is no volume-volume interaction. The weak or dilute interaction among adjacent bulkettes allows an exchange of energy and material at the surfaces of the bulkettes, just what characterizes diffusive flows that tend to equilibrate those presumably intensive properties among the bulkettes.[32] Yet, in such a weak interaction, the properties of each bulkette are statistically independent of the properties of adjacent (and distant) bulkettes—at least at equilibrium for that bulk matter (yet perhaps not at phase transition).

Now, say we measured some property of the bulk matter, such as temperature or pressure. In effect, we are averaging over the property for many bulkettes, and such an average for very many independent objects is rather sharply peaked (again, by the central limit theorem), as long as the variance of the probability distribution for any single bulkette is finite—which is just what we expect, except at phase transitions when there is correlation and statistical dependence among the bulkettes.

Recapitulation

In sum, that there is a temperature attributable to bulk matter—and more generally a separation of extensive and intensive properties—would seem to depend on two technical features, features physically characteristic of equilibrium: a single $N_S \to \infty$ limiting value for a parameter, and statistical independence of otherwise identical bulkettes from each other. As we shall see, these features play a central role in technical definition and proof of the thermodynamic limit for bulk matter.

More generally, an enormous amount of physics may be seen to inhere in what appear to be technical features. Physicists who have worked on a particular problem are often quite aware of how much physics adheres to the devices and techniques they employ for solving a problem.[33] None of what I say here would be surprising to them, at all. It is usually part of an oral tradition, conveyed in lectures, at the blackboard, or in asides. The conventions of textbooks and scholarly papers, however, often do not permit the author to display these physical connections on the way to providing a demonstration or a proof or a discovery. Thomas Kuhn's *Structure of Scientific Revolutions* has drawn very deep inferences about the processes of science, such as how textbooks canonize a way of thinking and practicing, from just this point.[34]

It may be useful to recall what we have been doing. We have seen how physicists provide a technical account of ordinary bulk matter composed of atoms, bound together by Coulomb forces. Such an account must show that atoms are stable, and that we might attribute an intensive variable such as temperature to such mater (technically, in the limit of an infinite number of similar systems). That limit must be expressed formally and technically in these mathematical accounts, and in its being so expressed, we discover some of what is essential about bulk matter.

In other words, in giving a technical account of everyday ordinary phenomena, we find ourselves coming to a different sense of what is obvious and ordinary and crucial about it. Moreover, the technical details in the proofs and accounts, details that might seem like mathematical and formal devices and artificial convenient assumptions (or *Ansätze*), can be shown to reflect deep and necessary features of the physics of a situation. A formalism set up to model some particular features of a situation often appears to have other deep facts built into it, facts that are made obvious only in the technical working out of the formalism.

THE THERMODYNAMIC LIMIT

The existence of a temperature for bulk matter, a measure of the average energy per degree of freedom, is a statement as well that bulk matter might possess what we might call pervasive properties, those size-independent intensive variables. The next task is to demonstrate how and when bulk matter also possesses properties linearly dependent on N, extensive variables. Given that the hydrogen atom, the multielectron atom, and matter composed of lots of atoms are each stable to collapse, matter composed of lots of atoms being bulky (having a size that goes at least as $N^{1/3}$), we need to ask whether some properties of bulk matter can scale with N (rather than being just bounded by N)—what is called the thermodynamic limit—and whether such matter is mechanically stable (so that when it is compressed, its density becomes greater).[35]

We shall first consider an archetype of such proofs, one that does not, on its face, apply to the Coulomb interaction between charged particles. The problem is to divide up bulk matter in such a way that it is seen to be composed of more-or-less independent bulkettes, and so some of its properties will scale with N. Put differently, the task is to build up matter from those bulkettes, combining them to form larger units, ad infinitum. We then consider how that archetype is adjusted so that the Coulomb force, with its infinite range, will still provide for extensive properties ($\approx N$) rather than properties dependent on N^2 (just the number of two-body interparticle interactions, were forces of very long range). In such an account, we find that we are saying something about how we might think of the constitution of matter, in terms of boxes (or balls) geometrically increasing in size and consisting of similar components, and as we get to larger and larger sizes the effect of N on the thermodynamic free energy becomes more and more linear. As we shall see in the next chapter, other accounts of crucial properties of matter suggest rather different constitutions, each of the constitutions to be taken as an aspect of bulk matter, a way we encounter and think about it.

Recall that, through a series of detailed model-dependent calculations, H. A. Kramers (about 1934–1936) came to realize that thermodynamics is a limit science, that thermodynamic properties (intensive variables, such as temperature and pressure, and extensive ones, such as entropy or free energy), and distinctive thermodynamic phenomena such as phase transitions, appear only in the infinite volume limit.[36] While, following Planck and J. W. Gibbs (1902), we might compute the partition function, Q (what Planck called "the sum over states," in German *Zustandsumme*, hence the conventional use of Z for the partition function), a measure of the number of ways of partitioning the energy of the system among its components, what we actually measure and is well defined is the free energy density, f: $\lim_{N \to \infty, N/V \text{ constant}} - 1/\beta N \ln Q$ (namely, $\exp - \beta N f = (\exp - \beta f)^{n_B N_B} = Q$, where n_B is the number of atoms in an independent bulkette, and N_B is the number of bulkettes), an infinite volume limit object. (We have seen a similar situation in the Darwin-Fowler derivation in which the temperature $[1/\beta]$ may be in effect "pulled out" as a good intensive variable, when the number of systems considered goes to infinity, the $N_S \to \infty$ limit. Here we might consider each of our bulkettes as a system and so identify N_S with N_B.)

But, having shown that matter is stable to collapse, just which sort of limit are we to take when we are concerned with extensive properties? What is the technical/mathematical construction of "the constitution of matter" (as Lieb and Lebowitz call it) that would enable us to demonstrate the existence of such limits? Put differently, just what is the meaning of a "linear law," as they refer to the thermodynamic limit? (Here I shall be following Lieb and Lebowitz's argument, an epitome and ingenious development of previous work.)[37]

As I have indicated, for these purposes matter's constitution might be archetypally modeled by a hierarchy of bulkettes, perhaps cubes, say each twice the dimensions of its predecessor—at each scale filling all of space; or space is built up out of larger and larger cubes—and so we might go to infinite volume.[38] (See figure 3.2.) Now, no bulkette has more than a limited number of atoms within it. The atoms are in effect impenetrable hard spheres. For recall that we already know, from a proof, that bulk matter is stable, having a lower bound of its energy proportional to N, and bulky, its linear dimensions going as $N^{1/3}$. The atoms' nonzero, finite size (radius a, $\infty > a > 0$), as we shall see, provides for the uniform convergence to the infinite volume limit of the series of approximations for the energy density or for the free energy density;[39] now each component bulkette (or subbulkette) is interacting with its neighbors at best through surface interactions, whose contribution to the energy density or free energy density proves increasingly negligible, so that the series converges, and so that there are extensive properties ($\approx N$) and in general no $N^{1+\epsilon}$, $\epsilon > 0$, properties, which might eventually lead to explosion.

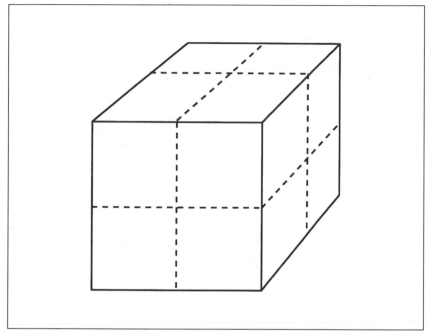

FIGURE 3.2   Hierarchical cubic decomposition of space.

Van Hove, and Yang and Lee, in their respective demonstrations of
the existence of a well-defined pressure in ordinary matter (that is, the
equivalence of the equation of state of matter in the grand canonical
ensemble to that in the canonical ensemble—that one might measure a
property of matter, such as pressure, in situ, rather than by isolating the
matter ahead of time), and of the impossibility of phase transitions
except in the infinite volume limit (if at all), used the apparatus of a
decomposition into bulkettes, composed of presumably stable atoms
(equal to hard spheres), and interacting through finite-range forces.[40]
Again, for the Coulomb interaction ($\approx 1/r$), the finite-range assumption is not at all obvious.

I should note that there are at least two kinds of decomposition in
these proofs.[41] The first allows the size of the component cubic bulkettes
to grow more slowly than does the volume, so that as one goes to the
infinite volume limit the number of components goes to infinity, but
more slowly. The second is a geometric series of nested cubes, each
twice the size of its predecessor, and therefore eight smaller cubes fit
into the next larger cube. As we shall see, when we use balls or spheres
as our bulkettes, we have a packing problem and we fill in the inter-

stices (and the interstices of interstices) with increasingly smaller balls (down to a minimum-sized one).

Again, the thermodynamic limit is a large-distance or large-r problem, unlike the stability to collapse I discussed earlier, which concerns the small-r possibilities. Put simply, if the range of an attractive interaction between particles is very large or "infinite," then we might expect that if we put pieces of matter together there might well be an explosion, the energy per particle rising so rapidly that, rather than a diffusive loss of excess energy through steady state heat conduction, there is shock wave and mechanical explosion. (If the interaction is repulsive, then it would be impossible to produce bulk matter in the first place. And if there were something like critical-mass effects as in radioactive matter, then adding on more matter allows even more matter for mutual interaction, nonlinearly increasing the energy of interaction.) With long-range interactions, bulkettes could not be independent of each other, and putting two in close proximity would produce a bulkette that is qualitatively different than its components. Practically, we could not hope that fluctuations in density would have little effect on thermodynamic properties in equilibrium (rather than, say, leading to explosion, which we surely do worry about for subcritical nuclear assemblies).

We shall want small but still sufficiently large bulkettes to be more or less separated and independent of each other, so that the energy per particle of bulk matter so composed is roughly constant. As we might infer from the Darwin-Fowler derivation, the statistical independence of the bulkettes should allow for sharply defined values of intensive variables and for the volume properties of the bulkettes just adding up; the fact that there is a unique temperature (or saddle point in the Darwin-Fowler derivation) suggests that the convergence to the infinite volume limit would be robust, namely, a uniform convergence, comparatively insensitive to details of the particular actual system or mode of convergence.

Can we so model bulk matter? Each of the proofs I shall discuss depends on finding a canonical way of decomposing matter into bulkettes, that way working out because atoms are of finite size, and so a limited number fit into a finite-sized bulkette—and hence there cannot be, along the way to the infinite volume limit, some bunching up of all the atoms within just one bulkette, or a few. And forces are of finite range, so that bulkette-bulkette interactions are surface- rather than volume-dependent. Such matter will have those extensive properties, its component atoms will not collapse, and the bulk will not explode.

As we approach the thermodynamic limit, it may be useful to recall that proofs of the stability of matter itself, that there is a linear lower bound of the energy, are quantum mechanical, with no need to resort to statistical mechanics as such. (The Thomas-Fermi model employs an average charge density, $n(r)$, but does not use statistical mechanics. There is no partition function. Thomas-Fermi theory is just a quantum mechanical theory.) As we might expect from its name, the thermodynamic limit is in fact statistical, and the partition function is the object of interest; that there is a linear limit for the energy is a statistical mechanical statement. The fluctuations here are not quantum mechanical, but refer to thermal fluctuations in particle number, for example.

The proofs that the energy and that the free energy possess thermodynamic limits are quite similar. In what follows, the free energy plays the central role. In each case, I show a linear N-dependent lower bound (I have already done so for the energy), and then an effectively decreasing sequence of estimates—and so prove the existence of a limit. Note, however, that interactions among the bulkettes contribute differently to the energy (they just add in) than they do to the partition function, $\Sigma \exp - \beta E$ (where they affect the combinatorics as well). In sum, to model and analyze the constitution of matter is to prove that a partition function (and so a free energy density) computed from that model possesses the desirable properties in the thermodynamic limit.

Proving a Thermodynamic Limit: The Free Energy[42]
Technically, one way of saying that matter is composed of bulkettes that are more or less independent is to say that the grand partition function, *GPF*—the partition function of systems when not only the energy but also the number of particles may vary—of some bulk matter (volume V) might be decomposed into the grand partition function of bulkettes of volume V/γ: $GPF(z,V) = \Sigma z^N Q_N(V) = [GPF(z,V/\gamma)]^\gamma$,[43] where the surface contributions to the grand partition function are comparatively small. And, in the infinite volume limit (V to infinity, V/γ constant and suitably large, so that each bulkette has lots of particles— noting that this kind of definition is just what is crucial and an invention), the above relationship holds with no surprises along the way.[44]

The proofs of the theorems of van Hove and of Yang and Lee, concerning the behavior of the partition function and of the grand partition function as the volume goes to infinity, along the way implement these features by several technical moves:[45] If the range of the interparticle forces is fixed and finite (and so volume independent), r_0, then each bulkette can (more or less) only interact with another bulkette along its surface. One sets up r_0-deep "corridors" between the bulkettes, let us say. (See figure 3.3. That one can do this means that there cannot be an explosion, since the interaction among bulkettes goes as $N^{2/3}$ rather than as $N^{1+\epsilon}$, $\epsilon > 0$.) As long as surface to volume ratios are of the usual sort, the contributions of these corridor interactions to the partition function are sufficiently small, in the sense that their contribution to the free energy density or to the grand canonical pressure ($P = (1/\beta V)\ln GPF$), each a logarithm of a partition function then divided by the volume, is negligible. The construction to prove this is by now conventional, but is a crucial invention—namely, how we might break up space into weakly interacting bulkettes: a geometric decomposition into smaller cubes, either an infinite number of such cubes or an infinite nesting of cubes.[46] Given this construction, and a finite range, then the contributions to the partition function (and so to the free energy density or to the grand canonical pressure) from the states in which corridor interactions are allowed stays sufficiently small and is bounded by a function that depends on r_0 alone.

Moreover, one wants to be sure nothing "interesting" happens on the way to infinite volume, that the limits converge nicely enough, hopefully uniformly. What might happen interestingly? There could be some weak $N^{1+\epsilon}$ effect, which when N is large enough starts becoming important. But I have precluded that by the finite-range assumption. Or perhaps all the particles might literally conglomerate into one bulkette, leaving all other bulkettes absent of particles. Now, we might just note that actual matter is bulky, and moreover it does not implode, and so this bunching up is impossible. What turns out to be crucial in the actual proofs, however, is that the particles are presumed to be hard spheres, of nonzero radius, a (namely, they are stable energetically); therefore, only a limited number of them could be packed into any bulkette.

The polynomial in z that is the grand partition function for a bulkette—$GPF = \Sigma z^n Q_n$—has a maximum degree, the maximum number of particles that could fit into a bulkette. The grand partition function for a bulkette or for a finite number of bulkettes is therefore bounded on a compact set for finite z (called the activity or fugacity). (Note that the activity is a measure of the [Gibbs free] energy needed, μ, to add one more particle to the system: $z = \exp \beta\mu$.) This boundedness turns out to be just what is needed to be sure of uniform convergence (namely, bounded derivatives) and so continuity.[47]

In sum, if r_0 were not finite, then interactions among bulkettes would not be bounded. If a were zero, then a single bulkette could have a very large fraction of the particles, and then fluctuation and macroscopic graininess would dominate the bulkiness of matter.

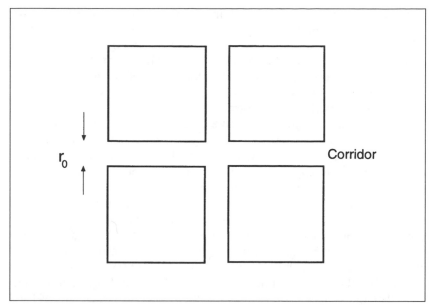

FIGURE 3.3   Corridors between cubes.

As a model for a proof of the thermodynamic limit for Coulomb forces, I have pointed to technical features of some of the proofs of the thermodynamic limit for finite-range forces: the capacity to bound the corridor contributions in a volume-independent way, and the boundedness of the partition function—which are crucial formally and mathematically. And then I suggest that those features reflect deep physical facts about matter, their absence implying a state of matter very different than the one we see everyday.

Still, we do not yet have the wherewithal to provide for a thermodynamic limit. Given the infinite range of the Coulomb force, how are we to apply the technology we have developed for finite-range forces?

Screening
For Coulomb forces, the thermodynamic limit is quite peculiar.[48] For example, were electrons positively charged, there would be no question about the stability of bulk matter: the excess charge would go to the surface, the minimum electrostatic energy would go as $N^2/N^{1/3}$ ($\approx e^2/r$), namely $N^{5/3}$, and there would be no thermodynamic limit: the energy per particle is not at all constant. Moreover, in general, only if the potential falls off as $r^{-(3+\epsilon)}$ would there be such a limit—except here ($1/r$, Coulomb) we are saved by the opposite signs of the charges of the electrons and the protons: "different parts of the system far from

each other must be approximately independent, despite the long range nature of the Coulomb force. The fundamental physical, or rather electrostatic, fact that underlies this is *screening* [in effect, there is an r_0 that is finite]; the distribution of the particles must be sufficiently neutral and isotropic locally so that according to Newton's theorem the electric potential far away will be zero. The problem is to express this idea in precise mathematical form."[49]

While "everyone knows" that the Coulomb force is shielded in bulk neutral matter (no net force at a distance, since in net neutral charged matter the distant effects of the positive and the negative charges cancel out), it is not at all apparent how what everyone knows might be used to show that such matter possesses a thermodynamic limit. The crucial insight is that the external effect of a sphere or a ball, isotropically filled with charge, is equivalent to having a point charge at the center ("Newton's [or Gauss's] theorem"). And if that net charge were (almost) zero, there would be (almost) no external force—in effect a finite r_0.[50] The task here is to pack space with such balls, and (as in van Hove and in Yang and Lee) the invention that is crucial is a mode of such packing —so that the spherical bulkettes or balls, at various scales, are of fixed uniform density, something that is automatic when we use cubes but that is not so obvious for spheres, since the leftover unpacked spaces are nontrivial. The development of a workable geometric construction —one that insures that the density of matter is constant, the infinite volume limit in practice—is as crucial here as was the earlier partitioning into cubes.[51] Since any but the smallest ball is filled with smaller balls, the partitioning looks something like figure 3.4. Again, such a partitioning is not only a mathematical device; it also shows how we might think of the constitution of matter, the neutral balls reminding us poignantly about the significance of screening in this realm.

Such a packing of space with neutral balls is not quite a claim that matter is actually composed of neutral balls, that matter is so constituted. It is a very useful assumption, one that is in fact not quite the case. But it is an assumption that can be relaxed enough so that the proof is quite believable as a model of actual matter. (One proves the limit for neutral balls and then shows that any surface charges due to nonneutrality that vanish in the infinite volume limit do not matter.) Namely, for each of the component balls, as long as the excess charge on each "neutral" ball is sufficiently small (its surface-charge density goes to zero) at each scale, and the net total charge on all balls is zero (macroscopic neutrality), all is OK. Hence, the dimensions of the characteristic charge fluctuations in actual matter have to be small compared to the size of the smallest ball considered. It can also be shown that these nonneutral configurations do not contribute much to

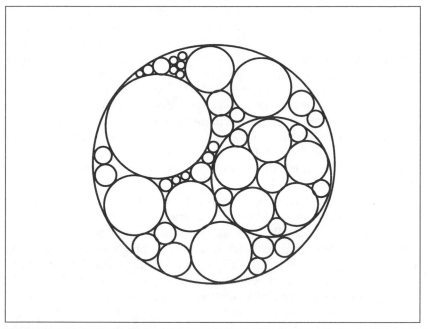

FIGURE 3.4   Filling space with balls.

the grand canonical pressure, the nonneutral parts being much like the corridor contributions in other constructions (as long as the surface-charge density goes to zero as the volume goes to infinity). It is a *façon de parler* to say that matter is constituted of neutral balls, justified by these physical facts.

The neutral-ball construction leads to a suitably convergent series of approximations to the partition function, and so it provides a constitution of matter. More precisely, first, it leads to a recursive sequence of estimates for the free energy density:[52] the free energy density of a ball at size l, f_l, is less than or equal to the free energy density computed by packing the neutral l-sized ball with smaller neutral balls (each of whose free energy density has also been bounded from above by its component balls' free energies), and then summing their contribution to the free energy: namely, $f_l \le \Sigma_{i=0,\ldots,l-1} f_i v_i$ (where v_i is the fraction of volume of the l-sized ball that is covered by the i-sized balls). Now, let us allow for almost neutral balls.

Second, in order to show that the residual Coulomb interaction energy of these almost neutral balls is not out of bounds and possesses a limit, one needs to pack space with l-sized neutral balls (each packed with $(l-1)$–sized and smaller almost neutral balls). What makes the proof go through is that, in computing a bound on the partition function for finite N and V, a sum or average for all configurations, one need consider only the average interaction among all the bulkettes, that is, "the *average* interdomain Coulomb energy in a canonical ensemble in which the Coulomb interaction is present in each subdomain but the. . . domains are independent of each other."[53] And it turns out that, for all overall neutral volumes, that average interdomain energy is zero, by the Newton-Gauss law. Even if fluctuations do give various balls small amounts of nonneutrality, that does not matter, since at each level one is demanding overall neutrality.[54] And when one goes to the next larger size ball, one can relax the neutrality assumption for the now smaller ball.

Crucially, the bulkettes must be statistically independent of each other, again what we would expect of bulk matter that is stable and in equilibrium.[55] Technically, that independence might be understood as an averaging that leads to $1/\sqrt{N}$ effects, damping charge fluctuations. More formally, $Q \geq (\exp - \beta U)\Pi_j Q_j$: the partition function for a finite system is greater than or equal to just $\exp - \beta$(Average Interaction Energy, U), representing the interactions among the bulkettes, times the product of the partition functions of each of the bulkettes that compose it.[56] (Note that, since the average energy is zero, the exponential equals one.) This inequality then defines a sequence of in-effect decreasing estimates for the free energy density as the volume gets larger ($f \leq \Sigma f_j$), and, given the stability of bulk matter (which leads to a bound on the partition function and so to a lower bound on the free energy, for which see the section below, "Limits Are Physical"), we are led to the proof of the existence of an infinite volume limit, the thermodynamic limit we are seeking.[57]

Now, say the actual matter itself were not electrically neutral. Would it still exhibit a thermodynamic limit? As I have suggested already, if the excess charge is small, it should have little effect. Moreover, the long-range character of the Coulomb force demands that, if there is to be a conventional thermodynamic limit, then the system must be effectively neutral—otherwise shape and size effects become important.

Technically, taking into account the amount of excess charge (over neutrality) in taking the thermodynamic limit: if the surface charge density goes to zero, then the limit is the same as for the neutral system; if the surface charge density goes to infinity, then there is no limit (as one might expect); and if it goes to a finite value, to the neutral free energy is added an electrostatic surface energy.[58]

In sum, the thermodynamic limit separates out N-dependence through the use of independent bulkettes. The proofs all depend on the hard-core or nonzero size of the atoms and the finite (that is, noninfinite) range of the interparticle forces. And screening does the finite ranging work for the Coulomb force, once we pack space with effectively neutral balls.

THERMODYNAMIC STABILITY AND MATHEMATICAL PHYSICS AS EMPIRICAL

These mathematical proofs are in effect philosophical analyses of thermodynamics, showing in various contexts the meaning of the infinite volume limit through the demands of technical constructions of the constitution of matter. The technical and mathematical features of the proofs depend upon three contingent physical facts, contingent in the sense that they need not be the case, but surely are the case in our everyday world: (1) The energy is bounded from below, namely, matter is stable, whether that is presumed by some hard-sphere repulsion or derived from the Pauli exclusion principle. The free energy, too, is bounded from below. (2) Bulkette-bulkette interaction terms are small; namely, they are surface terms, their contribution to the energy or the free energy densities being negligible. This smallness depends on the effective finiteness of the range of the interaction. And so space might be usefully partitioned into independent bulkettes—the remaining task being to find a partitioning that in some sense converges, that nicely reaches the infinite volume limit. (3) Hence, nothing interesting will happen, such as all the particles going to one bulkette (granularity), when one allows for larger- (or smaller-) scale bulkettes.[59] The latter fact is, essentially, the uniformity of convergence. Technically, at each stage in the approximation, on the way to the infinite volume limit, one can show that the free energy per unit volume, f_l, l being the scale, converges nicely enough for there to be uniform convergence: $|f_{2l} - f_l| < c/l$.[60]

The convergence to a limit (independent of the mode of sequencing or the shape of the volumes) can be shown to have rather deep and generic physical consequences: matter becomes denser when compressed, and its temperature is raised when heat is added to it (it is "thermodynamically stable"); and in general (except perhaps at phase transition), measurements of properties as a function of density or temperature for large samples will be a good approximation to the infinite volume limit.

For, even if there is a lower bound on the energy, so that matter is energetically stable, and even if there is a thermodynamic limit (asymptotic $1/N$ limit), we have yet to show there is such thermodynamic stability: that the system will not exhibit a negative compressibility, Δdensity$/\Delta$pressure < 0, so that pressing upon it would decrease its density; nor will it possess a negative specific heat, Δenergy/Δtemperature < 0, becoming hotter as it emits energy.[61] As W. Thirring puts it: "Systems of negative specific heat cannot coexist with other systems. They heat up [temperature going as N, say] and give off energy until they reach a state of positive specific heat [just what happens for gravitational systems in collapse]."[62]

Technically, it was Gibbs (1873–1878) who realized the convexity and concavity implications of entropy maximization. If the entropy, S_i, of two formerly isolated equilibrium systems stays the same or rises when they are put in contact and come to equilibrium, then

$$S_{12}(U_{12}) \geq S_1(\alpha U_1) + S_2((1 - \alpha)U_2),$$

where U is the internal energy, and $U_{12} = \alpha U_1 + (1 - \alpha)U_2$ (since U is conserved and extensive). But $S(\lambda U) = \lambda S(U)$, by the extensive character of the entropy. Hence, we have

$$S(\alpha U_1 + (1 - \alpha)U_2)) \geq \alpha S_1(U_1) + (1 - \alpha)S_2(U_2):$$

the definition of a concave function.[63] (Here, think of an upturned parabola. The value of the curve at a point intermediate between two points is always larger than the linearly weighted sum of their values. The straight line connecting the two points is always below the curve.)

A nonnegative specific heat can be shown to be a matter of the concavity of βf as a function of β for finite systems (where f is the free energy per particle), and the fact that a convergent series of concave functions has a concave limit as well. Similarly, a nonnegative compressibility is a convexity property, and with some work (to take into account the discreteness of N and the peculiarities of packing with balls) one shows that the limiting function is also convex.[64]

In each case, the free energies for finite systems do have the appropriate stability properties, and the fact that the free energies converge to a limit, that they possess an infinite volume limit, is crucial to the free energy's possessing those same properties in the infinite volume limit. And so bulk matter, an infinite volume limit object, might be modeled by this scheme.

Moreover, given these stability properties—as embodied in the concavity and convexity properties—"*the thermodynamic limit is uniform* on bounded (ρ [density], β) sets. This uniformity is sometimes overlooked as a basic desideratum of the thermodynamic limit. Without it one would have to fix ρ and β *precisely* in taking the limit—an impossible task experimentally. With it, it is sufficient to have merely an increasing sequence of systems such that $\rho_j \to \rho$, and $\beta_j \to \beta$."[65]

The thermodynamic properties we measure are in general not delicately dependent at all on the usual spatial fluctuations in temperature and density, upon which microscopic part of the system we measure them within, or upon the size of the macroscopic sample. In modeling this empirical fact, delicate limit properties must be fulfilled within the mathematical structure of the theory.[66]

In sum, these technical features of concavity and convexity, a consequence of the existence of a limit as well as of features of the finite-size free energies, which are formal embodiments of the physical properties we might expect, then lead to another technical feature, uniformity of convergence, which embodies another physical property we might also expect.

Contingent physical facts, here screening represented by those neutral balls, justify mathematical moves, here the convexity properties of f, needed to prove the stability of matter. Conversely, mathematical observations about the model (for example, that there exists a limit) may then lead to proofs of otherwise contingent significant physical facts (the thermodynamic stability of matter), which if they were not proven would considerably weaken our confidence in the model. Limit properties of models, as we shall see, are likely to be physically interesting.

Mathematical and Physical Argument

In general, I have been demonstrating just how mathematical argument and physical argument parallel each other (at any time there perhaps being leads or lags), how what is most crucial to making a mathematical argument go through is as crucial a feature for a corresponding physical argument. We might hope this to be the case, and surely it is an artifact of successful mathematical physics. (By the way, for most practicing physicists, physical arguments would appear to be distinguishable from mathematical ones, the former characterized by physical interactions, the latter by formal structures and manipulations.)

Physical properties can be made to correspond to mathematical properties, which lead mathematically to other mathematical properties, which then correspond to other physical properties. And we might hope that the first set of physical properties are eventually seen to lead physically to the second set. Mathematical moves have physical meanings.[67]

My claim here about mathematical models is a working hypothesis. For the practicing scientist, the physical world is indexed by those mathematical models:[68] the meaning of the physical ideas is often articulated through a mathematical model, one that shows the implications of our intuitive notions and so educates our intuitions. The mathematical model of the physics becomes the mode of interpreting what we empirically see when we fit data to formulas. I believe that physical models corresponding to mathematical objects affect some mathematicians similarly, physical arguments leading to mathematical ones.

In general, we are properly quite skeptical of a claim that properties we might deduce of a model, mathematically or physically, correspond to properties of an actual system. This is a contingent empirical fact. For example, a representative body might be taken to model the politics of a society. It is hard to countenance a belief that what happens in such a body corresponds, systematically, to phenomena in the larger society—although this would be an interesting program for a modeling endeavor.

My concern here is with the relationship of mathematical and physical argument. Some provisos: The distinction between mathematical and physical argument may not be so sharp, especially when we are actually doing a calculation. But it would seem that mathematical physics arguments are more intimately driven by mathematical structures and demands of rigor. Moreover, I do not wish to insist that every addition or replacement in a mathematical argument corresponds to something physical. Rather, significant movements in an argument so correspond (and, of course, this leaves "significant" open to question). Finally, a mathematical argument, once we understand the physics behind it, may no longer be seen as mathematical. Still, there are times, especially in mathematical physics, when the distinction between mathematical and physical argument is fairly clear: there is less modeling and more proving, so to speak.

I should confess that I take it that mathematics is as empirical as is physics, its objects of observation as much located in the world as are physical objects. (As indicated in the preface, this is a controversial point for philosophers. But practically, I suspect that many working mathematicians would be sympathetic.) A mathematician's observations of the world of mathematical objects lead to mathematical structures (much as a physicist's observations of physical objects lead to physical structures).[69] Moreover, I take it that both sets of observations are of the same world, each being different phenomenologies of aspects of that world.[70]

Mathematical Physics as Philosophical
Mathematical virtues are not always physical virtues, and their absence may be of some physical significance. There are situations when uniform convergence is not necessarily a given, and we might convince

ourselves that it ought not be. So, for example, at phase transitions the absence of uniform convergence to the infinite volume limit indicates that more than one state is possible at the transition point (that ice and water are present together, in arbitrary ratios, at the freezing point).

Or consider a system consisting of charged particles, but with no explicit constraint on the neutrality of the system—an "ordinary" system. As I indicated above,

> The thermodynamic limiting free energy per unit volume exists for both the ordinary and the neutral grand canonical ensembles and are independent of domain shape for reasonable domains. Moreover, they are equal to each other and to the neutral canonical free energy per unit volume.
>
> . . . systems that are not charge neutral make a vanishingly small contribution to the grand canonical free energy. While this is quite reasonable physically, it does raise an interesting point about nonuniform convergence because the ordinary and neutral partition functions are definitely not equal if we switch off the charge before passing to the thermodynamic limit, whereas they are equal if the limits are taken in the reverse order.[71]

Again, the nonneutral systems make a vanishing contribution to the grand canonical partition function.[72] But this fact is not robust to alterations in how we take our limits. Physicists know that noncommutativity of operators or limits can lead to significant physics; the question is just how does noncommutativity appear and what is its significance in each case.

My point, again, is that what seem like technical points, such as the order of limits, ones that have curious further technical consequences (such as nonuniformity of convergence), correspond to physical situations in which those curious consequences are quite reasonable. (In chapter 5, I shall examine the definition of a phase transition and of permanent magnetization, in which the order of limits is crucial.) And were those curious consequences absent, then the mathematical model would fail as a model of the physical system. In this sense, mathematical physics is again philosophical, showing just what is essential in models of the physical world. Moreover, it is usually good practice for the physicist to conflate models of the world and the world itself.

Another example: One can show quite generally that two of the following imply the third: stability against implosion (in effect, hard spheres); stability against explosion (in effect, a finite range of interaction, so that one does not release too much energy ($\approx N^{2/3}$) in putting two chunks of matter together); and thermodynamic stability (in effect, uniform convergence).[73] Note that the stability of matter (which I have defined in terms of stability against implosion) does not imply thermo-dynamic stability—unless there is stability against explosion (the linear law). The program I sketched in chapter 2 for demonstrating the stability of matter is not only a description of a proof and an analysis of what we might mean by the stability of matter; here it is justified as a necessary program, in that it is formally consistent internally.

Technically and mathematically, what can be shown is that two of the following imply the third for certain reasonable functions: homo-geneity $(h(\alpha x) = \alpha h(x))$, subadditivity $(h(x + y) \leq h(x) + h(y))$, and convexity $(h(\alpha x + (1 - \alpha)y)) \leq \alpha h(x) + (1 - \alpha)h(y))$—corresponding to stability against implosion, stability against explosion, and thermody-namic stability, respectively. The mathematical formulation, suitably interpreted for the physical situation, makes precise just what is re-quired for there to be a model of stable bulk matter, and in this sense the mathematical model is philosophical. A set of mathematical-physi-cal results analyze the constitution of matter as we know it—pointing out just what is essential about it if it is to exist as we know it as physicists.

The failure of stability for gravitation is just what allows for our lives in this universe. One might argue that there is a sort of thermody-namic limit for gravity, $\lim_{N \to \infty} N^{-7/3} E(N)$, for fermions (actually, $N \approx 10^{57}$ for a star), and this is true in that the properties of stars are in this sense N-independent. (Note that $E(N)$ goes as N^3 for bosons in gravity.) But that does not guarantee thermodynamic stability; there is relativistic gravitational instability and collapse, leading to a region of negative specific heat for these systems (as nuclear reactions become significant). That gravity is unstable to collapse provides the universe with its stars, our sun, and our livelihoods.

That we have everyday matter is just as subtle. Coulomb forces between charged particles, one kind of which, say the negative ones, are fermions, whose charges can cancel each other (unlike their masses, as in gravitation), allow for a thermodynamic limit and thermodynamic stability. Again, given that the Coulomb force is between all pairs of particles, one might expect an energy proportional to N^2. But, in fact, the existence of $\lim_{N \to \infty} E(N)/N$ is a consequence not only of screen-ing but also of the fermionic character of electrons. Recall that, if only bosons are involved, then we do not have a limit, for $E(N)$ goes as $N^{7/5}$.[74]

Limits Are Physical
As I have indicated already, while a proof that a limit exists is presum-ably a mathematical proof, in the end it had better be physical. Mani-

festly, one needs certain properties of a sequence of presumably better and better models of the physics, $\{M_i\}$ (where i might refer to the ball size, 28^i), to achieve convergence to a limit, $M_i \to M$, where M is taken to be a good-enough physical model (and, here, is the infinite volume limit). So one needs either finite range forces or average charge neutrality for the Coulomb force, so that interbulkette forces prove negligible in the limit, and the contribution of surface terms proves vanishingly small. One designs the $\{M_i\}$ so that this is possible. But that design is possible only if the physics presumably allows for such negligibility, in the limit. It does not allow such for gravity: stars collapse. I mention this example to emphasize that in some cases the physics itself should not allow for certain kinds of modeling—since that model could not be valid.

Again, the existence of the thermodynamic limit is proved, first by bounding from below the free energy per particle, at constant density —for each of the models, M_i. This requires that each of the models, M_i, actually have a constant density or at least a density that converges to a limit; hence, all the effort expended in constructing a sequence of packings of space by balls that provides for this. Second, one employs the boundedness to show that the sequence of free energies, $\{f_i\}$— corresponding to the $\{M_i\}$, which is in some effective sense decreasing, namely, $|f_i - f_{i-1}| \to 0$—implies the existence of a limit.[75]

That bounding from below for the energy depends on the fact that atoms do not chemically bind with each other within the Thomas-Fermi model of multielectron atoms, so that the energy of Thomas-Fermi matter is just the sum of the energies of its components, and on the fact that the energy of the Thomas-Fermi atom is a lower bound for actual atoms, so that the lower bound on the energy of bulk matter is proportional to the number of atoms.[76] As for the free energy, the bound is provided for by the Thomas-Fermi free energy bound plus a bound due to treating these atoms as composing an ideal gas, which latter bound is proportional to the number of atoms.[77]

The effectively decreasing nature of that sequence depends not only on small interbulkette forces, so that the bulkettes might be treated individually, but also on the design of those decreasing-in-proportion-to-volume interstitial spaces between the composing bulkettes at each scale (the corridors for the cubes; the interstices of the packing of balls, smaller balls packing larger ones).

All of these features are physical, and the actual physical world— whatever the mathematics is—might not provide for them within such a model. (Of course, there may be another proof altogether that does not depend on these features.) Multielectron atoms just might bind with each other significantly in our model (as they surely do to some extent in actual matter); the division of the free energy of bulk matter into that of individual independent atoms and that of an ideal gas might not be possible; and there might be no such sequence, $\{M_i\}$, given large enough surface forces.

The existence of a thermodynamic limit, and so presumably the existence of the limit P of certain properties P_i of the M_i (pressure, free energy per particle,. . .), says that the model, $\{M_i\}$, might well correspond to the physical world.[78] For we know that the physical world has such properties as P, and these are well defined in our everyday version of the infinite volume limit (10^{23} atoms). As we shall see, there are times when convergence or at least uniform convergence does not occur in a good model. These situations, as in phase transitions, are just those where one does not expect there to be such convergence, such a well-defined P. For, again, there is something like a multivalued limit, to be oxymoronic, as in the specific volume in a first-order phase transition (the water-to-ice ratio is undetermined at the freezing point) or when a discontinuity is present.

A model that may be employed to prove the existence of a thermo-dynamic limit, when we take that limit to model everyday matter, testifies to the possibility that that model might be a good model of the physical world; and if it is so testified, the model might even be taken to tell us about what is essential about the physical world and our models of it (effectively hard spheres, effectively finite-range forces), so display-ing what I have called the philosophical character of mathematical physics.

The kind of convergence we expect (to the thermodynamic limit, say, of the P_i) is also physical. *Pointwise* convergence says that "nothing happens" on the way, $\{M_i\}$, to a good model, M. Pointwise convergence might be defined by a Cauchy criterion—given ϵ', then there exists an n and for all $s > 0$, $|h_{n+s} - h_n| < \epsilon'$, namely, the difference between approximations beyond a certain n, $n(\epsilon')$, remains small. Or it might be defined in terms of a bounded monotone sequence, as I did earlier. In any case, there are no surprises on the way to M—although continuous functions do not necessarily converge to continuous functions under this regime (nor do analytic functions necessarily converge to analytic func-tions).

Uniform convergence assures us there is nothing very delicate about such a convergence, in the sense that there is a continuous range of parameters or variables (a "region") that characterize each of the M_i, for which convergence occurs in the same fashion, and so, within the region, convergence of continuous functions is in fact to continuous functions. Here convergence might be said to be robust, and, as indicated in the earlier quote, this corresponds nicely to the fact that the properties of everyday matter are never fixed precisely by us—yet matter seems roughly the same over a range of parameters (except at phase transition). Uniformity of convergence would seem to be the norm for most of the everyday physical world. Put differently, with uniform convergence the resulting objects, such as the free energy, are justifiably subject to those partial differential equations of mathematical physics (they have nice derivatives).

Asymptotic convergence is to a particular form, in the case of a thermodynamic limit to a linear relationship, $E \approx N$. That relationship is not just a lower bound. One can have a lower bound without a linear relationship, as we saw for bosons. And that linear relationship is just what characterizes extensive properties of matter, one of the foundations of thermodynamics.[79]

In general, a sequence of models, $\{M_i\}$, had better be a convergent approximation if it is about the physical world with its well-defined properties (with perhaps notable exceptions at phase transition). And limits are not only of mathematical interest. Rather, they are just what is needed if the model is to be about the physical world. The kind of those limits is physically meaningful as well: at a point, over a region, and/or to a given form. For example, tangents are limits, at least in one formulation, and those limits had better exist. Otherwise we would not identify tangents with the rather well-defined property we measure as velocity, and insofar as those limits are uniform in a region—namely, bounded derivatives—we might expect the path to be a smooth one, which is often the case in nature.

In sum, Kramers's thermodynamic limit turns out to require a suitable partitioning of an infinite volume that then leads to an effectively decreasing sequence of estimates for the free energy, one that is bounded from below—for then there is a limit. Moreover, the stability of matter does not necessarily imply the existence of a thermodynamic limit (that infinite volume limit), nor does such a limit necessarily imply thermodynamic stability (just what we expect from ordinary extensive or bulky matter). In the next two chapters, I explore how we might describe situations when bulkettes are not independent, that is, phase transitions, examining just how we might constitute such matter mathematically, and I explore the physical meanings of these different constitutions.

4

Formalism: Constituting Bulk Matter

SOLUTIONS TO THE ISING MODEL, AND DUALITY IN THOSE SOLUTIONS

*The geometric constitution of matter • Preview and prospect • The
arithmetic constitution of matter and transfer matrices: duality, the central
limit theorem, the physical constitution of matter • The holographic
constitution of matter • The topological / combinatoric / polymer
constitution of matter: duality, enumerating the polygons—an analogy with
quantum mechanics • Mathematical physics as philosophical—again*

> The most successful elaboration of technique in statistical mechanics
> exists in connection with the Ising model. . . . these techniques are rather
> far removed from either the basic starting assumptions or the final results.
> . . . [W]e believe it unwise to suppress successful techniques when
> discussing a subject in mathematical physics, for in the last analysis a
> subject grows with the techniques available to handle it. In the present
> instance the physical content of the techniques is not well understood,
> particularly as they relate to second-order phase transitions. However, the
> physical information must be buried somewhere in the formalism and will
> perhaps be found sometime in the future.

G. Wannier, *Statistical Physics*

In this chapter, I shall examine some of the various ways mathematical
physics has conceived of everyday bulk matter, specifically, such matter's
constitution, how it is put together out of its constituents; thus I will be
philosophically saying what is crucial about bulk matter, and thus
connecting deep mathematical features with deep physical features with
everyday features: here, apparently technical formalism is actually physical.[1] I show how the mathematical devices employed along the way
have physical content, in that they point to important physical mechanisms and interactions and symmetries, and in that the precise form of
those techniques and devices can be seen to express the particular
physics of the model. This is a common informal theme in the actual
doing of physics, pointed out in oral presentations, perhaps offhandedly,

but is rarely mentioned in the published papers. It is a theme not unfamiliar to lay persons from how geometry's role in general relativity is often presented, for example.[2] My purpose here is to take seriously the physical content of mathematical devices, the most technical of devices being quite philosophical in their implications. For the devices point to what is arguably most essential about bulk matter.

Moreover, we might hope that the various ways of conceiving of everyday bulk matter could be aspects of a whole, whether those aspects are connected by physical argument or by mathematical argument and proof. Especially in chapter 7, I suggest that the variety of mathematical devices employed share certain features that point to crucial physics in accounts of the constitution of matter: what would seem to be a pervasive and necessary presence of antisymmetric functions and fermions.

Rather more speculatively, one might argue that, since the various mathematical structures and devices are drawn from empirical observations of the everyday world,[3] it is no more surprising that many are roughly equivalent (their application to physical problems only suggesting which mathematical forms might be formally shown to be equivalent within mathematics), than that various physical constitutions of matter are also seen as equivalent aspects of that matter (their mathematical representations perhaps suggesting which physical models might be shown within physics to be equivalent).[4]

Conversely, one might well complain about "the depressing and puzzling fact that the introduction of new methods has, in fact, produced very few new results,"[5] although new methods often do illuminate old results. A more general and recurrent theme in this vein is to show how all previous solutions reduce to the current new method.[6] For example, "In the remainder of this chapter, we shall show that all these approaches could, in fact, have been derived from a single basic treatment, the evaluation of traces or 'vacuum-to-vacuum expectations' of products of Dirac type matrices by means of Wick's theorem."[7] I first review the mathematical devices and constitutions employed in demonstrating the stability of matter in the infinite volume limit. Then I examine a number of additional models of mathematically and physically constituting matter, devices employed for the various solutions to the two-dimensional Ising model, a lattice of spins, σ_i, each spin interacting with just its nearest neighbors, in a particularly simple way, the energy or Hamiltonian, H, equals $\Sigma - J\sigma_i\sigma_{i+1}$ ($\sigma = \pm 1$), so that the preferred or ground state is all spins aligned with each other (for the energy is then lowest). I emphasize dimensionality here because the solutions and constitutions I shall present are exact as solutions only in one and two dimensions (figure 4.1).[8]

Over the past fifty years, we have come to appreciate the particular topological features of the two-dimensional lattice with nearest-neighbor interactions that allow for exact solutions (and its field theory analog, in which space-time is that lattice), given the techniques we have been able to muster.[9] In three dimensions, various approximative schemes can give quite precise results even if they are not formally exact. So, for example, a series approximation in terms of the combinatorics of the number of successively interacting spins, exact in three dimensions at $T = 0$ and $T = \infty$, can be effectively extrapolated to the critical temperature (technically, by means of Padé approximants for

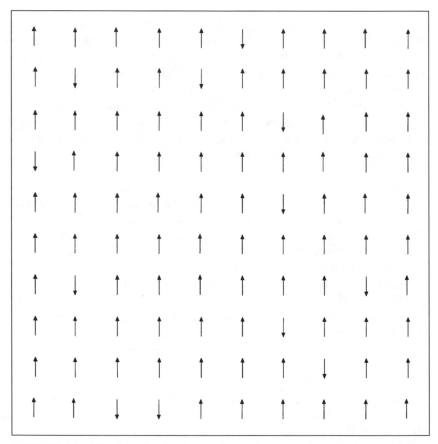

FIGURE 4.1   The Ising lattice in two dimensions, at a low temperature.

the term ratios, which enable the singular behavior to the taken into account).

By a solution of the model, I mean an account of the lattice's having a phase transition and a critical temperature, T_c, and of its thermodynamic properties—its free energy, its specific heat, and its spontaneous magnetization and correlation functions, and especially their behavior near the critical point (namely, the critical exponents, α_i, characterizing that behavior, $|T - T_c|^{\alpha_i}$)—especially in the infinite volume limit.[10] In order to account for these properties, most solutions must solve for either the partition function or an effective Hamiltonian, while some solve directly for the correlations among spins. The partition function is a measure, as a function of temperature and density, say, of the probability of all the possible states a system might find itself in (its volume in phase space), of all the ways of partitioning its components given certain constraints (typically, on the average energy, which is a measure of temperature). And the effective Hamiltonian is a measure of the energy of a system of many microscopic components expressed in terms of the macroscopic properties of that system. In principle, and often in practice, one can then deduce the thermodynamic properties from the partition function or the effective Hamiltonian. Of course, in actual practice, various different paths are also employed, as I shall discuss in chapters 5–7.

The Geometric Constitution of Matter

As we have seen so far, one way of analyzing bulk matter is in terms of a hierarchy of nested components, with two sorts of nesting: incommensurable levels—nuclei, atoms, molecules, bulkettes—each stable at its level (where the technical definition of stability might be rather different at each level);[11] and commensurable, similar levels, as in those cubes or balls, for cubes proportionately dimensioned by, say, 2^n $(1, 2, 4, 8, 16, \ldots)$. (For balls, it was $(28)^n$.) Such an analysis might be called reductionist and geometric, respectively, the latter as in a geometric series, recalling the composition of space in terms of l-sized cubes, each containing cubes $l/2$ in size (or contained in cubes $2l$ in size), so the dimension at each level, $2^n l$, would be a constant multiple of the adjacent level—and so eight cubes of size l would fit into a cube of size $2l$ or four squares of size l would fit into a square of size $2l$ (figure 4.2).[12]

Macroscopic extensive quantities, such as the free energy per unit volume or per spin, are defined in terms of an infinite volume limit, density held constant—the free energy being measured for successively

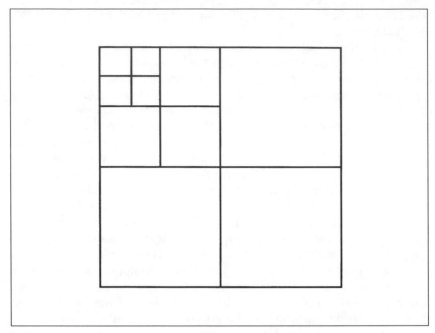

FIGURE 4.2   Hierarchical decomposition of two-dimensional space.

larger and larger systems. (Technically,

$$\beta f = - \lim_{\substack{n \to \infty \\ \text{constant density}}} (\ln Q_n)/(2^n l)^3,$$

where the size $2^n l$ object consists of eight $2^{n-1} l$-sized objects, and where Q_n is the partition function for a $2^n l$-sized object.) Again, it is worth noting that we are concerned with the density of the logarithm of the partition function and not that of the partition function itself, for, as Kramers pointed out, that free energy density is the physically relevant quantity in the infinite volume limit.

There are several interesting features of this reductionist and geometric analysis.

1. There is a smallest size, a, that is considered, namely a hydrogenic atom, where stability is against collapse; and there is a slightly larger size, the multielectron atom, where again stability is against collapse: namely, $a > 0$. Bulkettes, the geometric objects, are much larger than this size.

2. Combining the geometric components, at each level, turns out to produce no big surprises, for our purposes here—although bulkettes

themselves with their truly negligible volume-volume interactions with each other and their crystalline structure are emergent objects. A recurrent theme is that bulkette-bulkette interactions are surface terms; namely, the range of interaction, r_0, is much less than $2^n l$ (more precisely, since $r_0 < \infty$, there exists an n such that $r_0 \ll 2^n l$). More generally, interactions are sufficiently modest at each scale (from the atomic scale where there is "no binding" among atoms, to bulkette scales where interbulkette surface terms are small compared to intrabulkette volume terms) that they have no substantial effect, in the infinite volume limit, on the energy or the free energy densities.

Close to the critical point, the surface terms are in fact significant, if the blocks are small compared to the correlation length.[13] Close enough to the critical point, that length is effectively infinite and the geometric constitution has a very different sort of infinite volume limit, one that measures those critical exponents, α_i.

Technically, recall that infinite volume limits are shown to exist because the limit sequence for the free energy density is proven to be effectively monotone and bounded.[14] It is bounded because, for example, atoms do not bind among each other in the Thomas-Fermi approximation for a multielectron atom (and that sets a free energy bound proportional to N), and the bound on the free energy of an ideal gas also goes as N. This geometric construction of cubes works to provide an effectively monotone sequence because the contribution of the surface terms (to the energy per particle, or to the free energy per particle) is increasingly vanishingly small.

In particular, stability provides, in effect, for a hard-sphere repulsion among atoms, so there can be no bunching up of all the atoms in any single bulkette, which then provides the lower bound. And the range of interaction is in effect finite, and so one can separate or isolate each of the bulkettes from its neighbors, treating them disjointly. The finite-range device is an empty corridor between bulkettes (the corridor's depth is greater than the range), or the screening provided by the Coulomb interaction, the effect of such a device on the free energy density by construction turning out to be negligible in the infinite volume limit.

3. At each scale or level, one deals with just a single cube or ball (albeit composed of a number of next-smaller cubes or all-smaller balls). Space filled with matter is to be built up by enlarging that single cube so that it is now, say, twice as large, ad infinitum.

Alternatively, one might start with matter present throughout all of space, space now subdivided into an infinity of bulkettes. The first division being atomic, at their scale, then perhaps combining their effects into those of cubes of size l (where l is the interatomic distance), and then combining l-sized cubes into $2l$-sized cubes,. . . into a sequence of geometrically larger cubes—and the sequence's infinite volume limit ($2^n l$, $n \to \infty$) is presumably the macroscopic bulk effects we are interested in (if such a limit exists, mathematically—namely, that the constitution works). One is thinning out or averaging out details at scale l, to produce a lattice at scale $2l$, and so one is approaching macroscopic properties. (Obviously, and we shall need this shortly, one might have, say, $3l$ rather than $2l$, and so 27 l-cubes in one $3l$-cube.)

This latter constitution is just that of the "block-spin renormalization group" approach to solving the Ising model—where those cubes are the blocks of spins.[15] (See figure 4.3.) The renormalization group refers to the fact that if $H = \sum - J\sigma_i\sigma_{i+1}$ and $Q = \sum_{\{\sigma_r\}}\exp - \beta E(\{\sigma_r\})$, where $E(\{\sigma_r\})$ is the energy of configuration $\{\sigma_r\}$, and we are summing

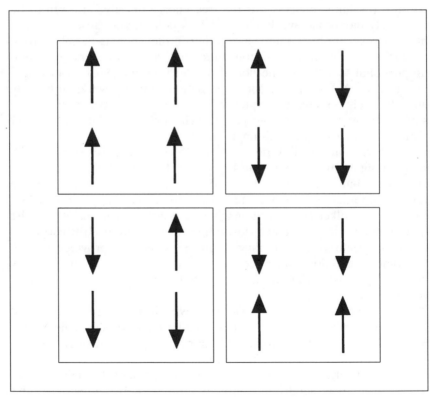

FIGURE 4.3   Block spins.

over all of them, then $H' = \Sigma - J'\sigma_i'\sigma_{i+1}'$, where the prime indicates a block of spins and σ' is the spin we assign to a block, then J' or $\beta J'$ is determined so that $Q' = \Sigma_{\{\sigma_r'\}}\exp - \beta H'$ is equal to Q (with perhaps some scale factor). That is, in computing the partition function, we may impute a ("renormalized") constant, J', to the interaction between the individual blocks (rather than between the individual spins, J), so that the partition function computed as if the blocks were now individual spins is roughly equal to the original partition function, and so that an imputed or effective interaction (or Hamiltonian) may be adequately represented as a spin-spin interaction, $-J'\sigma_i'\sigma_{i+1}'$, much like that between individual atoms, but now the interaction constant, J', is modified as the block size increases—in fact, it scales up or down ("renormalizes"), becoming larger or smaller, depending on whether we are below or above the critical temperature.[16] And we may iterate this process.

Of course, it is not apparent ahead of time whether such a constitution can actually work, and in fact various not-so-nearest-neighbor block-spin interactions must be included. Yet one finds that $J(l)$, where l is the current scale, does scale up or down. (Note, as well, that the Coulomb interaction energy that concerned us in chapters 2 and 3 is much, much larger than the magnetic interactions between the actual spins. In this Ising model case, there is a natural smallest length or cutoff (the interatomic distance), so stability questions in the constitution of this matter may be put aside.)[17]

Technically, that modification of the interaction constant is a group transformation: something like $J(l + m) = J(l)J(m)$, so that $J(l)$ is perhaps like $\exp(\alpha l)$, where $J(l)$ is the block-block interaction at block size l. Note that $\lim_{l \to \infty} J(l)$ is zero or infinity, except if $J(l) = 1$, at presumably the critical temperature—so that the limiting interaction is either negligible or absolute, or all scales would seem to matter equally.

This limit is not quite what we have so far taken as the infinite volume limit. Rather, we scale up and so incorporate the whole lattice, but the interaction also changes (except at the critical temperature). The infinite volume limit here concerns a rate of convergence, namely, the critical exponents.

For example, and a bit more technically, consider a two-dimensional lattice, scaled by factors of three (to allow for a straightforward majority-wins determination of the spin we assign to each block). Along the way to computing the partition function, we compute the probability of each configuration for two nine-spin blocks (ignoring other interactions outside the blocks), for each configuration assigning a spin to each of the blocks equal to the majority of its nine spins, either up or down.

Add up the probabilities for all block pairs that look alike at the larger $3l$ scale: namely, add all the probabilities of all the pairs that are, say, up-up by majority rule, although they may be composed of different arrangements and different degrees of majorization. So we have computed the probabilities, $p_{\uparrow\uparrow}$, $p_{\uparrow\downarrow}$, $p_{\downarrow\uparrow}$, and $p_{\downarrow\downarrow}$, for configurations at the $3l$ scale—and these are, it turns out, different than the probabilities at the l scale. Finally, impute an interaction $\beta J(3l)$ by assuming that the $3l$-probabilities are accounted for by the usual formula for the canonical ensemble, $\exp \beta J \sigma_i \sigma_{i+1}$ (suitably normalized), but now the spins, σ_i, are the block spins (by majority rule) at the $3l$ scale. (One can do this because the probabilities for the parallel spins are equal, as are the probabilities for the antiparallel spins, and $p_{\uparrow\uparrow} + p_{\uparrow\downarrow} = 1/2$: so $\beta J(3l)$ is determined.) On then iterates this process, now with blocks (or superblocks) of size $3 \times 3l, \ldots$, and so forth.

In sum, the imputed or effective interactions of blocks with each other (taken as individual block spins)—which we have imputed from the probabilities of particular configurations of atoms (recall that at each level, except the atomic level, there are no direct block-block interactions in this model)—have different strengths than atomic spin-spin interactions. As we scale up, we are in effect revising the temperature-interaction product, βJ. Again, it turns out that diagonal and not-so-nearest-neighbor terms also become important. And it is not apparent ahead of time that the $\beta J(3^n l)$ will have nice behavior as $n \to \infty$.

As the block size increases, in this model the effective interaction strength of superblocks in fact does converge to either zero or infinity, except at the critical point, which is a fixed point.[18] We might get a feel for how this occurs: Imagine that we had a lattice of spins filling all of space in one particular configuration. It is quite likely that for most configurations the up and down spins are fairly randomly distributed among each other. Avoiding comparatively rare configurations, such as all spins to the left of the axis are up and to the right are down, say, we go from a single-spin "block" to a block of nine spins, and to each block, we assigned the spin of the majority of those nine spins. Let us now go, sequentially, to larger and larger blocks. As the averaging of averages proceeds, we might expect that either almost all the larger block spins are up or almost all are down (if there were some, perhaps small, original inequality of the two kinds of spins), or they are rather randomly distributed (were the ratio of up to down spins originally equal).

For very small inequalities (typical of configurations just below or above the critical point)—where to statistically significantly distinguish the inequality or equality we would have to have lots of spins or very large blocks—the rate of convergence is very slow.[19] At the critical point, it is in fact infinitely slow, averaging out to neither order nor disorder.

Now, if there were net equal spins, but also clusterings of ups and downs originally (as we might expect, even randomly, and this is surely the case near the critical point), those clusterings would persist until the size of the block were larger than the cluster size. But if there were, say, reasonably probable clustering at all scales, there would be no scale at which clustering would average out, even if there were equal spins up and down—and this is just what happens at the critical point. Put differently, as the block size increases, blocks eventually become more statistically independent of each other and the central limit theorem applies more appropriately (or the blocks become more dependent on each other)—if there is no phase transition. But, in principle, we cannot know just what is the case until we arrive at the infinite volume limit.[20]

In going to the infinite volume limit geometrically, in this way, the system either goes to total order or to total disorder, in effect, to zero and infinite temperature, respectively (infinite or zero interaction, respectively)—except at the critical point. Let us call this a "duality," connecting high-temperature behavior to low-temperature behavior. The reason for the name will become apparent shortly.

It may be helpful to recall some of the experimental facts that made it plausible to think geometrically about the constitution of matter in this context. It was known already from Onsager's exact solution that the Ising model exhibited a logarithmic specific heat, $c_v \approx -\ln t$, $t = |T - T_c|/T_c$, near the critical point, and that actual matter exhibited similar behavior.[21] Density fluctuations were known to similarly peak, causing anomalous scattering of light (critical opalescence). More generally, properties near the critical point were of the form t^α. Moreover, the critical exponents, the various αs, seemed to be "universal," independent of the detailed properties of the matter being considered (although T_c surely was so dependent), dependent on more general features such as the spatial dimensionality of the matter.

One mode of explaining just these features was to account for the thermodynamic properties in terms of blocks of matter whose size was geometrically increasing. The distinctive feature of the geometric constitution is its explicit concern with behavior near the critical point (in terms of t), especially with correlations and fluctuations near that point.[22] (As we shall see, other constitutions are more focused on high and low temperatures.)

PREVIEW AND PROSPECT

There are two sorts of duality that shall concern us. Let me preview what we shall discover, presuming on the reader's having scanned the rest of the chapter already. In the first sort, what I call geometric duality, I address the interaction between spins (the coupling constant), which as I indicated goes effectively to zero or to infinity, now between blocks of spins, as we scale up to the infinite volume limit, except at the critical point. The interaction between spins is directly related to the bulk magnetization of the lattice, which is a measure of the correlation of spins with each other. The "geometric" and "holographic" constitutions are of this sort. The fact that the lattice and its interactions can be divided into two interlaced parts (it is said to be bipartite) is, it would seem, almost crucial for the constitutions' being practically workable. Moreover, besides leading to exact solutions, these constitutions lead rather directly to methods of numerical approximation for the bulk properties near the critical point, in the infinite volume limit, for even more complicated situations. And some of the most perspicuous of solutions of the Ising model, directly calculating correlations, depend on working with this sort of geometric duality and with the interactions or couplings.[23] As we shall see, one focuses on an effective Hamiltonian rather than on a partition function.

The second sort of duality is arithmetic, and it appears in the "arithmetic" and "topological" constitutions. High-temperature configurations of the lattice (configurations that dominate at high temperatures, namely, rather random arrangements of spins) are directly connected to low-temperature configurations (rather orderly arrangements): adding a bit of order to the former is "the same" as adding a bit of disorder to the latter. One pays attention to the fact that a lattice is composed of bonds between atoms, whether one expresses this in terms of the actual bonds themselves or in terms of the "transfer matrices" that take account of the contribution to the partition function by the interactions at each of the bonds. Besides the exact solutions, those various modes of solution lead to perturbative expansions, in the number of bonds considered, which lead to powerful series approximations for thermodynamic properties. Here, one works in the space of actual spin configurations, rather than in the space of the coupling constants as for geometric duality.

Not only are these two kinds of duality united by a parameter, k, which is a measure of order in the system, but they can be shown to be mathematically quite similar transformations, picking out different symmetries of the lattice.

I have been using *the constitution of matter* to mean the computation of its partition function or its effective Hamiltonian, that constitution presumably indicating ways bulk matter might actually be put together. Now, as I have just indicated, there are other ways of constituting matter than through geometrically growing bulkettes. One might arithmetically add on one atom at a time, the infinite volume limit (of the free energy per particle) being $N \to \infty$, where again "adding on" literally places one more atom onto the lattice and computes the change in the partition function, that additive change being a multiplicative factor—since for the partition function we are dealing with an exponential in the interaction, $\exp - \beta H$ (and $\exp - \beta(H + \delta) = \exp - \beta H \times \exp - \beta \delta$).

Or, reminding us of the geometric block-spin technique, one might constitute matter in terms of the spatial repetitiveness of its constitution. One might start out at fine resolution and go to rougher and rougher resolution, a kind of hologram. (Technically, we are in Fourier-transform or momentum space, while the block spins are in configuration or ordinary space. In effect, the spatial Fourier transform goes from a minimum wavelength to an infinite wavelength, the wave numbers go from a maximum to zero, the infinite volume limit being the sum of all the Fourier components up to zero wave number.) Or matter might be seen as disturbances in a uniform continuum or an ether, those disturbances being either the likeness of neighboring points in a random continuum or the unlikeness of neighboring points in a constant continuum, in effect a polymer. And the infinite volume limit would take into account all such possible polymers, adding in their configurations as contributions to the partition function. This latter is in effect a topological constitution of matter.

Each of these modes of constituting bulk matter and its properties involve different mathematical technologies: for example, (uniformly) convergent sequences of models and group theory, linear operators in a Hilbert space, semigroups and Fourier transforms, and combinatorial topology. We might well expect that each mode allows us to think of the constitution of matter in a different way, each highlighting different physics of matter. Moreover, since they are all meant to model bulk matter, here two-dimensional Ising matter, we might expect that their mathematical technologies might be shown to be in some sense equivalent, at least in the infinite volume limit.

So far, our consideration of the constitution of matter and of the existence of an infinite volume limit has centered on matters of stability and of the existence of extensive thermodynamic properties such as the free energy. As I have already indicated, another feature of bulk matter is its capacity for rigidity.[24] Literally, it can be rigid, as is a stick or a bar. More generally, it exhibits orderliness over macroscopic distances, as in a permanent magnet, the spins of its iron atoms more or less aligned. Anderson describes such "generalized rigidity" as a consequence of the fact that the energy is minimized if there is an orderliness throughout the sample, in the same but otherwise arbitrary direction—what is called a broken symmetry. Concretely, one cannot change the magnetization of an iron bar a little piece at a time; the energetics, or the "infinite" force from the rest of the bar (the rest of the iron's atomic spins), precludes this—in the infinite volume limit. Hence, features at point x are directly related to features at point y. So, for example, if the available energy is low enough (the temperature is low enough), the direction of magnetization should be roughly the same throughout a homogeneous sample, or a rod is mechanically rigid (so that moving the stick at x has implications for what happens to the stick at y).[25] Such strong correlations are meant to capture our intuitive sense of rigidity. The problem then becomes one of showing how matter's constitution—in other words, our mathematical models of that constitution—might allow for that rigidity.

We know that bulk matter of a particular composition may not be very rigid under certain conditions, say, if it is a fluid, yet that same matter may become rather more rigid—it freezes—through a phase transition. And so the model of the constitution of matter must show not only how bulk matter might be rigid, but also how it makes a transition to rigidity. So, for example, the block-spin geometric analysis of matter is employed to show how, if the temperature is low enough, matter constituted in this way will exhibit rigidity—namely, a permanent magnetic moment throughout a bulk sample without an externally applied magnetic field: larger and larger blocks are correlated with each other, and so there is greater order.

Thermodynamically, phase transitions are taken to be indicated by discontinuities in the specific heat or its derivatives. But, in statistical mechanics, what is interesting and needful of explanation is just how generalized rigidity sets in as well (at the same temperature, something we do not know how to prove generically, but can show in particular models). Not only do we want to compute the partition function, show that it exhibits the suitable zeros or singularities at phase transition, and show that the correlation functions become nonzero (so there is rigidity)—we also want to have a detailed picture or model of what the molecules are doing at that point and as order sets in or as it dissipates.

I shall now shift considerations of the constitution of matter from questions of stability (for the only proofs of the stability of real Coulombic matter that I am aware of are geometric)[26] to matters of rigidity and order and correlation. As we shall see, in the arithmetic case, we model rigidity as the propagation of order, as one more atom is added on; in the holographic case, we model rigidity as interactions and so correlations at larger and larger scales; and in the topological case, we model rigidity in terms of the connectedness of one part of matter to another, and the multiplicity of such connections.

Crystalline structure is nature's most manifest testimony to long-range order and rigidity, and is the foundation for the stability of matter.[27] The question I am now asking is how and whether such a structure or lattice can support as well a long-range order that reflects a local order—the permanent magnetization of a bulk sample, reflecting the local neighboring interaction of spins.

THE ARITHMETIC CONSTITUTION OF MATTER AND TRANSFER MATRICES

We might imagine building up bulk matter atom by atom, one after another, adding on one more, or one more row, or one more layer at a time—extending these additions to infinity, and so arithmetically constituting matter in the infinite volume limit. The task is to take this notion and embody it in a technically powerful means of constituting matter, that is, a means of computing its partition function.[28]

The fact that there is a definite spacing between atoms, and that the Ising model interaction is restricted to nearest neighbors, means that issues concerning stability may be put aside. The lower bound of the energy is linearly proportional to the number of atoms.[29] What remains open is whether a solution to the Ising model along arithmetic lines possesses an infinite volume limit—or better put, how it can be made to possess such a limit. For, if the Ising model were to be a model of ordinary matter, there had better be an infinite volume limit, since we know that such Ising matter (if it is to model actual matter) must have a well-defined bulk permanent magnetization, for example, although it might well be zero in some cases. Early general considerations of a two-dimensional lattice by Peierls (1936) and more specific considerations by Kramers and Wannier (1941) suggested that the arithmetic model possesses a transition point or critical point, and Onsager's (1944) solution, as such, showed not only that there could be an infinite volume limit at all temperatures, but just how the critical point was approached.[30] Note that, in this case, we are taking a $\lim_{N \to \infty}$, a linear progression, rather than a $\lim_{\ln N \to \infty}$, as in the geometric case. The constant-density assumption is built into the fixed spacing and the fixed interaction of the spins or atoms.

The task here is to give an account of the collective behavior of many, very simply interacting spins. Arithmetically constituting or building up bulk matter becomes a matter of the multiplication of a matrix, V, N times.[31] V is called the transfer matrix, for it measures the change in the partition function from adding on one more atom, transferring the interaction down the line, so to speak. (There should be no problem distinguishing when V is used for the volume, from when it is used for the transfer matrix.) In effect, each factor of V adds on one more atom, measuring the contribution to the partition function of its interactions with its neighbors. (Again, since the partition function is a sum of $\exp - \beta E$, multiplication will add in interactions.) And a sum of the eigenvalues of that matrix, $V^N - \sum_i \Lambda_i^N = \mathrm{Tr}\, V^N$ —is the partition function. As $N \to \infty$, $\Lambda_{\max}$ dominates the sum.

More technically, the partition function, a measure of the number of different ways of partitioning or arranging the component spins, is $Q = \sum_{\text{all spin configurations} = \{\sigma_r\}} \exp(-\beta E)$, where, schematically, the energy, E, of a spin configuration, $\{\sigma_r\}$, is $\sum_i - J\sigma_i \cdot \sigma_{i+1}$, assuming that σ_N interacts with σ_1, as it does in circular boundary conditions in one dimension.

Technically, since $Q = \sum_{\{\sigma_r\}} \exp(-\beta \sum_i - J\sigma_i \cdot \sigma_{i+1})$, and $\exp \sum = \prod \exp$, then $Q = \sum_{\{\sigma_r\}} \prod_i \exp(\beta J\sigma_i \cdot \sigma_{i+1})$, which (since it is not difficult to show that the sum of all spin configurations cancels out the intermediate sums over each of the spins) reduces to $Q = \mathrm{Tr}\, V^N = \sum_i \Lambda_i^N$, where V is a matrix that represents the value of $\exp - \beta E$ for each of the values of the two adjacent spins (both up, both down, one up and one down, and one down and one up);[32] N is the number of atoms; and Λ_i are the eigenvalues of V.

Before going farther, let us estimate the partition function at its extremes of low and high temperatures, for a two-dimensional square lattice. At low temperatures, there are only two states that should be significant, all up and all down. For each, the energy is $-2JN$. Hence the partition function is $2 \times \exp \beta 2JN$, and the partition function per spin (the Nth root) is $\exp 2K$, for large N, where $K = \beta J$. For high temperatures, note that the number of addends in the partition function is about 2^N and so $Q = 2^N \langle \exp - \beta E \rangle$, where $\langle \ \rangle$ stands for an average value or an expectation. The expectation should be about $2\cosh(\beta J(2N)^{1/2})$, by assuming in the computation of the energy that we have something like a random walk of spin flips, and then adding up the contributions from both the positive and negative walk endpoints or energies. The partition function is now dominated by the 2^N factor, and the partition function per spin is equal to two.

The question remains, is there an infinite volume limit here in this model? Does $-1/\beta N \ln Q_N$, the free energy per particle, possess a limit? Does the exponent on N in the above analysis actually behave nicely? If not, this model cannot be about the physics. Does this model have a chance of representing nature?

Just as the free energy density, not the partition function, is the physical quantity of interest in the infinite volume limit, so it turns out that it is the few largest eigenvalues of the transfer matrix, not the matrix itself, that are of interest. More precisely, one is interested in the infinite volume limit of the free energy per particle, f; technically, $f = -\lim_{N \to \infty}(\ln Q_N)/\beta N$, and so $f = -1/\beta N \ln(\sum_i \Lambda_i^N)$. If there is an appropriately bounded largest eigenvalue, there is not too much degeneracy ($< 2^N$), and the eigenvalues are all positive (all of which must be demonstrated, each property connected to deep physical facts, such as the existence of a ground state for an analogous system), then the largest eigenvalue term, that is, Λ_{max}^N, dominates all the other summands—in the limit of infinite N.[33] Then $f = -(1/\beta N)N \ln \Lambda_{max}$; the exponent on Λ_{max} is canceled by the normalization—just the infinite volume limit we want.[34]

This model of the arithmetic constitution of matter is exactly solvable in one dimension, and in two dimensions if there are only nearest-neighbor interactions and no external magnetic field.[35] Moreover, that interaction must be truly local; it does not change as N increases. No other formulation or constitution is any more exact, although some are suggestive of good approximative schemes for the three-dimensional or magnetic-field cases, and of just why these situations prove intractable to exact solution.[36]

Onsager's strategy for solving the two-dimensional Ising model is fairly straightforward, if not easily discerned (see the appendix). The formulation of the two-dimensional Ising problem is already set by Kramers and Wannier, Montroll, Lassettre and Howe, and Ashkin and Lamb, and it would seem that all were talking to Onsager at the time. After writing the transfer matrix as exponentials in sums of his spin-spin operators $(s_i s_{i+1})$ for the horizontal interactions, and sums of his spin-flip operators (C_j) for the vertical interactions, his A and B operators, respectively (pp. 120–121 of his paper), Onsager notes that $A \leftrightarrow B$ under a dual transformation (p. 123). He also notes that s_i and C_i anticommute, and form a quaternion algebra (pp. 121 ff.). Onsager then explores the spectrum of eigenvalues of A and of B separately, since in effect they are the low-temperature and high-temperature operators (pp. 124 ff.). A's good eigenvectors are highly ordered arrangements of spins; B's are when all possible configurations of spins are equally likely (since β is then small and the exponential in the partition function is approximately one). He then proceeds to diagonalize them together, and so diagonalizes V (pp. 126 ff.).

Onsager's problem is analogous to the standard quantum mechanics problem of diagonalizing the energy or the Hamiltonian, $p^2/2m + V(x)$, which is composed of one part, the kinetic energy, diagonal in the momentum, p, and one part, the potential energy, which is diagonal in x or a function of x.[37] (Just as p and x do not commute, as operators, so here s and C do not commute, for the same index.) Note that at low temperatures the kinetic energy is expected to be low. And so, conversely, the potential energy should correspond to A, the order operator aligning the spins. The kinetic energy corresponds to B, the disorder operator, becoming relatively more significant as the temperature is raised.

To diagonalize A and B simultaneously, Onsager creates a set of operators that produce rows of spins with specified flip positions (for example, ↑↑↑↑↑↑↓↓↓↓↓↓↑↑↑↑↑↑ ...), his P_{ab} (pp. 125 ff.), through which he can then express A and B as sums of Ps (for example, $A = \Sigma P_{a, a+1}$). While A and B do not commute, they can each be expressed as a sum of operators that for the most part do commute (and when they do not, the commutation relations of the P_{ab} are familiar [p. 126]). If we now let $B = -A_0$ and $A = A_1$ and generate an algebra of spin configurations by taking commutators of the A_i, leading to Onsager's A_i and G_k, those commutation relations look like convolutions, namely $[A_k, A_l] = 4G_{k-l}$ and $[G_m, A_k] = 2A_{k+m} - 2A_{k-m}$ (p. 127). (These are grounded not in the translation symmetry of the lattice, but in duality, which is a form of Fourier transform.)[38] That suggests a Fourier transform on the A_i and G_k to linear combinations of operators X_r and Y_s, and to Z_t, respectively, which act like Pauli spin operators, not commuting only when $r = s = t$ (Kaufman's commuting plane rotations) (p. 128). (This is a standard strategy for free fields, going to momentum eigenstates, and in effect the potential energy or A operator has disappeared.) A and B are then sums of the X_r, Y_r, and Z_r, and each Fourier component (essentially a periodic arrangement of spin flips) is then separately diagonalizable (by rotating each to and in its plane of rotation).

Onsager points out that the largest eigenvalue (the value of the partition function) of the all positive transfer matrix, V, is associated with an all positive, spatially even eigenvector (pp. 130–136). It is then straightforward to guess the eigenvectors of the X_r and X_r^* (disorder and order, respectively), and so the eigenvalues of A and B (sums of the eigenvalues of the X_r) and the eigenvalues of V (exponentials of the eigenvalues of the X_r). Onsager then computes the partition function, using elliptic integrals (pp. 138–141, 148–149). The partition function per spin was approximately but correctly surmised by Montroll

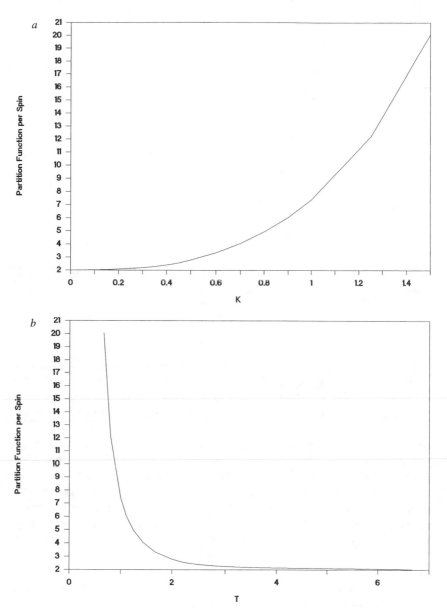

FIGURE 4.4   The partition function per spin (a, b), and the free energy per spin (c) for the Ising model in two dimensions. T is in $1/K$ units.

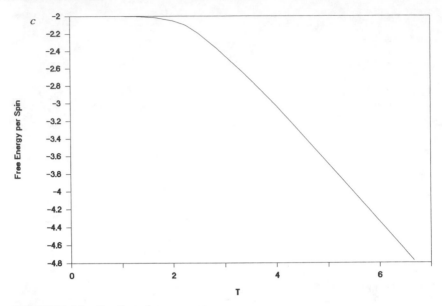

FIGURE 4.4   Continued.

and by Kramers and Wannier, at high temperatures being 2, and at low temperatures being $\exp 2J/k_B T$ (as we might expect from the above estimate; see figure 4.4). For the square lattice, that partition function per spin at the critical point was estimated to be 2.5335. Onsager's exact value is $\sqrt{2}\exp 2G/\pi$, where G is Catalan's constant, $= 1^{-2} - 3^{-2} + 5^{-2} - 7^{-2} + \cdots \approx 0.92$ (the integral that defines G, $\int_0^{\pi/4} \ln\cot x\, dx$, is within an integer factor of 8 the crucial definite integral of the energies (at $k = 1$) of the quasiparticles I shall discuss shortly).

Kramers and Wannier, and Montroll, compute the eigenvectors explicitly for small $n \times \infty$ lattices. They point out that, for low temperatures, the first excited state is just one inverted spin, and so an energy shift of $4J$ (in a row). At the critical point, the eigenvector is a "resonance" of the almost ordered and almost disordered eigenvectors characteristic of low and high temperatures.[39]

Onsager gives a beautiful analysis of the physics (pp. 135 ff.), which we shall return to regularly in what follows. For example, he shows how the eigenvectors of the transfer matrix might be seen as a consequence of the Z_r introducing greater and greater disorder into the orderly low-temperature eigenvector as the temperature is raised (pp. 135–137).

Technically, the maximum eigenvector is $\exp i/2 \sum \delta^*_{2r-1} Z_{2r-1} \times \chi^*_0$, where the Zs here act as disorder operators and χ^*_0 is the orderly zero-temperature ground-state vector. Onsager then shows how the amplitudes of the disorder operators, δ^*, grow as the temperature rises, becoming significantly nonzero for long wavelengths only when near the critical temperature, and then they increase quite rapidly (see figure 7.12).[40] Onsager explains all of this using lovely diagrams, some of which we shall encounter in chapters 6–8. In effect, Onsager has given an account of how the permanent magnetization disappears as the temperature becomes larger.

(As we shall see, an explanation in terms of particles will be particularly illuminating about the infinity of the specific heat at the critical point. It also suggests that the degeneracy of the eigenvalues at low temperatures is due to the fact that it is rather easy to excite the first excited state, $|0\rangle = |\uparrow\rangle - |\downarrow\rangle$, from the necessarily even ground state, $|0\rangle = |\uparrow\rangle + |\downarrow\rangle$, an equally likely mixture of all up or all down spins. This is not the case above the critical point.)

The limitations I have mentioned of Onsager's exact solution—two dimensions, no magnetic field, nearest-neighbor interactions—are perhaps no more an indictment than is saying that someone has exactly solved Maxwell's equations for electromagnetism for only certain geometries (and hence the history of special functions). Moreover, exact solutions, or at least formal approximative schemes, may in their formalism show us physics that an effective and sometimes jerry-built approximative scheme may miss (while a more ad hoc approximative scheme or a computer simulation may show us phenomena that cannot be found through exact solutions). Here is Baxter on exact solutions.

> It's worth specifying that by the phrase *exactly solved* we don't mean *rigorously solved*. In fact *exactly* applies to the results, while *rigorously* applies to the way you get it. Exact results can be obtained by an applied mathematician derivation which is not properly a rigorous proof. For instance one can multiply and diagonalize infinite matrices without demonstrating that matrix products are convergent, believing that this is true and that the results are exactly correct.
>
> Exact results can also be obtained by a conjecture.[41]

The mathematical device of a transfer matrix was proposed more than fifty years ago as a way of solving the Ising problem.[42] Manifestly, it builds up the lattice arithmetically—again in the sense of computing the partition function. There was, perhaps, no reason to believe then that this was more than a device, an important device nonetheless. Yet it turns out that such building up atom by atom is directly analogous to building up the trajectory of a particle step by step, as in the calculus and in Newton's second law: Δvelocity = acceleration $\times$ Δtime = Force/ mass $\times$ Δtime; and distance = Σ velocity $\times$ Δtime, and velocity = Σ Δvelocity. Adding on one more atom corresponds to adding on one more time step, the infinite volume limit corresponding to bulk matter or the path of the particle, respectively (where the step size here does not go to zero, unlike the usual presentations for differential equations).[43] What was once a technical device is now rather more familiar and natural. And just as there are conserved quantities in the Newtonian case, typically energy and momentum, here too are such quantities (what I shall call k, for example), constant as we build up a lattice, as we constitute matter.

Moreover, we have here a formal analogy of the summands of the partition function in statistical mechanics with the evolution operator in quantum mechanics,[44] suggested by $\exp - \beta H_{sm} \approx \exp - i H_{qm} t / \hbar$, where H_{sm} is the Hamiltonian (or the energy) for the statistical mechanical problem and H_{qm} is a Hamiltonian for a different, quantum mechanical, problem, and $\hbar$ is Planck's constant$/2\pi$. (t here stands for time, not reduced temperature.) The real statistical mechanical exponent corresponds to the imaginary exponent in quantum mechanics: quantum mechanics is statistical mechanics in imaginary time (with due allowance for matters of boundary conditions). The infinite volume limit is analogized to particles before and after they interact with each other, what are called in scattering theory free in- and out-states, or ground states before and after the Ising interaction is turned on. In each "after" case, all interactions have been accounted for.[45] In effect, what we are saying is that if we build up the world arithmetically, there is, it would seem, one way to do this, whether that buildup is in spatial increments (bulk matter) or in temporal ones (trajectories).

Furthermore, and technically, the partition function—

$$\sum \exp(-\beta E) = \sum\prod \exp \beta J \sigma_i \sigma_{i+1} = \operatorname{Tr} V^N,$$

where $H_{sm} = \Sigma - J\sigma_i\sigma_{i+1}$—can be made analogous to the quantum mechanical evolution operator that takes a system from time 0 to time t, $U(t,0), = \exp - i H_{qm} t / \hbar$, namely,

$$U(t,0) = \sum_{\text{intermediate states}} \langle 0 | \exp - i H_{qm} \, \Delta t / \hbar | \Delta t \rangle$$

$$\times \langle \Delta t | \exp - i H_{qm} \, \Delta t / \hbar | 2 \, \Delta t \rangle \ldots \langle (N-1)\Delta t | \exp - i H_{qm} \Delta t / \hbar | t \rangle,$$

$$N\Delta t = t,$$

which again leads to a trace, here of $\exp - i H_{qm} \, \Delta t / \hbar$.[46] Hence, if $Q_N = \operatorname{Tr} V^N$ and $V = V(H_{sm})$, where H_{sm} is the statistical mechanical

Hamiltonian, then $V(H_{sm})$ corresponds to $\exp - iH_{qm}\Delta t/\hbar$ and so $\ln V(H_{sm}) \approx -H_{qm}t$ (where β absorbs the i and the $\hbar$). More precisely, $V^N(H_{sm}) \approx \exp - H_{qm}t$. By "corresponds to," I mean that a solution to a statistical mechanical problem with the Hamiltonian H_{sm} corresponds to a solution to a quantum mechanical problem with the Hamiltonian H_{qm} (again, pace correct handling of boundary conditions), $Q(\beta)$ and $U(t)$ being the corresponding objects being calculated, and $-1/\beta \ln \Lambda_{max}$ corresponding to the ground-state energy for the quantum mechanical problem. It is now clearer why $N \to \infty$ corresponds to out-states (namely, $t = +\infty$).[47]

Parisi puts it all nicely: "By introducing the transfer matrix we have reduced the evaluation of the partition function of a linear chain [or a higher dimensional lattice] to the computation of the largest eigenvalue of a linear operator, this result allows us to use many powerful theorems of analysis, as we have just seen. The transfer matrix is a bridge between two apparently unrelated fields, probability theory and linear operators on Hilbert spaces."[48] But this mathematical statement turns out to be highly physical as well, about an equivalence in another arena. For what Parisi here calls probability theory is also in effect formally the path-integral formulation for the transition amplitude in quantum mechanics (the Feynman-Kac and the Dirac-Feynman formulas), and what he calls linear operators is just the Schrödinger equation.[49] And the evolution operator (or Heisenberg representation) links them. Bulk matter is again seen to be analogous to the final states in a scattering problem, when all the interactions have been taken account of, here as a sum of two-body interactions. Bulk matter is like a free particle, which of course it is, since presumably all interactions have been accounted for in the infinite volume limit of the arithmetic constitution of its components.

Schultz, Mattis, and Lieb have pressed this latter point most forcefully.[50] They showed how Ising matter can be thought of as a system of noninteracting, independent quasiparticles, each with a well-defined energy or mass, truly an arithmetic constitution. In effect, a two-dimensional statistical mechanics problem is made into a one-dimensional lattice problem (the row particles), governed by something that looks like a quantum mechanical Hamiltonian with Pauli spin matrices at the lattice sites.[51] And "time" is the second dimension (hence the well-defined masses). Those row-particles are the normal modes of the system, corresponding to the Fourier components in the Onsager solution, row configurations of periodic up-down flips, $\uparrow \downarrow$, uniform between flips (either $\uparrow \uparrow \uparrow \uparrow \uparrow$ or $\downarrow \downarrow \downarrow \downarrow \downarrow$), standing waves of wavelength $2\pi/q$, along a row (let us say).[52] (The energies of these row states or particles are shifted due to column interactions, the kinetic energy or B term, which are in effect perturbations on the row interactions, A.)[53] The effective Hamiltonian is $\Sigma \epsilon_q N_q +$ constant, where ϵ_q is the energy of these quasiparticles and N_q is their number, which, since they turn out to be fermions, is either zero or one. The transfer matrix is exp $- \beta \Sigma \epsilon_q (N_q - 1/2)$. The largest eigenvalue (the ground state of the system) occurs when all the N_q are zero, the sum of all the energies. Once one knows the energy spectrum of the particles, the various properties of the system, including the infinity of the specific heat, can be readily deduced, for which see chapter 6. It should be noted that the T_c spectrum and the $T \neq T_c$ spectra look very much like the acoustic and optical modes of crystal lattice vibrations, respectively.[54] (See figure 4.5.)

More generally, the technology associated with quantum field theory has proved powerful in this realm. The right objects to be added up, the right free particles, the right degrees of freedom, turn out to be fermions, particles described by antisymmetric functions, particles having two states (as if they had spin $1/2$)—it would seem reflecting the yes-no character of whether adjacent spins are parallel or not, or, as we shall see, "that not more than one side of a polygon may be drawn between any pair of neighboring lattice points."[55]

In the various exact solutions, one may employ Onsager's quaternion algebra, or fermion pair-creation operators (pairing positive and negative wave numbers, from the theory of superconductivity), or anticommuting fermion emission and absorbtion operators to begin and end a bond, or more general anticommuting variables, to diagonalize the transfer matrix and find its eigenvectors.[56] In each method, an exact solution is possible, since the Ising model can be made to correspond to a noninteracting field. If spin interactions cross over (so that the lattice is no longer planar), as in crossing diagonal interactions, or in higher than two dimensions, or if there is a magnetic field present, there is no exact solution, because the analogous fermion fields are interacting.

Duality

In this section, I have chosen to be didactic and historical, rather than saying something like, duality, an early formal observation by Kramers and Wannier, really is a topological and homological notion, or, perhaps, is really about harmonic analysis on an abelian group, or a field theoretic notion expressing. . .[57] The process through which we come to a deeper notion, one that may take decades to achieve, is worth recalling, very much for the physical insight it provides.

Kramers and Wannier made a crucial technical observation in a 1941 paper, connecting a high-temperature or disordered system to a

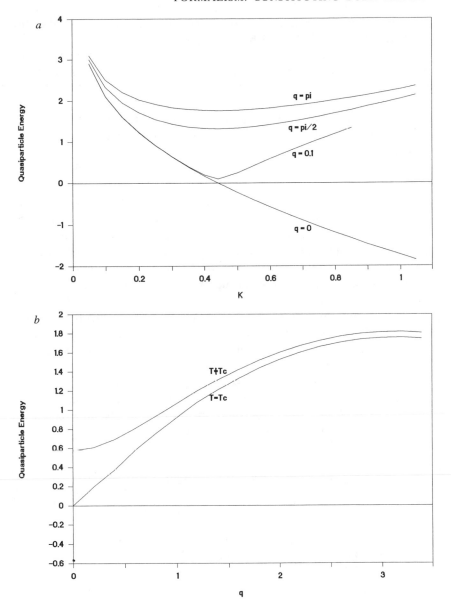

FIGURE 4.5 ϵ_q, the energies of the quasiparticles in $k_B T$ units for the two-dimensional Ising model. a, $\epsilon_q(K, q)$ for (from the top) $q = \pi$, $\pi/2$, and 0.1, and for the $q = 0$ odd state (which is degenerate with the $q = 0^+$ state for $K < K_c$); b, $\epsilon_q(K, q)$ for (from the top) $T \neq T_c$ and $T = T_c$. Adapted in part from Kaufman and Onsager, "Crystal Statistics," fig. 3.

low-temperature or ordered system: A row of spins, each spin either up or down, might be indicated by ↑↑↓↓↑↓↑↓. Now let us transform this row into another one, where if two adjacent spins are alike we write a zero, and if different a one (again, comparing the last spin with the first): let us call this transformation D_1. Then ↑↑↓↓↑↓↑↓ → 01011111. Now do a 2s complement, which we shall call D_2, so that 01011111 becomes 10100000. And then go back to the ↑ and ↓ form, so that 10100000 ← ↑↓↓↑↑↑↑↑ —namely, $D_1^{-1}D_2D_1$ (reading the operators right to left).[58]

Note that ↑↑↓↓↑↓↑↓ has equal number of up and down spins, while ↑↓↓↑↑↑↑↑ is six up to two down. We went from a configuration of spins with no particular order to one that is quite orderly. Kramers and Wannier realized that through this transformation every orderly configuration could be transformed one to one, into a disorderly configuration, and insofar as a predominance of disorderly configurations signifies high temperature, and orderly ones a low temperature, we have here a temperature-inverting transformation that is strictly formal.[59]

The crucial point for Kramers and Wannier is that this transformation can be expressed formally on the transfer matrix itself by a matrix transformation, hence on the partition function—and in so doing one relates the partition function at a low temperature to the partition function at a high temperature. If $Q = \Lambda_{max}^N$, and if $K = J/k_BT$ and if K^*, the dual coupling, is defined by $\exp(-2K^*) = \tanh K$, or $\sinh 2K^* \sinh 2K = 1$; so that K small means that K^* is large, connecting high temperatures to low temperatures, respectively; then Kramers and Wannier find

$$\Lambda_{max}(K)/\cosh(2K) = \Lambda_{max}(K^*)/\cosh(2K^*):$$

$$Q(K) = \{\cosh(2K)/\cosh(2K^*)\}^N Q(K^*).$$

(Or, as we shall see, $Q(\beta_{low}) = k^{-N/2}Q(\beta_{high})$, where $1/k = \sinh^2(2J\beta_{low})$. More generally, define $1/k = \sinh 2K_1 \sinh 2K_2$ for an asymmetric lattice.)

Technically, Kramers and Wannier showed specifically how the transfer matrix for a high-temperature system, which is equally likely to add on the next spin parallel to its predecessor or opposed in direction, may be transformed by a unitary matrix and a matrix transpose (in effect the transformations performed above, by the way rotating the lattice) into the transfer matrix for a low-temperature system, one that preferentially adds on one more spin oriented like its predecessor. These formal manipulations interchange the statistical weights for low and high temperatures.

In Kramers and Wannier's notation, the transfer matrix indices, i and j, are the decimal values of the binary spin representation of a row (not quite what I did above), say $0 = \uparrow$ and $1 = \downarrow$, of configuration i and configuration j, so that spin configuration 00011101 corresponds to $i = 29$. The transfer matrix is a $2^n \times 2^n$ matrix, referring to the 2^n spin configurations possible for n spins in a row, the index referring to a particular configuration. So, it is perhaps not hard to see that matrix manipulations, interchanging rows and columns and the like, could exchange the weights for orderly and disorderly configurations.

Given the matrix transformations, the eigenvalues of these two transfer matrices are simply related, and so the partition function for a high-temperature system may be expressed in terms of the partition function for a low-temperature system. At some intermediate point ($\sinh 2K = 1$) this duality transformation turns a system into itself, and this point is taken as the critical temperature for the Ising model.

Nowadays, the duality relationship is usually expressed at the level of the formal analogy of the high- and low-temperature power series expansions for the partition functions (for which, see below).[60] Onsager (1944) showed how this transformation might be thought of in more general topological ways—a high-temperature lattice is the same as a low-temperature topological-dual lattice—as long as, in two dimensions, there are no crossed bonds of interaction among the spins.[61] (Or we might think in terms of the interchangeability of Onsager's A and B.) The dual lattice is defined as a lattice formed from the original lattice by exchanging vertices and faces, by placing perpendicular bisectors on the original's bonds (figure 4.6); so, for example, a rectangular lattice (defined as a square lattice in which the horizontal and vertical couplings are unequal, $K_1 \neq K_2$) goes into a rectangular lattice where $\{K_1, K_2\} \rightarrow \{K_2^*, K_1^*\}$, while a triangular lattice is converted to a hexagonal one (figure 4.7a).

In sum, Kramers and Wannier show how formal features of the transfer matrix, inferred from the equivalents of the D_1 and D_2 transformations, then led to physical insights about the structure of Ising matter. Onsager's achievement was developing an algebraic formalism for these duality transformations,[62] one that would allow for a computation of the partition function, the algebra generating the full variety of spin configurations.

More recently, Baxter has made the most of apparently formal features of the (diagonal-to-diagonal) transfer matrix in order to solve

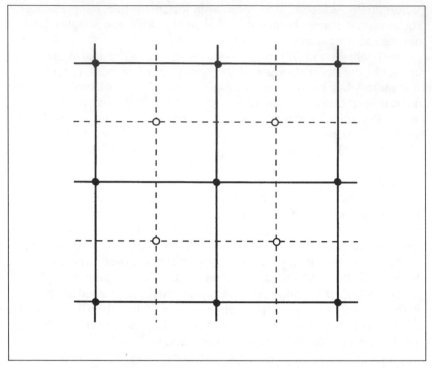

FIGURE 4.6 Dual (*dotted lines*) of a square lattice. Adapted from Baxter, *Exactly Solved Models*, fig. 6.1.

the Ising model, working only with its algebraic features—commutativity, invertibility, and symmetry (as he calls them)—without any specific representation for the spins.[63] But each of these features may be seen not only mathematically and formally, but also physically, in terms of the interactions of spins and how the lattice is built up by such an arithmetic procedure: for commutativity, in effect, there is near the critical point a directionally uniform correlation length for the lattice, ξ ($1/\xi \approx |\ln k|$), as a measure of fluctuation and so of temperature ($\ln k \approx t$, the reduced temperature); for invertibility, in effect, there is duality; and for symmetry, essentially this is a two-state spin system, its Boltzmann weights (exp $-\beta E$) invariant to various spin-flip transformations.[64] Conversely, given such intuitions of what is crucial about Ising matter, we might show why we might expect algebraic features such as commutativity.[65] For example, since the correlation length depends only on k, as discovered by Onsager; and the eigenvectors of the transfer matrix express the orderliness of the lattice and depend only on k; and

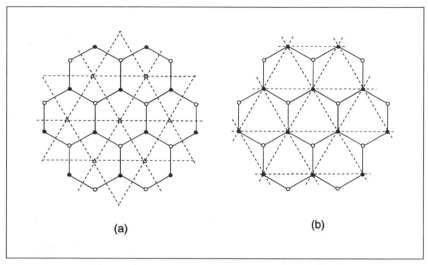

FIGURE 4.7 Duality (*a*) and star-triangle transformation (*b*) on a honeycomb lattice. Adapted from Baxter, *Exactly Solved Models*, fig. 6.4.

shared eigenvectors are characteristic of commuting matrices: we might expect that, if two transfer matrices had the same value of k, they would commute.

I should note that Baxter and Enting eventually show that one might solve the Ising model paying attention only to formal symmetries of the Ising lattice itself: duality, and a consistency condition for the interaction constants of a lattice.[66] The latter is called by Onsager the star-triangle relation (for which see chapters 6 and 7), referring to the hexagonal or honcycomb lattice, which may be seen to be composed of "stars," Y, and the triangular lattice, topological duals of each other. (The star-triangle relation allows us to express the partition function of the hexagon in terms of the partition function of a specific triangle, both at the same low or high temperature. Applying duality, we can then relate a high-temperature hexagon to a low-temperature hexagon. If the three couplings are equal, then it is the same hexagon.)[67] The star-triangle relation may be seen to be a requirement that the lattice possess a uniform correlation length, since it is about k invariance (figure 4.7).

I take it from such demonstrations, and many more that will follow, that what is crucial in the constitution of Ising matter are just these symmetry features: the star-triangle relation and duality. The star and the triangle lattices are intimately connected by a parameter, k, which is a measure of fluctuation, a surrogate measure of temperature. Or, as I shall discuss in chapters 6 and 7, in general in a physical system, absorption and scattering are intimately connected (as Einstein showed in the fluctuation-dissipation relations). As for duality, as Onsager put it, we may "associate order in a cold state with disorder in a hot one,"[68] so that disorderly configurations in a cold state correspond to orderly configurations in a hot one. Practically, this means that we may be able to solve problems with large couplings by going to the dual, so having small couplings that allow for perturbation theory approximations. The various devices employed for solving the Ising model may be seen as means for converting a highly coupled system into a weakly coupled one (in effect a "solid" into a "gas"), a fairly standard strategy for doing physics.

More technically, in the original work of Kramers and Wannier, as I indicated above, the product $\sinh 2K^* \sinh 2K = 1$ is used to define the dual temperature. Onsager, in computing the integrals that are the energy and the specific heat, shows how the product $\sinh 2K_1 \sinh 2K_2$ $(= 1/k)$ proves crucial, where K_1 and K_2 measure the horizontal and vertical interaction strengths in a lattice $(K_1 = J_{\text{horizontal}}/k_B T)$.[69] k will turn out to be both the "modulus" of an elliptic function, as Onsager originally introduces it, and a measure of correlation along the diagonal. A duality transformation converts $k \to 1/k$, although this is not explicitly mentioned at that time. (A large k, $k > 1$, means a high temperature, a small k, $k < 1$, a low temperature. $k = 1$ is the critical point.)[70] And the star-triangle relation or transformation is k preserving. Baxter's achievement was taking these k symmetries and showing explicitly how they were just about sufficient (given sufficient analyticity of the eigenvalues) to determine the full solution to the Ising model. k preserving leads to commutativity, and k inverting leads to the invertibility of the transfer matrix.[71]

Moreover, since $K \to K^*$ (or $\{K_1, K_2\} \to \{K_2^*, K_1^*\}$) corresponds to $k \to 1/k$, we can now restate the Kramers and Wannier result as $k^{-N/2} Q_{\text{dual}}(1/k) = Q(k)$. This is just what we might expect from what is called a modular transformation of an elliptic modular function of modulus k. Also, if $w = \tanh K$, then $w^* = (1 - w)/(1 + w)$, a conformal mapping. These observations, reformulations that were not employed by Kramers and Wannier or by Onsager, will be seen to preview much of what occurs in this area over the subsequent fifty years. But it is clear that Kramers and Wannier were aware of the importance of a variable like k (their κ), and of course Onsager defined k.

One might say that by expressing the constitution of matter arithmetically, physicists have gained enormous insight into major symmetries—and so the crucial physical features and conserved quantities—of bulk Ising matter, the mathematical technologies being ways of revealing, pointing to, and expressing those symmetries.[72] Insofar as those symmetries are poignantly understood, then one can exactly solve the Ising model in two dimensions with greater and greater perspicuousness. But it is also true that it took a great deal of collective effort before these facts about k were appreciated.

The Central Limit Theorem

It may be useful at this point to again bring up an analogy with the central limit theorem of statistics, a description of the asymptotic shape of normed sums of random variables $(1/\sqrt{N} \sum_i^N X_i)$—which I suggest is something like the thermodynamic limit or an infinite volume limit. Instead of building up bulk matter atom by atom, one "builds up" a random variable, S_N, in terms of a telescoping sum of its purported components: $S_N = \sum^N X_i$ (note that these Xs have nothing to do with Onsager's Xs). Again, the central limit theorem says that if the X_i are identically distributed, mean-zero, variance-one, independent random variables, then $\lim_{N \to \infty} S_N/\sqrt{N}\,(= (1/\sqrt{N})\sum^N X_i)$ possesses an asymptotic limit, the Gaussian with mean zero and variance one. As importantly, the theorem holds up fairly well if the distributions are not so identical and if there is some statistical dependence among the X_i.

Here the central or asymptotic limit (the Gaussian, $1/\sqrt{N}$) is analogized to the thermodynamic or asymptotic limit $(1/N)$, so that a normed arithmetic sum possesses an asymptotic form. The fact that the limit in the central limit theorem is fairly robust is much as we would hope for the constitution of ordinary matter as well, where small deviations in the composition or interaction strength would in general have small or negligible effects. As for the analogy to thermodynamic stability, I suspect that the corresponding feature for random variables is that the variance of the normed sum, the variance of the asymptotic Gaussian, should converge to a constant, approximately one, even if there is some dependence among the variables; and that variance would become larger if we increased the variance of the component X_i.[73] Otherwise, we would have the situation where increasing the variance of the components would decrease the variance of the normed sum—less accurate measurements leading to a more accurate average measurement.[74] I take this as the correspondence, since for thermodynamic stability we expect that the specific heat is nonnegative for a similar reason—otherwise, a system would spontaneously heat up (at least until its specific heat became nonnegative).[75]

There is one other interesting correspondence. At the phase-transition point for these thermodynamic systems, the specific heat or its derivatives are infinite—because there is correlation among the atoms at all scales, and there is in effect no small bulkettes (namely, surface terms are nonvanishing). As we shall see, the number of excited degrees of freedom for the two-dimensional Ising lattice is infinite (their energies are close to zero). Similarly, here, the variance of the normed limit distribution grows as N if there is sufficient statistical dependence among the X_i, and hence is infinite in the infinite volume limit.[76]

The Physical Constitution of Matter

As I have indicated, these models of the constitution of matter are not only mathematical devices and formalism. In actual physical circumstances, we might imagine building up bulk matter dynamically and arithmetically, atoms one by one building up or freezing out onto a substrate, the substrate rearranging itself so that its crystalline structure becomes better established. (Actually, freezing is a cooperative phenomenon of all the molecules, while we might eventually believe that adding on one more atom to a crystal is a local increment.) Or we might imagine atoms aggregating into bulkettes (by whatever means, including arithmetic), but then those bulkettes aggregating among themselves at each scale. Since these models are designed to account for the gross features of bulk matter, they are not likely to be correct in detail about how this dynamical physical constitution takes place. But a detailed model of the dynamics of that physical constitution might be expected to fit under the umbrella of one of these formal models of the constitution of matter.

In any case, whether we are being geometric or arithmetic, we might expect emergent phenomena at various scales or particle numbers, so that mere multiplication or addition might have surprising effects. The claim, however, is that there is some scale or particle number at which there are no new surprises as we scale up or increase the number of particles, while maintaining a constant density: the free energy density approaches its limit, or a fairly sharp phase transition appears, even if it is only truly sharp in the infinite volume limit.[77] (Recall that, in the arithmetic composition of the Ising model, N cancels out of the free energy per particle because the largest eigenvalue [whether degenerate or not] dominates the sum, since the relative contribution of the next largest eigenvalue goes as $(\Lambda_{next}/\Lambda_{max})^N \approx 0$.) In fact, that scale or particle number is in practice not so very large: this is the infinite volume limit in actuality.

Moreover, one expects that there is some separability of scales or numbers, namely, that whatever the details at one scale or size, they are, for the next significant larger scale, epitomized by a very small number of variables. The microscopic details are for the most part averaged away. (Hence, there are effective field theories, which mask these microscopic details, and there is, for example, a chemistry independent of elementary particle physics.)[78]

In general, at the infinite volume limit, only a very small number of features of all the smaller scales or sizes will play a role in characterizing bulk matter. Those features are likely to be epitomes, representative of in-effect emergent, global features, such as dimensionality and topology, or temperature and pressure, whose presence is already embedded, albeit implicitly, in the rules (of topological connectedness, for example) we use to build up or analyze the bulk matter. So, for example, the transfer matrix presumes there is a single temperature assignable to the lattice, and dimensionality is built into the transfer matrix's construction. A constitution of matter, a model, initially builds in the possibility of particular emergent features, in detail (although we might have surprising discoveries as well). So, for example, models of a phase transition that do not have the possibility of a failure of continuity or analyticity (in the infinite volume limit) cannot work for many purposes.[79]

Insofar as there is an infinite volume limit, there is in a deep sense a sharp separation between the fundamental properties of the presumed components of bulk matter and the properties of bulk matter itself. And so there may well be a variety of very different sorts of model-dependent components, yet they all lead to just one phenomenological bulk matter.

THE HOLOGRAPHIC CONSTITUTION OF MATTER

A hologram captures an image whole, rather than in spatial units as in an ordinary photograph. Each part of the negative provides the entire image, albeit in reduced resolution. And the more of a hologram we have, the more detail at finer and finer spatial resolution we have, again unlike an ordinary photograph where the more we have, the more of the whole picture we get. A hologram may be said to capture an image at a hierarchy of scales or degrees of spatial resolution.

Now, imagine if we were to constitute bulk matter—namely, to compute the effective Hamiltonian (at bulk scale)—in terms of varying degrees of nearness or neighboringness of atoms to each other, the physical world seen at varying scales or degrees of spatial resolution: in effect, we compute the partition function at each scale separately, by device implicitly incorporating interscale configurations such as rearrangements at scale $2l$ with those at scale l—and thereby systematically take into account all the configurations of the system. (The geometric constitution does this by means of its bulkettes or blocks, in which non–nearest-neighbor interactions do become significant. Here we stay in the realm of the actual individual spins.)

Inverting the hierarchy of a hologram of low resolution to high, we start off with configurations and interactions between nearest neighbors, the only spins in the Ising model that interact directly; then, we consider next-nearest neighbors, which only in-effect interact with each other, through their direct interactions with their mutual nearest neighbors; and so forth, until we have constituted matter as a hierarchical sequence of spin configurations, and of effective interactions or Hamiltonians between atoms and their "neighbors," at the various scales. (A large-scale neighbor would in reality be quite remote.)

At any single scale, we might well expect that, since the distant neighbors are nth hand, the effective interactions with them is weaker than those with nearest—but since there are more such farthest than nearest neighbors (so that in two dimensions the number of neighbors at distance l is proportional to l), the effect on the partition function of all those farthest neighbors (that is, the configuration or partitionings they allow) might be very substantial. (Recall that, when direct interactions are presumed to be of infinite range, such systems are not energetically stable, in the infinite volume limit: Coulomb forces work only because there is screening.) Conversely, the direct interactions with nearest neighbors are strongest (and effective interactions with comparatively near neighbors are still strong). But, in each such case, the number of configurations that contribute to the partition function is quite small, since a comparatively small number of particles are involved. The crucial technical question becomes how and whether the account of matter in terms of varying degrees of nearest neighbors converges to a limit. Does such an account provide for phase transitions and for rigidity?

At this point, it may be useful to emphasize that these descriptions of the constitution of matter are effectively realized in detailed mathematical derivations that do display the crucial phenomena. While I have been relatively nontechnical in my exposition, it is meant not to be hand waving. Rather, I want to provide some enlightenment about what is going on in those derivations, connecting the technical apparatus with the physical models.[80] (It will prove rewarding the consult the original papers.) It is a redescription of experts' work that, hopefully, they will recognize as unexceptionable and unexceptional. This is what one might say to a student or a colleague before or after launching into a formal derivation. Namely, it is motivation.

Such motivation is motivation for something, and without that something it would not be worth much. So, for example, Kenneth Wilson provides motivation for the holographic constitution (as well as for the geometric one) by analogizing the infinite volume limit (what he calls the "statistical continuum limit") to the continuum limit characteristic of the differential calculus. The partial differential equations of mathematical physics, such as Maxwell's equations, express the fact that interactions are local through those well-defined continuum limits (derivatives), as well as local source terms. Similarly, Wilson argues, no matter how you formulate the geometric/holographic constitution, the kind of decimation (see below), the shapes of the bulkettes—in the end, the infinite volume limit ought to be the same. That is, in the limit of the largest scales, the properties of matter ought to be invariant to how you took that limit.[81] Such motivation is set up so that the reader can appreciate an analogy to the calculus, one that makes sense of Wilson's renormalization-group calculational technology.

Again, the holographic constitution—which, for the partition function, means enumerating all the possible configurations and their contributions, $\exp - \beta E$, to it—starts at the scale of nearest neighbors. As we go up in scale, instead of contributions due to bulkettes composed of

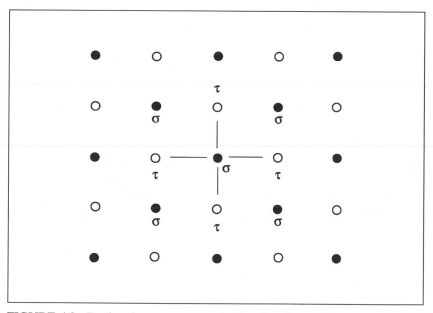

FIGURE 4.8   Decimation on a square lattice. See figure 4.7b for a similar process in the star-triangle relation. τ and σ are the fixed and unfixed spins, respectively.

bulkettes, or those due to individual atoms, we cumulate contributions due to configurations of atomic interactions at each relevant scale.[82] All but nearest-neighbor interactions at the atomic scale are induced or in-effect interactions. As we shall see, this scheme would seem to demand that the lattice and its interactions be divisible into two interlaced parts—essentially the light and dark circles in figure 4.8—so that there are only light-dark interactions (and no light-light or dark-dark) ones.

The decimation method of Kadanoff takes the above insights and constitutes matter holographically as follows: Compute the partition function as $Q = \Sigma_{\text{all other spins}} \Sigma_{\text{neighboring spins}} \exp - \beta H_0$, where H_0 is the original nearest-neighbor interaction. The inner sum has few possibilities and is readily done, if one fixes the spins on the light diagonals (namely, leaving them as variables, τ_r)—that is, one actually sums over the dark circles adjacent to each light one, σ_s—because there are no light-light interactions. Then the contribution to the partition function from each unfixed spin (namely, the sum of the exponentiated interaction energies for all its light-dark configurations) is formally identical with and reasonably separable from the contributions from the other unfixed spins. To anticipate our discussion, formally

$$Q = \sum_{\{\tau, \sigma\}} \exp - \beta H_0 = \sum_{\{\tau\}} \sum_{\{\sigma\}} \exp - \beta H_0$$

$$= \sum_{\{\tau\}} (2 \cosh K\tau_i \tau_{i+1})^{N/2} = \sum_{\{\tau\}} \exp - \beta_1 H_1,$$

where $H_1 = \Sigma(A + B\tau_i \tau_j + C\tau_k \tau_l \tau_m \tau_n)$.[83] (The cosh is from the sum of $\exp - \beta E$ and $\exp + \beta E$, and the $N/2$ shows that we have summed over half the spins.)

What would be a straightforward problem in adding up independent random variables becomes a much more difficult problem when we are adding up exponentiated random variables—and the technical apparatus that we shall encounter (transfer matrices or decimation or Pfaffians or renormalization groups) can be seen as ways of dealing with this fact.[84]

One can formally represent that inner sum as an exponential in the leftover or fixed spins ($Q = \Sigma_{\text{leftover spins}} \exp - \beta_1 H_1$, where $H_1 = \Sigma_{\text{neighboring spins to the fixed spins}} \exp - \beta H_0$), and so pick out an effective Hamiltonian, H_1 (for it has already added in or summed over or integrated out the effects of the unfixed spins)—namely, the effective Hamiltonian being whatever is in the exponential. Formally, H_1 more or less looks like H_0, except that now there are not only "neighboring" spin-spin interactions, of the τs ($B\tau\tau$, above), but also next-nearest-neighbors interactions (the C term), and so forth, at a scale that is $\sqrt{2}$ as large as H_0's. The original σ spins have been summed over, the dark sites erased, so to speak. And B is different from J. Moreover, β_1 is different from β; it is said to be renormalized. One then fixed spins on the next superdiagonals (so that one is scaling up), in effect a new set of σs and τs; computes the effective Hamiltonian; and so forth. Each scale is $\sqrt{2}$ the size of the predecessor; and, again, as scales go up, implicit not-so-nearest-neighbor and diagonal interactions appear in the effective Hamiltonian.

It seems that what formally allows this procedure to work is the bipartite character of the lattice and the curious fact that $\cosh K\tau\tau = \exp - \beta(A + B\tau\tau + C\tau\tau\tau\tau)$. This would seem to depend on the fact that the τs have just two values, $+1$ and -1, and that the lattice is partition-function invariant (the sum of the Boltzmann weights) to not only a total spin reversal, but to other spin-inversion symmetries. Moreover, as we go to larger scales, the cosh-to-exp transformation, which defines the effective Hamiltonian, remains comparatively tractable. Implicit complicated spin interactions may be safely neglected.

So far, all we have is a formal device or a scheme for adding up or integrating out contributions to the partition function (through the modified effective Hamiltonian) at small scales. However, it turns out that, much as in the geometric constitution, the behavior of the interaction or couplings, A_i, B_i, C_i, and β_i, in the effective Hamiltonian as a function of scale, $\sqrt{2}^i$, can tell a great deal about the properties of the system near the critical point.

The nontrivial task is to show that such a constitution of matter, such a cumulation procedure, might actually be effectively implemented, theoretically and practically and numerically.[85] There must be a sequence of comparatively simple effective Hamiltonians (in effect, not too many τs in any one term). And they must converge to an infinite volume limit in a nice way, in that their coupling constants (β and A, B, and C above, for example) are well behaved (perhaps converging to zero or to infinity). Now, we might expect that such a sequence ought to converge, for at some wavelength the effective Hamiltonian is much like that of a spin bulkette interacting very weakly (or very strongly) with other bulkettes (except at the critical temperature).

In actual models, it is shown that, as the wavelength approaches infinity, the "nearest-neighbor" effective interaction or couplings (at the current scale or wavelength—again, having taken into account or integrated out or summed over all smaller scales) go either to zero, or to infinity, or to a fixed point. In the limit, either there is no rigidity (the coupling is zero, just what we would like for the stability proofs) or there is perfect rigidity (the coupling is infinite, and this system is not stable in that it exhibits $N^{1+\epsilon}$ effects), or there is a critical point for which all scales matter and for which the coupling of nearest neighbors at each scale does not converge to zero or infinity, yet perhaps converges to a limit rather than cycling. We are interested in the rate of convergence with respect to scale so as to determine the exponents of the reduced temperature, t, the αs.

What is of course remarkable is that there might be such limits and fixed points, and this is, presumably, a mathematical fact and a physical fact about our models, presumably nicely in accord with empirical observations of critical-point behavior. But it is still a quite remarkable fact, one that needs to be demonstrated, that the way the coupling constants scale up or down as we go to larger wavelengths is related to ways macroscopic variables scale as we move away from the critical point.

Again, the holographic constitution of matter, insofar as we might actually and practically employ it, seems to depend on a number of features. Interactions between atoms have to be local, between nearby neighbors. Otherwise, procedures such as decimation will not readily converge (although that procedure may still safely involve second- and third-nearest-neighbor interactions, still quite finite neighbors). The Ising model stipulates such locality; it is a desideratum for the proof of the stability of real matter, a desideratum that requires much ingenuity (those almost neutral balls) and physics (screening of the Coulomb force) to be achieved.[86]

A second curious feature of decimation is the apparent desideratum that the lattice and its interactions be bipartite, so that it can be repeatedly divided up into two interlaced parts, one of whose spins is fixed, while we then may sum more readily over the other's spins. We see a similar phenomenon in the diagonal-to-diagonal transfer matrix of Onsager and of Baxter, so for example the lattice is built up as shown in figure 4.9.

The transfer matrix, V_a, builds up one layer, V_b the next, and $V_a V_b$ is the arithmetic mode of the constitution of matter. We see a similar phenomenon in the star-triangle transformation, where, by dividing the hexagonal lattice into two parts, we create a triangular lattice (figure 4.7b).

These observations suggest how crucial is the bipartite character of the lattice for any of the modes of constituting matter through solutions of the two-dimensional Ising model.[87] Namely, as I indicated above, these constitutions of matter are exactly solvable for a planar lattice (or an orientable graph) in which interactions do not involve crossovers in two-dimensions.[88] Onsager points out that a duality transformation is possible only if there is such planarity: the duality of Kramers and

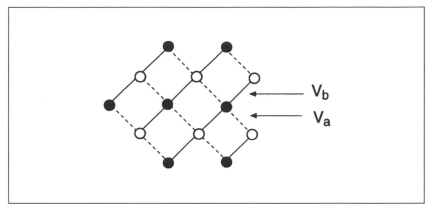

FIGURE 4.9   Diagonal-to-diagonal transfer matrix, $V_a V_b$, on a bipartite rectangular lattice, with different horizontal (*solid line*) and vertical (*dotted line*) interactions. Adapted from Baxter, *Exactly Solved Models*, fig. 7.1.

Wannier, which "*always* converts order into disorder and vice versa," was "but special applications of a transformation which applies to all orientable graphs."[89] The solution to the Ising model then proposed by Onsager, an algebraic arithmetic one, did not manifestly employ this topological fact (although that solution was motivated by a duality transformation, and it is built into the Fourier transform he performs). Baxter and Enting use duality and the star-triangle relation directly (k being the crucial variable), in order to perform scale transformations (in effect a holographic constitution) on the correlation function of spins and so to solve the Ising model.[90]

Duality here is much as it is for the geometric solution. In the holographic constitution of matter, duality shows itself as we integrate out or add up the contributions of configurations at larger and larger scales. At each point in the process, there is a residual spin-spin interaction, but now of more and more widely separated "neighboring" spins. For temperatures below the critical point, that residual interaction becomes stronger and stronger, implying that spins far apart are well correlated with each other. For temperatures above the critical point, that interaction becomes weaker and weaker. As we go to larger and larger scales, the lattice either goes in effect to zero or infinite temperature (since the apparent interaction strength is $\approx J/k_B T$). At the critical point, those interaction strengths become finite nonzero constants, in effect saying that the lattice looks the same at all scales.[91]

I should note that, in the actual numerical calculations of the infinite volume limit within the holographic constitution of matter, one ends up dealing explicitly with only a very finite lattice of perhaps a few hundred atoms. It is, however, a contingent fact, as far as I can tell, that a not-so-large-N approximation employed in any particular model of the constitution of matter actually works effectively. We might say that, if more than a few decimations were needed to get accurate results (comparable to experiment, let us say), we would not have come to appreciate the power of the holographic constitution. (Of course, the exact number of this "few" depends on our ingenuity and our computational resources.) However, the presumed group property of the coupling constants allows one to get much closer to the critical point than numerical approximation would seem to allow.[92] We might well say physics is doing lots of the large-N numerical work, the group property reflecting the self-similarity of the system at the various scales near the critical point.

As we shall see later, certain functional forms (modular or elliptic functions), in which k plays a central role, are just the right forms to extrapolate from low or high temperature toward the critical temperature. Put differently, if a constitution and its mathematical formulation gets at the deep physics that informs a system, presumably fewer terms in an asymptotic expansion are needed for good results, so N need not be so large to get effectively close to the infinite volume limit.

THE TOPOLOGICAL / COMBINATORIC / POLYMERIC CONSTITUTION OF MATTER

The localized units or atoms interacting with their neighbors that we have used to constitute Ising matter have been quite varied. For the geometric and arithmetic constitutions, those atoms are localized in ordinary space, in one case bulkettes interacting among themselves, in the other individual atoms interacting with a substrate that is presumably quite large already, the added energy and contribution to the partition function of one more atom eventually being a constant marginal add-on or multiplicative factor. In the holographic constitution, those atoms are localized by their scale of interaction (in what physicists call momentum space, or Fourier-transform space), and again at each scale they interact more-or-less locally with each other, that interaction becoming trivial (infinite or zero) as we scale up, unless we are at the critical point.

Now, very differently, we might well think of matter as a polymer, namely, a set of intertwined closed chains of linked atoms, chains weakly interacting among each other.[93] Matter so constituted would be more rigid if those polymers were longer, more intertwined, or shared some bonds. We might reasonably expect that, if the temperature were high, the closed chains would be quite short in length, while, if the temperature were low, the chains would be much longer—since at lower temperatures, any single link is less likely to be broken by the comparatively smaller random thermal energies $(k_B T)$. Were we to enumerate all the possible chains, weight them by their likelihood of being stable (longer chains in general being less stable than shorter ones), and find a way to add their contribution to the partition function, we might provide ourselves with a constitution of matter. Again, the issues are, How are we to actually do such addition? Will such an account converge to a limit? Will that limit be recognizable?

The partition function for a two-dimensional Ising model can be shown to be a weighted sum of the number of length-l simple closed polygons and their superposition (which have no shared sides) that can be drawn on the lattice or its dual. One expands $\exp \beta J \sigma \sigma$ to $\cosh K \times (1 + \tanh K \sigma \sigma)$, and the products of the second term for all the adjacent spin pairs do the required work of defining the polygons. In effect, the partition function is a generating-function for the number of these polygons of length l, $Q = \Sigma P_l w^l$, where $w = \tanh K$.[94] (Note that, in effect, a side of a polygon corresponds to all the possible arrangements of each pair of spins that make up a bond: $\uparrow \uparrow, \uparrow \downarrow, \downarrow \uparrow, \downarrow \downarrow$; each polygon corresponds to a term representing certain spatial configurations in an expansion of the partition function.) That weight in the high-temperature expansion, $(\tanh J/k_B T)^l$, is small (approximately zero) for higher temperatures, longer lengths, and weaker bonds. It approaches one for lower temperatures, shorter lengths, and stronger bonds. The reverse is true for the weight, $(\exp - 2J/k_B T)^l$ ($= \tanh^l K^*$), the change in the partition function due to l-pair flipped spins, in a low-temperature expansion on the dual lattice (and hence the K^*). Changing the temperature shifts the intermediate point between polygons that count for a great deal $(P_l w^l > 0)$ and those that do not count for much $(P_l w^l \approx 0)$ in the partition function.[95] (It should be kept in mind that P_l is topological, and is temperature independent.)

Such a constitution of matter becomes of practical interest if we can analytically compute the coefficients, P_l. Or, if not, at least we might hope to fairly accurately and rapidly compute the coefficients of such series numerically, and so actually do the high- and low-temperature expansions. Given enough terms in the series, we might analytically continue the approximations close to the critical point, from either the high or the low temperatures.

Formally, it would be desirable if we could more or less automatically and systematically enumerate all the possible polygons. For most situations, a great deal of art is needed to actually do this. But, in fact, following an observation of van der Waerden, it was shown by Kac and

Ward how two-dimensional Ising matter might be constituted explicitly and automatically in terms of such polygons, using the device of a matrix, which they call A, constructed so as to enumerate the various polygons and weight them appropriately by means of its determinant. A particular A_{ij} represents a link or bond between two sites, and the products that appear in the determinant, $A_{1,5}A_{2,6}\ldots$, represent particular polygons. (We are now explicitly in the realm of mathematical partition functions and generating functions, as in chapter 3, and, as we shall see, these polygons are in effect random walks.)

However, the expressed purpose of Kac and Ward's endeavor was "to understand how the algebraic method [of Onsager, what I have called the arithmetic constitution]. . . performs the actual counting [of the configurations of closed polygons]."[96] (Onsager's quaternion algebra and his Lie algebra of operators, $\{A_i, G_k\}$, where Onsager's A operator and Kac and Ward's matrix A are not directly related, can be seen to generate all the various polygons, and so might be called an algebra of loops. Onsager's P_{ab} draw the sides of the polygons.)[97]

Ironically, or at least as a sign of free trade among the constitutions of matter, Kac and Ward say, "[T]he exact formulas (for finite lattices) of Onsager and Kaufman [the algebraic arithmetic method] should provide a clue to the proper method of counting and thus lead to an elementary combinatorial approach to the [Ising] problem" (p. 1332)— the idea being that one might be able to guess the form of the combinatorial solution (the generating function for the number of polygons of length l) from the arithmetic, algebraic, analytic solution of the Ising model for small finite lattices (reminiscent of Onsager's guessing the general arithmetic solution from the solution for narrow lattices).[98] Onsager was aware of the topological aspects of his mode of solution, in particular how thermal duality was simply expressed on a lattice geometrically dual to the original one. Put differently, the arithmetic constitution of matter and the topological are taken by their proponents to be intimately connected and suggestive for each other.

By the way, it is surely a curious fact that the exact formulas for finite-sized lattices are sums of all four elliptic theta functions (each of which can be expressed as an infinite product).[99] The partition functions for numerical partitions are of a similar functional form. As we shall see, this is a quite freighted fact.

Duality

While it might make sense to speak of polymers and polygons, and mathematically this turns out to be a powerful technology for constituting matter, just what are these closed polygons and polymers as physical objects? (I am speaking in two-dimensions for the moment.) Again, there are two sorts of models: the high- and the low-temperature expansions. In the high-temperature expansion, the polygons may represent the polymers themselves, here quite short: the set of lines or links between spins, representing reasonably strong interaction between a pair of spins, where *reasonably strong* is meant to say that the interaction dominates significantly over thermal noise. Recall that, as the temperature becomes larger and the thermal noise becomes more significant, we would expect that the relevant polymers would become somewhat shorter—the probability of any single link in the chain being strong enough to dominate over noise going down, and so the probability of a chain being closed going down even more (presumably something like p^l, where, roughly, p is the probability of a single link being stronger than thermal noise).

The polygons, rather than being taken as a chain polymer, might also be seen as polygons of enclosure of like spins, in effect droplets or inclusions, what is emphasized in the low-temperature expansion on the dual lattice. If the temperature is very low, then most of the spins will point in one direction, the exceptions will be infrequent, and the size of any single enclosure will be quite small. (Note that the dominant direction of spin is taken to have no enclosure, going to infinity. See figure 5.3.) Matter at very low temperatures is constituted by small inclusions of uniformly oriented spins surrounded by a sea of oppositely oriented spins. And again the probability of an inclusion of perimeter l is $(1-p)^l$, where $(1-p)$ is the probability that adjacent spins are not like each other (if they were like, then the boundary would not be drawn between those spins). Note that, by definition of the dual lattice, boundaries of inclusion are drawn on that dual lattice.

More technically, the polygons of inclusion are derived from thinking of a lattice in terms of like and unlike neighbors, re-marking the lattice so that at each bond or edge one indicates that likeness or unlikeness, and noting that one has a new lattice, again the dual, whose polygons are polygons of inclusion.

In any case, my discussion of the actual physics of this topological constitution of matter has been motivated not only by a desire to show that the mathematical mode of constitution is physical. Rather, I have been exposing the duality properties of the topological constitution. To each small polygon of inclusion at low temperatures, on the dual lattice, there corresponds a small polygon of interaction at high temperatures, on the original lattice. Polygons of inclusion, if they are mostly small, represent highly orderly states; polygons of interaction, if they are mostly small, represent highly disordered states. Hence, we expect high- and low-temperature constitutions of matter to be formally connected.

Note as well that, if the geometric and holographic constitutions and their geometric duality focus on behavior near the critical point, where scaling and self-similarity are the dominant phenomena, then the high- and low-temperature expansions and their arithmetic duality would seem to have a vary hard time being accurate approximations as they approached the critical point, where more and more lengthy and complex polygons become important. The regularities of those theta functions and of k, and of that modular transformation we encountered earlier, would seem to be behind the success of the polymer constitution.

What for Kramers and Wannier was a symmetry of the transfer matrix, here becomes both a topological and a physical fact. What was formalism, or in Onsager's hand an observation leading to a formalism, here becomes the physical justification of a constitution of matter.

Enumerating the Polygons: An Analogy with Quantum Mechanics

As I have indicated already, the partition function, as a weighted sum over all states of a system, is formally much like the Feynman formula for the probability amplitude in quantum mechanics, as a weighted sum over all paths or histories of the system.[100] The task in each case is to be able to estimate those sums in a perspicuous fashion. The polygonal expansion is much like the Feynman graph expansion, for the partition function and the probability amplitude, respectively. (They are both an asymptotic series.) Ideally, one wants a device to automatically enumerate the polygons or the Feynman graphs, and thereby perhaps develop a methodical way of summing such polygons and graphs. The Kac and Ward determinant is just such a device.[101]

Kac and Ward's paper is a wonderful example of an ingenious construction meant to achieve a certain end, worth reading by any budding scientist. It should be noted that the determinantal method soon yielded to a more firmly grounded method, one with more physics directly attached to it. But now it was the antisymmetric determinant, the Pfaffian, of another matrix (again meant to count all the polygons), that Pfaffian, like the determinant, being a multilinear form on the elements of that matrix. If the lattice is bipartite or planar, then it can be shown that such a method is "obvious," both for topological and quantum field theoretic reasons.[102] In chapters 6–8, I shall examine this approach, employed by Hurst and Green when they used quantum field theoretic methods to enumerate all the polygons on the lattice, the crucial fact being that in the enumeration there was at most just one link between sites. Therefore, one is led to link-operators whose product is zero if in the enumeration there are two joins between the same sites; namely, they are antisymmetric or fermion-like operators.

What at first might seem like a device for solving a physical problem, the expansion in terms of polygons, is seen to be a model of how matter might in general be constituted out of its parts, as a polymer, its parts having been chosen to do that constitutive work. Such a constitution is seen to be much like how physicists constitute dynamical processes, such as particle scattering. The matrix forms (and the determinants and Pfaffians) can be shown to be physical as well:[103] Statistical mechanics on a lattice is a model of quantum field theory.

MATHEMATICAL PHYSICS AS PHILOSOPHICAL—AGAIN

Each of the various ways of solving the Ising model draws on different mathematical technologies in order to constitute matter. Those technologies emphasize different features of matter's constitution.

We might summarize our considerations in terms of constitution, mathematical technology, conception of matter, and manifestation of duality, respectively—geometric: block-spin transformation and the renormalization group, bulkettes, and fixed point of a transformation; arithmetic: transfer matrices and analysis of linear operators, rigidity (transmitting a force down the line), and symmetry of the transfer matrix; holographic: semigroups and correlation functions, orderliness in terms of linear scales, and fixed points of transformations; and topological: topology and graph theory, linked and clustered atoms, and topological duality and star-triangle transformations.

Each mode of constitution suggests just what is essential about bulk matter: local interactions, a capacity to propagate order, effective interactions at all scales, and high interlinkages of the lattice. The orderliness and rigidity we associate with bulk matter, and its curious behavior at the critical point, would seem to depend on these essential features. *This* is what matter is, at least in its most general constitution: bulkettes, transfer matrices, scale invariance, and polygons/polymers. And "this" is understood in a formal technical way that shows why each feature is essential, its absence being disastrous for the model. Each feature makes possible a solution to the Ising model, with convergence to an infinite volume limit, so that the model is a candidate for representing nature. Just as philosophical analysis may show precisely what we are up to when we engage in ordinary life, so here mathematical analysis shows precisely what we are up to when we speak of bulk matter. And, it should be recalled, such an analysis then makes it possible for us to solve very difficult practical problems, showing how matter is constituted. Moreover, deep features of matter, such as duality, show themselves in different ways; those features enable each model or constitution to actually work as a means for calculation.

While there has been so far only one mode of proving the stability of matter, the geometric constitution, I have delineated here other constitutions of matter that have been employed to demonstrate the existence of a phase transition in the two-dimensional Ising model.[104] Each of these methods of constituting matter might be said to depend in the end on some technical feature: uniformity of convergence of a sequence, the commutation relations of matrices, the existence of a group (actually a semigroup) of transformation among Hamiltonians, and the employment of antisymmetric link operators. One would like to believe that these features could be shown to be "the same" in some sense—mathematically and physically. Similarly, given the pervasive significance of duality (and, as we shall see, the star-triangle relation), one would like to believe that its various appearances might also be connected physically and mathematically.[105] Finally, one might hope that the mathematical technologies of the modes of constituting matter could be shown to be formally equivalent, and the physical features also shown to be physically related.[106]

Matter as constituted by mathematical physics and so understood philosophically is still that everyday matter we encounter ordinarily, or so we might hope that to be the case. We might also hope that, since the mathematical accounts refer to that everyday matter, they are intimately connected. Now, were we to take mathematics as an empirical science, in that it studies the general properties of various objects (sets, spaces, polygons),[107] we might hope that various expressions of those general properties would be in some precise sense equivalent— that sense of equivalence being itself of some interest.

One last proviso: The constitutions of matter I have been describing are exactly solvable in only two dimensions with only certain kinds of nearest-neighbor interactions. The inability to generalize to three dimensions in an exact way is by no means a casual fact but surely reflects the topology of the two-dimensional lattice (versus that of the three-dimensional one).[108] Yet, even if these are toy models for three-dimensional problems, for my purposes they do illustrate that the constitution of matter is a philosophical problem analyzed through mathematical device. In any case, no theoretical physicist would allow the lack of an exact solution to get in the way of a model that seems productive and even true.

5

Analysis

GENERIC, FORMAL, MODEL-INDEPENDENT ACCOUNTS OF THE CONSTITUTION OF MATTER AS PHILOSOPHICAL

Marking a phase transition: formally defining a phase transition, the Lee-Yang model, modeling discontinuity • Phase transitions as marked by changes in symmetry • Phase transitions as marked by fixed points • The indicators of phase transitions • Formally defining a bulk quantity—the spontaneous magnetization

In chapters 2 and 3, I reviewed an apparently rigorous account of bulk matter, one that indicated that, in the infinite volume limit, bulk matter would be energetically stable, exhibit thermodynamic properties, and be thermodynamically stable as well—significantly, that we might successfully model bulk matter, as such. Other than questions of thermodynamic stability (that the specific heat and the compressibility are both nonnegative), and that the pressure, for example, is well defined, perhaps the most interesting question concerns the possibility of phase transitions in such bulk matter—namely, the sharp onset of long-range order within such matter: freezing into a crystalline solid or becoming permanently magnetizable. I then in chapter 4 described several more particular ways of constituting matter, each way being a solution of the two-dimensional Ising model. Each constitution and solution exhibits a sharp phase transition, indicated by a singularity in the specific heat, and by the onset of long-range order at that same point. Notably, the specific heat exhibits a logarithmic peak or singularity ($c_v(T) \approx -\ln|T/T_c - 1|$), and one can describe the behavior of the long-range order in terms of a function that scales up or down, dependent on the reduced temperature ($\approx |T/T_c - 1|^\alpha$).

From Kramers's detailed calculations of particular model systems (1934–1936), he came to appreciate that phase transitions appear only in the infinite volume limit: thermodynamics is a "limit science."[1] Then Kramers and Wannier (1941) indicated how duality—a systematic partition-function correspondence between orderly configurations at low temperatures and disorderly ones at high temperatures—would point to an intermediate temperature (rather than more than one) invariant to, or a fixed point of, that correspondence. Such duality, in an extended form, is just what I noted in chapter 4 for each of the particular models of the constitution of matter I discussed. In each model, or in each solution, the phase transition exhibits the expected sharpness or discontinuity—in the specific heat, say—only in the infinite volume limit.

In the arithmetic and topological formulations, which are usually presented as algebraic and exact and formal until one does a final approximation to an integral, the various solutions provided in the literature over the last fifty years arrive at Onsager's integral or some equivalent form for the free energy; then they rely on Onsager's original and exhaustive analysis of its behavior. Onsager's original paper, as well as Kramers and Wannier's, exhibited the maximum of the specific heat as a function of size explicitly ($\approx \ln N$), and Onsager as well estimated the shift in the critical temperature (the peak in the specific heat) for finite width ($n \times \infty$) lattices: $\approx n^{-2} \ln n$.[2] In the geometric and holographic solutions, which are differently algebraic and are often presented more directly in numerically approximate versions, the solutions most directly demonstrate the scaling behavior near the critical point, namely, the critical exponents.[3]

We might desire as generic (or model independent, even Ising model independent, say) and rigorous an account of phase transitions as we supplied for the stability of bulk matter, so that we might discover just what is crucial about any model of matter that allows for such sharp transitions and the long-range order or rigidity that, it would seem, often accompanies it. As we shall see, it turns out that a rigorous, formal, generic, mathematical definition of a phase transition is perhaps closer to experimental practice than is the result of any substantive model-dependent detailed calculation: the formal definition exhibits the discontinuities or infinities directly. Moreover, such a formal definition allows us to ask questions about whether a phase transition might occur within any particular kind of model, again perhaps more readily than if we had a detailed particular calculation. As we shall see, for such a definition to work, we need to go to the limit of infinite volumes, what is more or less approximated in actual practice ($10^{23} \approx \infty$) but, most important, what actual practice tries to or is presumed to emulate. There is, in general, a subtle interplay of the formal models and the particular ones: "Of course these [formal] considerations of Yang and Lee are not a complete theory of the condensation phenomenon. They show only a possible (and I think very likely) *mathematical* 'mechanism,' and they show how subtle and difficult the real theory will be."[4]

So, in analyzing experiments on phase transitions, physicists impute discontinuities to data sets that might well allow for smoothed curves. But, in the infinite volume limit, and we presume for good reason we are essentially there, our expectation is for discontinuity. Such an imputation of discontinuity is not to be justified by a calculation with the correct finite N, which solution is always smooth and analytic. We might show how a particular large finite N leads to a discontinuous-appearing curve, but a more generic account justifies that appearance. (I should note that in experimental studies of finite samples, when the number of atoms is much much less than Avogadro's number, such discontinuity is not expected, or seen, or presumed.) Actual experiments —those actual messy situations in which one is trying to get some sort of result that holds up—initially do not manifestly accord with the laws of physics and the expectations of theory. But they can be legitimately made to appear to do so—and this is good experimental practice.[5]

I shall also consider a rather more specific problem, the technical definition of the spontaneous magnetization, such as in a permanent magnet, to show how our way of thinking about a phenomenon is influenced by what we can calculate.[6] There is an obvious definition of the spontaneous magnetization; however, it does not readily lead to doable calculations. Instead, somewhat different technical definitions do lead to practical calculations, different modes of calculation demanding different definitions—although, and this is crucial, all lead to the same formula for the spontaneous magnetization. Along the way, seemingly problematic definitions that work in some calculations are shown to work because of the peculiar or contingent physics of the model system being studied. I take it that one task of a theoretical science is to find technical definitions of ordinary phenomena that are true to those phenomena, both experimentally and theoretically, and that are computationally tractable as well.[7]

MARKING A PHASE TRANSITION

An orderly phase of matter in equilibrium is stable in the sense that for long times it retains its order. Yet we also know from theoretical calculation that a physical system that is in equilibrium, one that consists even of a very large number of molecules, can, over very very long times, spontaneously go from one seemingly stable state to another.[8] (Of course, if it is not in equilibrium, the transition can be quite rapid—think of metastable, supercooled water.) Since our observations are always for times that are short compared to long times, and very short compared to very long times, observed spontaneous transitions between orderly states play a limited role in actual experimentation in bulk matter.[9]

If we are far from the infinite volume limit and have a very small system, we are in a very different realm. For "[a] detailed computation shows that we need to wait for a time proportional to $\exp N^{(D-1)/3}$ [where N is the number of atoms and D is the dimension] (an incredibly large time!) to see the spontaneous reverse of the magnetization."[10] For one-dimensional systems (which, for localized interactions, cannot maintain a spontaneous magnetization) or for small systems, the time is perhaps not so small.

Moreover, "[i]n order to reproduce the experimental situation [of a phase transition, in a calculation] we must first send the volume to infinity, and only later the observation time to infinity: The two limits do not commute."[11] For, if we "send the observation time to infinity" first (and let the volume be quite small), all sorts of transitions between states are possible—especially for quite small systems, where the expected time for transition is small compared to ordinary everyday observation times. Were we then to go to the infinite volume limit, the final state would be indefinite (dependent on the nature of that short-time transition, what physicists call boundary conditions). Put differently, writing in operator terms (and hence to be read right to left), $\lim_{N \to \infty} \lim_{t \to \infty}$ does not possess a well-defined value, while $\lim_{t \to \infty} \lim_{N \to \infty}$ may possess a well-defined value.

Again, ordinary matter in thermodynamic equilibrium just sits there and makes no such arbitrary spontaneous changes (with any substantial probability), and in general possesses no such indefinite form—except, perhaps, at phase-transition points, or in a spontaneously broken symmetry, as in the orientation of a crystal. Also, what I mean here by ordinary stable matter is matter that is observed for reasonably long but not too long times, and that is bulky enough so that it more or less fulfills the infinite volume limit. This, of course, corresponds to much actual experimental and most everyday practice.[12]

As we shall see, noninterchangeability of measurements and non-commutativity of mathematical operators is one of the phenomenological and technical features that indicate that we have a system in which phase transitions are possible within a range of density and temperature (and, for first-order transitions, the coexistence of different phases, ice and water, for example).

Ordinarily, there is a lack of sensitivity of measurements to just how we do them, corresponding to a mathematical commutativity between limits or between limits and derivatives. But at some points this is not the case.

For example, consider measuring the density before or after going to the infinite volume limit. By definition, we might measure the density as $\mathbb{IN} \ln GPF$ (again in operator notation, reading operators right to left; GPF = grand partition function), first measuring the number of molecules in a specific volume ($\mathbb{N}$) and then seeing how that density changes, hopefully converging to a limiting value, as we go to the infinite volume limit ($\mathbb{I}$). (Formally, $\mathbb{N} = \partial/\partial \ln z$, where z is the activity, and $\mathbb{I} = \lim_{V \to \infty} 1/V$.) Or, commuting the operators, we might measure it as $\mathbb{NI} \ln GPF$, taking the infinite volume limit first and then counting the number of atoms in a specified volume. It would seem that it ought not matter in which order we do this: $\mathbb{NI} = ? \mathbb{IN}$.

Now, $\mathbb{IN}$ is not always well specified; it does not have a unique limiting value at phase transition since fluctuations in density or N are then very large, or if two phases coexist their relative aliquots are not specified. On the way to $V = \infty$, one has fluctuations that arbitrarily determine the limit or the outcome.[13] On the other hand, by its definition, $\mathbb{NI}$ would appear to be always well specified. But taking the infinite volume limit at a phase-transition point is still subject to fluctuations at all scales (which, therefore, do not average out at any finite scale), and so $\mathbb{NI}$ is not well specified there. At phase transitions, both $\mathbb{IN}$ and $\mathbb{NI}$ are not well defined, and they need not be equal.

Technically, as we shall see, Pressure $= 1/\beta \, \mathbb{I} \ln GPF$, and this pressure can be proven to always exist and be well defined as a limit; $\mathbb{N}$ is a derivative of the pressure, P, a derivative of the well-defined $\mathbb{I} \ln GPF$. Yet $\mathbb{NI}$ is not well specified if there is no limiting value for the derivative of the pressure when the volume becomes infinite. The pressure need not be a smooth function, possessing such a derivative, if the infinite volume limit that defines it is not a uniformly convergent process. At phase transition, the infinite volume limit is not a uniformly convergent process.[14] Therefore, $\mathbb{NI}$ is also not likely to be well defined.

Taking the infinite volume limit first and measuring the density second, $\mathbb{NI}$, presumes on the uniform convergence of the pressure—and at phase transition that is not the case. Measuring the density first and then going to the infinite volume limit, $\mathbb{IN}$, presumes on the uniqueness of the limits. Again, phase transitions are marked by the lack of commutativity of measurements or their representative mathematical operators, since the measurements here depend on conditions that are in effect arbitrary and the operators are not well defined at these points.

Before going further, I should note that I have deliberately not presented the most recent formulations of these problems. For example, there are phase transitions that are continuous. Rather than be concerned with infinite volume limits, much of recent research is concerned with proper definitions of states and phases, which in effect achieves much the same goals.[15] I shall review some of this in chapter 8. Here I examine a particular moment in the development of these formulations.

Formally Defining a Phase Transition

Besides the noncommutativity of operators in particular models, we might provide three sorts of more generic, formal accounts of marking a phase transition: zeros of the partition function, changes in symmetry, and fixed points of group transformations.

In one account, due to Yang and Lee, the real zeros of the grand partition function $(GPF(z_c) = 0,\ z = \exp - 2B/k_BT$, considered as a complex variable, where B is the magnetic field) or of the partition function $(Q\,(B,T) = 0)$, are taken as marking a phase transition. Those zeros imply singularities in the thermodynamic functions since those functions depend on $\ln GPF$ or $\ln Q$ ($\ln 0 = -\infty$). And those real zeros might appear only in the infinite volume limit—those zeros indicating discontinuities in the density (or the specific volume) or in the spontaneous magnetization, or in their derivatives. Hence, first- and second-order phase transitions are known by discontinuities in derivatives of suitable functions, such as the free energy, the more conventional thermodynamic way of marking a phase transition.

Technically, for example, the grand partition function is a polynomial sum of positive terms $(GPF = \Sigma z^n Q_n)$ and therefore cannot have real zeros for finite systems. But for infinite systems, the infinite volume limit, there can be such zeros, at least as limit points of finite-N complex zeros.[16]

For example, in a magnetic system whose magnetic moment is proportional to the volume, V, the roots of the partition function in terms of an external applied magnetic field, as that field approaches zero from above, are the unreal and curious values of the external field, $(2n + 1)i\pi/2\beta V$, the smallest root being $i\pi/2\beta V$ from the real axis, and obviously that root is just "at" the real axis, $B = 0$, for V infinite. (The density of roots of the partition function along the imaginary-field axis is $V\beta/\pi$, and that density can be related to the thermodynamic functions that describe the system.)[17]

One example that we shall need in chapter 6: Technically, consider a system with a very large or infinite degeneracy of zero (or lowest) energy states. Or, better put, there is a dense continuum of excitations from the ground state rather than a gap (which dense continuum can be shown to be the case for the Ising model at its critical point, in the infinite volume limit; see figure 4.5). Then the partition function is dominated by these almost-zero energy states ($\exp - \beta\epsilon \approx 1 \gg \exp - \beta E$, where $E > 0$) and the grand partition function, $\Sigma^N z^n Q_n$, for Q_n that grows slowly enough with n, has very nice properties. Namely, for $Q_n \approx$ constant or $\approx \exp n$, the zeros of the grand partition function lie on a circle centered at the origin, and they get closer to the positive real line as we go to the infinite volume limit ($N \to \infty$).[18] One can show that a similar result applies to the exact solution to the two-dimensional Ising model; that is, the partition function is expressible in terms of its zeros in the complex B or tanh K plane.[19]

A second indicator of a phase transition is, as already mentioned, a change in spatial symmetry. We know that stable bulk matter is orderly, in that it might be isotropic as is a gas or be crystalline as are many solids. Phase transitions are almost always accompanied by an abrupt change in that orderliness. Within a model, the mathematical possibility of such changes in spatial symmetry, namely, the possibility of a transition to a state of lower free energy, corresponds to the possibility of there being another phase of matter present in the actual world.

A third formal indicator of phase transitions is just that fixed point (or, say a fixed line) of a generalized duality transformation between a system at a high temperature and one at a low temperature, or between a system at finite temperature ($T < T_c$ or $T > T_c$) and one at $T = 0$ or ∞. Note that, within an arithmetic or topological constitution, that formal transformation links excitations of the Ising lattice from $T = 0$ to deexcitations from $T = \infty$. Within a geometric or holographic constitution of matter, that formal transformation corresponds to actual similarity and scaling phenomena exhibited by a system as it gets close to the transition point.

In each of these formal definitions of phase transition—zeros of the partition function, changes in symmetry and orderliness, and fixed points—a sharp transition is possible: the transition temperature is well defined, and the emergence of new orderliness is abrupt. Moreover, these formal features are indicators for phenomena we should look for: an abrupt change in specific volume or density, an actual change in symmetry, or similarity or scaling of properties as measured by critical exponents and correlation functions. The mathematical form analyzes and justifies our everyday intuition of a phase transition, one built up from a variety of observations and experiments, and gives it precise meaning. Again, technical features, such as uniform convergence (or the lack of it), symmetries, and fixed points, are in fact the physics.

THE LEE-YANG MODEL

Without an infinity of particles, there can be no phase transition—according to the now canonical Lee-Yang theory (1952), following the earlier observation of Kramers.[20] Yang and Lee observe that the grand partition function is an analytic function in z (the activity) if N is finite —namely, it is a finite polynomial in z, in fact with nonnegative coefficients, Q_n. They also show how and why we might think of a single phase of physical matter as a region in complex-valued activity space. Such a region includes a segment of the positive real line and is without zeros in the partition function in that region.

Activity is a measure of the Gibbs free energy required or released, μ, to add one more particle to a phase: $z = \exp \beta\mu$. $z = 1$ corresponds to zero free energy, and hence we might expect something like a second-order phase transition at that point. And so the grand partition function ought to be zero, which is no mean achievement for such a polynomial.

Such a phenomenology is also no mean abstraction. Yang and Lee justify it by means of actual models, such as the lattice gas (a gas of atoms, atoms being present or absent at the vertices of a spatial lattice), and by formal results. The lattice gas has a first-order transition, but has the same form for its grand partition function as the partition function for an Ising model in a magnetic field. And, if $z = \exp - 2B/k_BT$, and $|z| = 1$ for the zeros (as they prove), then B is complex for finite N, as we saw earlier. (Note that the zeros for other variables may be identified differently, so that, for K, $|\sinh 2K| = 1$.) They prove that the infinite volume limit of the logarithm of the grand partition function always exists (it is just the pressure times β), for nonzero activity (Lee-Yang theorem 1). The also prove that that limit is uniformly approached as N goes to infinity—just what we might naively hope for in describing an actual homogeneous physical phase—as long as there are no positive real roots of the grand partition function (no phase transitions), as a function of the activity, within the phase region (and for which there surely are no such real roots for finite N) (Lee-Yang theorem 2).[21] In other words, and to reiterate once more, no phase transition is possible at all unless N is infinite—by their definition of a phase transition and by a theorem concerning convergence properties.

While the Lee-Yang theorems suggest that a finite-N theory will naturally work well when we go to the infinite volume limit, it also allows for the breakdown of that limit. For, again, a phase transition is indicated in this account by a zero of the partition function on the real line (or at least a limit point of zeros); and it is just in that case that the uniformity of convergence fails, and so we might expect some new phenomenon to occur in the infinite volume limit.

In general, everyday phenomena and technical features in an abstract model are meant to correspond to each other, each influencing how we think about the other. The abstract model allows for a philosophical analysis of what we mean by those everyday phenomena—saying what is essential or crucial to such phenomena and how. And those phenomena, conceived of in terms of that model, change how we think about that formal mathematical account, for it becomes interpreted in detail through examples and cases.

In light of the Lee-Yang account, let us return to the question of the relationship of $\mathbb{NI}$ to $\mathbb{IN}$. We might first measure the number the particles in a finite subsystem, and its volume—and so compute the density—and then go to larger and larger subsystems, expecting a convergence of the measured densities to an infinite volume limit ($\mathbb{IN}$). This limit cannot be unique if there are very large fluctuations, or if two phases coexist (say, a gas and a liquid), at constant pressure and temperature, and so the density is not determinate. And that is what we shall discover at the phase-transition point, where fluctuations in the number of particles in any volume are large. Mathematically, $\mathbb{N} \ln GPF = z\partial/\partial z \ln GPF = z/GPF \partial/\partial z GPF$. Since the grand partition function has a zero at the phase-transition point in the infinite volume limit (by the Lee-Yang account), once we go to the infinite volume limit (applying $\mathbb{I}$) we are manifestly in trouble.

Now, say we reverse the order of the limits or operations ($\mathbb{NI}$). Taking the infinite volume limit first, experimentally we might then measure the pressure, P, of the system (which is always possible, by Lee-Yang theorem 1).[22] Then we might note how that pressure changes as we change the density, and from that impute the value of the activity, z. In effect, we are measuring the activity ($z = \exp \mu\beta$) by changing the Gibbs potential, μ, or by changing the number of particles, since z^N is the crucial measure here. (Since the density $\rho = \partial P/\partial\mu$ at constant temperature, the density is $\mathbb{N}\beta P$ ($=z\partial/\partial z \beta P$).) This works well as long as P is continuous and smooth in z, which can be shown to be the case for finite volumes.[23]

If we assumed that the pressure we were to measure for finite volumes (which is a smooth function of ρ or z) is uniformly convergent in the infinite volume limit (we assume the applicability of Lee-Yang theorem 2), then there could not be discontinuities in the derivative of P, in the infinite volume limit, and hence the density ought to be well defined in the limit. However, as I have mentioned, in actuality the density may not be well defined at phase transition (fluctuations, or two phase coexisting)—just at that infinite volume limit. What we presumed we could measure (all the way to the infinite volume limit) is in fact not well defined there. The presumption of uniform convergence is erroneous just at the phase-transition point; without uniformity of convergence, smooth functions can converge to nonsmooth ones. So the derivative, $\partial \beta P / \partial z$, may well exhibit discontinuities, reflecting the indeterminateness of the density at a phase transition. Hence, this order of measurement, of taking limits and derivatives, is problematic, since it would seem to insist on the existence of a limit (namely, the derivative of βP), the existence of a well-defined density, one that in fact does not exist.

Modeling Discontinuity

My interest in this formal theory of Lee and Yang is not only definitional and scholastic. The theory corresponds to practice in two ways. It provides an account of the peculiarities of certain prescriptions for performing a measurement, namely, how and when can $\mathbb{N} = \mathbb{N}$? (Answer: if and when there are no zeros of the grand partition function in the region.) So, in computing the density or the specific volume as $1/v = \lim_{V \to \infty}[(1/V)z\partial / \partial z \ln GPF(z,V)]$, "[t]he limit $V \to \infty$ must be understood in the strict mathematical sense. In particular, the order of the operations $\lim_{V \to \infty}$ and $z\partial / \partial z$ may not be freely interchanged."[24] If they were always interchangeable, then one would not get the singularity in the free energy or its derivatives (such as the specific heat) in the infinite volume limit, a discontinuity that indicates a phase transition.

More precisely, an infinite sequence of analytic functions need not converge to an analytic function, if the convergence is not uniform.[25] Were the volume infinite, then the polynomial in z that defines the grand partition function, *GPF*, would be infinite in degree, and that would allow for unsmoothness in the polynomial and so for discontinuities in its derivative in that limit.[26]

Second, in actual experimental practice, where $N < \infty$, even if in practice one expects that Avogadro's number (6×10^{23}) is effectively infinite, a phase transition is recognized by a finite change in one property such as a density or a magnetization for an infinitesimal change in another property such as temperature or magnetic field—as in water freezing or in a magnet becoming permanently magnetized. Yet, within an exact substantive theory for finite volume or finite N, we could not be sure we are identifying a phase transition, as such; for the formal continuity of the pressure-volume curve is guaranteed by the analyticity in the activity for finite N. For finite N, any potential infinite volume limit unsmoothness or discontinuity is rounded off or smeared.[27]

> For any finite value of the total volume V, no matter how large, we cannot easily recognize a phase transition unless the equation of state can be explicitly calculated. Suppose, for example, that the system under

consideration actually exhibits a first-order transition. For a finite value of the total volume, no matter how large, the pressure cannot be strictly constant in the transition region [a sign of a phase transition] because it is an analytic function of v [the specific volume, the inverse density, otherwise it would be constant everywhere]. Nevertheless, $\partial P/\partial v$ might be extremely small in some range of v, e.g., of the order of 10^{-23} atm/cc—or inversely, a small change in pressure changes the specific volume dramatically. On a macroscopic scale we would regard the pressure as constant, but there is no way to see this from the partition function unless the function $P(v)$ has been explicitly calculated.[28]

(An explicit calculation for finite N requires our judgment (one we might readily make in practice) of when there is in a curve a suitably discontinuous or unsmooth change.) Or,

> In a real macroscopic system of finite extent the internal energy will change a finite amount (in units of k_B) when the temperature changes by 10^{-23} Kelvin [where $\Delta E = Nk_B\,\Delta T \approx k_B$], and for all practical purposes we are in the presence of a discontinuity [in the magnetization]. In order to have a mathematical discontinuity we must go to the infinite-volume limit.[29]

The question is, just where is this limit to present itself in a formal theory?

Again, we need the formal criteria and information about limits because what we mean by a phase transition as a phenomenon is perhaps not best captured by an exact finite-N calculation, but rather by formal discontinuities in a generic function at the infinite volume limit. Similarly, when we do an experiment, what we are looking for empirically are jumps or discontinuities in the data, ones that cannot be smoothed over—for the closer we look the sharper is the jump, or so we tell ourselves. Actually, of course, if we are inclined to see smoothness, we can almost always draw a smooth curve and get away with it. It is only in a very curious sense, but actually quite common if we are so inclined, that we see discontinuous jumps.[30] (Or perhaps it is the other way around, that it is curious that we see smooth curves.) Jumps and kinks in curves are phenomena themselves, their shape being archetypes we prepare ourselves to discover. It is the purpose of a formal theory such as Yang and Lee's to warrant such archetypal forms and such preparation for the experimenter. It is the purpose of models such as the Ising model to tell us the exact form of these jumps (such as "logarithmic singularity").

Recalling the discussion at the end of chapter 3, I again emphasize the significance of the kind of convergence to the infinite volume limit. Within the Lee-Yang account of phase transitions, uniformity of convergence is crucial to interchanging limits and derivatives, and so to a claim about the meaning in some situation of the absence of real zeros in the partition function in the infinite volume limit—namely, there is no phase transition in that situation. More precisely, for uniform convergence, we say that a function (of the activity), $F_N(z)$, approaches its limit within ϵ for N greater than $N(\epsilon)$, for all activities in the range, independent of activity. One N works for each ϵ, so to speak, over a range or region of activities—that is, over a single physical phase. There is no delicate dependence of the limit on these intensive properties, no surprises to appear along the way to the infinite volume limit (such as, for first-order transitions, the coexistence of different phases with different specific volumes)—just what we might mean by a single stable physical phase of matter.[31]

Uniformity of convergence means there are no surprises, no jumps, no kinks, as we get to the infinite volume limit. Recall that, in my discussion of the thermodynamic limit in chapter 3, uniformity of convergence plays a similar role, but in that case the convergence is of the free energy as a function of β and ρ, as $N \to \infty$. ($N(\epsilon)$ is independent of β or ρ, for a range of β and ρ.) By the way, only the average density, ρ, is given in the stability of matter constitution, not the composition, so that two phases of different densities might be coexistent in the thermodynamic limit—as they surely are some of the time.[32]

If the grand partition function were only pointwise convergent to the infinite volume limit, small changes in activity might well require very different Ns for the same ϵ (or, better, different ϵs for the same N). In effect, there would be no sense of a single phase. For an uncertainty in z (an uncertainty that is surely the case in the measured world) might have enormous consequences—when there is no phase transition expected: the phase (= the partition function over a range of activities) would not come along "together" in the infinite volume limit; the phase would be seen to break apart.

At phase-transition points, there will be such a breaking apart, as well as a delicate dependence on intensive variables along the way. Those points are marked by real zeros of the partition function—when there is no uniform convergence and so no application of the Lee-Yang theorems and their interpretation of a uniform phase. Technical features, here uniformity of convergence, analyze what we might mean phenomenologically, here by a phase of matter.

Lee and Yang provide a poignant electrostatic analogy that makes the formalism of their model of phase transitions intuitive to well-trained practicing physicists. Namely, the locus of the zeros of the grand partition function as a function of z—and so discontinuities in the equations of state, say for the specific volume—correspond to a charge sheet, across which there is a discontinuity of the electric field.[33]

More specifically, it can be shown (for a hard-sphere attractive potential) that the zeros of the partition function in the activity z lie on the unit circle in the complex plane.[34] (And if $z = \exp - 2B/k_B T$, the zeros in the B plane lie on the imaginary axis.)[35]

Consider a unit-radius, infinitely long cylinder with charge density, $g(\theta)$, proportional to the density of zeros of the partition function as a function of $z (= \exp i\theta)$, and from which $g(\theta)$ one can recover that partition function (see figure 5.1). To model the discontinuity, we

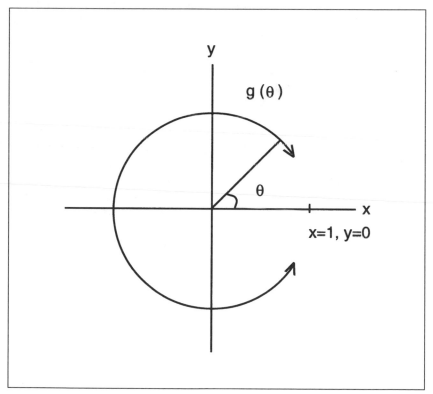

FIGURE 5.1   Electrostatic analogy for zeros of the partition function.

consider the horizontal electric field, E_x, along the real axis (for, by symmetry, $E_y(y = 0) = 0$). That field is smooth across $z = x = 1$ as long as there is no charge on the sheet at that point. As the volume goes to infinity, $g(\theta)$ may close in on the x-axis (it "pinches" the real axis), and so the electric field develops a discontinuity across the charge sheet at $x = 1$. That discontinuity is just the relevant one for the grand partition function. What might seem like a formal mathematical device for modeling phase transitions recalls the intuitions of any physicist who has taken an intermediate course in electricity and magnetism.[36]

Moreover, that the activity might be complex is less of a mathematical curiosity or formalism, when in the electrostatic analogy it represents off-the-real-axis points $(y \neq 0)$, points that may have nonzero values of E_y.[37]

More generally, complex values of the activity, z, might be interpreted as reflecting either a complex β, and hence disequilibrium, as I have suggested in the discussion of the Darwin-Fowler method, or a complex μ (where $z = \exp \beta\mu$, μ being the Gibbs potential per particle).[38] Now, from the lattice gas analogy, $\mu = \alpha J - 2B$, where α is the coordination number (4 for a square); it is essentially the energy change due to a spin flip. For a complex μ, either the spin-spin interaction in the Ising model, J, has a width or a lifetime and hence no matter what the temperature there will be relaxation of any ordered state, or the external magnetic field is complex, again implying a relaxation or oscillatory phenomenon.[39]

In the case of the Ising model, it is perhaps interesting to display the zeroing of the partition function more directly. In general, the $GPF = C' \times \mathscr{P}(z)$, where C' is a constant and $\mathscr{P}$ is a polynomial in z with the constant coefficient equal to one. Correspondingly, the Ising model partition function may be expressed as $Q_N = C^N \mathscr{P}_N$, where $C = C(K)$.[40] So, for example, the high-temperature expansion is

$$Q_N = (2\cosh^2 K)^N \left\{ 1/2^N \sum_{\{\sigma_i\}} \prod (1 + \sigma\sigma w) \right\},$$

where $w = \tanh K$. The bracketed expression is $\mathscr{P}_N$, here a function of w. We might also express Q in terms of the partition function per spin, $Q_N^{1/N}$. If $Q_N^{1/N}/C = \mathscr{P}_N^{1/N} < 1$ as $N \to \infty$, then we have a zero of the partition function, and we would expect it to otherwise be greater than or equal to one. Employing Onsager's asymptotic estimate for the partition function per spin (figure 4.4), we may then compute $Q_N^{1/N}$ for N infinite, so to speak. As we can see in figure 5.2, the normalized partition function per spin reaches its minimum at the critical point.

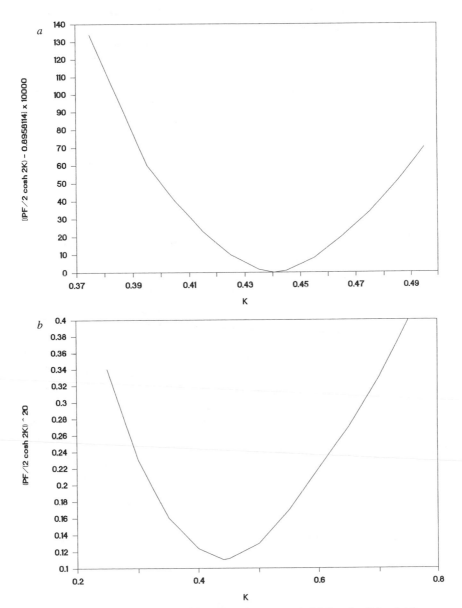

FIGURE 5.2   Normalized partition function per spin(s) for the Ising lattice, showing the possibility of zeros of the partition function: *a*, for one spin (notice ordinate); *b*, for twenty spins.

The minimum is not just at one (or slightly less than one) because we have employed the asymptotic estimate (for which see note 34).

What I have tried to show in this section is that formal mathematical analysis and device is not only laden with physical content, but that experimental practice is perhaps best modeled by that formal abstraction. Conventional physical intuitions can be recalled to make that formalism seem not only natural but straightforward and physical. The physics is not just one fact but a set of linkages to other facts and phenomena. Each analogy and model builds up the set of linkages, so that a formalism is attached to what a physicist already knows as physical.[41]

PHASE TRANSITIONS AS MARKED BY CHANGES IN SYMMETRY

As we have seen, matter achieves a thermodynamic limit because volume interactions between bulkettes are negligible (and the residual surface interactions are small). And so we might constitute matter from smaller into larger and then larger bulkettes—whose free energy densities are in general decreasing with increased bulkette size and so push us to the thermodynamic limit, since that free energy density is bounded from below.

The following analytic deconstruction shows how actual matter (at equilibrium, minimizing the free energy density) is a limit of its being divided into bulkettes. "[T]he insertion of a hard wall, infinite potentials on the boundaries of the subregions [namely, the bulkettes] only decreases the partition function [since there are fewer possible states, and surface energies are more and more negligible]; the further restriction of a definite particle number to each subregion further reduces the partition function."[42] Namely, $f_{\text{actual matter}} \leq f_{\text{bulkettes}}$ (since $Q_{\text{actual matter}} \geq Q_{\text{bulkettes}}$).

Now it is the thermodynamic limit, seen in a geometric analysis, that suggests that, when the kinetic energy is low enough (when the temperature is low enough), we might expect a regular array of atoms in bulk matter, namely, orderliness.

Now we imagine putting a much larger array together of $N = mn$ atoms [n atoms in each one of these m minimal-sized stable units or bulkettes]. An attempt at a low-energy configuration may be made by simply piling all the boxes of n atoms together and removing the interior walls. We can expect that small harmonic readjustments will further improve the energy; but our basic argument is that the energy of misfit between the surfaces of the small pieces is a surface energy, of order $n^{2/3}$, while the energy necessary to change the interior configuration in an essential way will be $\approx n$, so that there will be a cell size, n, beyond which it will not pay to modify the internal configurations; this *then* gives us a regular array.[43]

So Anderson argues, "nonrigorously,"[44] that in general among the lowest energy configurations there is at least one regular lattice (except when there is a high multiplicity of ground states, namely, frustration). The trick is to convince oneself of the notion of that stable smallest unit of size n—the ideal being that any essential change from that ground state will change the energy by an amount of the order n.[45]

Lenard and Dyson put the issue in a similar way.

> It is a remarkable fact that our proof of the stability of matter. . . boils down in the end to an estimate of the binding energy of a single electron in a periodic Coulomb potential. We conjecture that this appearance of the periodic Coulomb potential at the kernel of the proof is not acciden- tal. After all, the ground states of most forms of matter are crystals in which electrons are actually moving in periodic Coulomb potentials. The essence of a proof of the stability of matter should be a demonstration that an aperiodic arrangement of particles cannot give greater binding than a periodic arrangement.[46]

We know that at high enough temperatures there will be no such regular or periodic spacing, for kinetic energies will completely domi- nate interatomic forces, and we will have a gas. Therefore, we argue, as we go to lower temperatures there might well eventually be a symmetry change to a more regular lattice. And, in general, such symmetry change cannot be gradual, in the infinite volume limit; hence there is a phase transition that is sharp.[47]

The essence of any particular, detailed model of such a phase transition is that at some temperature there are two configurations of comparable (free) energy, one of which is more highly ordered than is the other. One might even say, and it seems that one can get away with it, that a hot crystal about to melt is like a cool liquid about to freeze—in principle smoothing over the surely sharp transition. In this case, the free energy is assumed to be (and in reality is surely not) an analytic function of the density and its degree of orderly modulation, and one searches for that onset of order.[48] For a second-order transi- tion, the free energies of the two differently ordered configurations are the same at the critical point (the two eigenvalues of the transfer matrix, representing those two configurations, are degenerate at that point).[49]

In sum, the possibility of stable bulk matter allows for the possibility of changes in the symmetry of that bulk matter at some temperature low enough so that interatomic forces dominate random ones (the coupling constant takes over from the temperature), so that potential energies dominate kinetic energies. At those lower temperatures, there is an orderliness, a regularity in space, let us say. Moreover, it is very expensive energetically to move or change just one of the component atoms of that orderly bulk matter—hence there is rigidity in the system.[50] A system that is microscopically asymmetrical is, in general, a rather rare fluctuation; while a system that is macroscopically asymmetrical has cut out so much of phase space it is in fact quite stable.[51]

PHASE TRANSITIONS AS MARKED BY FIXED POINTS

In my discussion of the constitutions of matter in chapter 4, I displayed in a variety of ways how fixed points of transformations that connect a system and its partition function at one temperature to the system and its partition function at another temperature, pick out a distinctive temperature—taken to signify a critical point and a phase transition. These transformation depend on two physical, or better put, structural, facts about these systems: there is an arithmetic (and algebraic and structural) duality between high- and low-temperature systems; and, geometrically, if we have a system at any particular temperature, we may then proceed to average out its structure, decreasing the resolution so to speak, and then it tends to either extreme order or disorder, except at an intermediate temperature in which its structure is scale invariant and so never averages out.

The crucial observation was made by Peierls (1936), who noted that in, say, two-dimensional systems we might well find a transition point.[52] At low temperatures (and hence low average total interaction energies, $\approx k_B T$, and low entropies), there is little room for differences among interacting particles (if the interaction energy is proportional to those differences). And so we might expect a fairly uniform system. In two dimensions, such different parts are separated into finite-area, finite-perimeter blocks of like spins—the crucial fact in the argument (see figure 5.3). At high temperatures, we would find the reverse. In order that there be a high average energy, there would have to be lots of differences among neighbors, and hence, following the above argument, there would be disorder in the sense that the numbers of up and down spins would be roughly equal and they would be randomly distributed among each other. (We might expect the average energy to go as $\sqrt{N}$, as for a random walk, although the maximum energy goes as N.) Somewhere in the middle, there has to be an order-disorder transition. (Again, one has to convince oneself there is just one such point, and that it is a sharp transition, the latter being the argument about the nature of changes in symmetry.)

Put differently, and more physically, duality is a restatement of the contingent fact that a pair of dissimilar spins at low (zero) temperature is equivalent as an excitation to a pair of similar spins at high (infinite) temperature—and the energies of excitation from their respective ground states of order and disorder are complements of each other (in

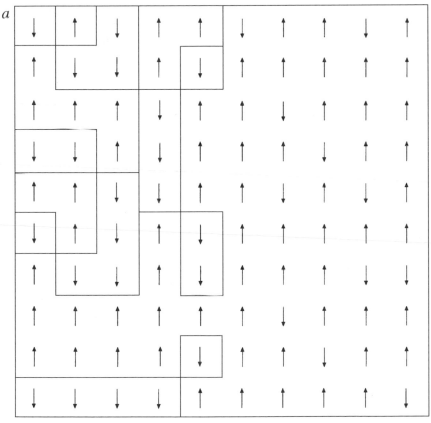

FIGURE 5.3 Polygons of enclosure on the dual lattice and the possibility of a phase transition (Peierls's argument), from figures 1.1 (*a*) and 4.1 (*b*).

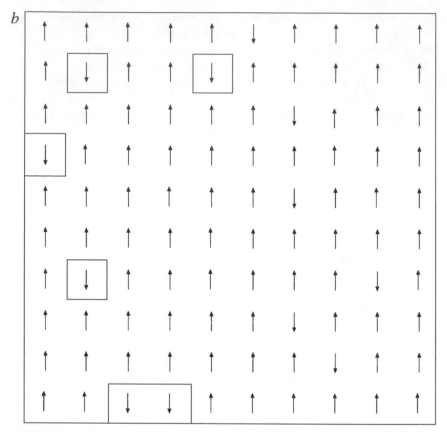

FIGURE 5.3   Continued

k_BT units). Each pair is a perturbation on a simply computable system, the totally ordered and the totally disordered systems (and this is reflected in the employment of low- and high-temperature expansions in the Ising model). And we might expect that the set of low-temperature excitations, when fully activated at a high-enough low temperature (all the various configurations of dissimilar adjacent pairs), would meet, in the middle, the set of high-temperature deexcitations (the set of configurations of similar pairs), when fully activated at a low-enough high temperature. In that middle, at the critical temperature, infinitely many of these states are mutually degenerate or equal in energy; it is the exact ratios or spectrum of their amplitudes that defines the critical point.

The second crucial observation was made in discussions of the geometric and holographic constitutions of matter.[53] A recursive transformation that averaged out details, and so could iteratively convert a system at any temperature into one at $T = 0$ or ∞, might well possess a fixed point, and the system at that fixed point was self-similar at all scales (as indicated by critical opalescence, for example), so that averaging would have no effect.

In sum, the physical facts of arithmetic duality and of geometric duality lead to mathematical fixed points as indicators of phase transitions.

THE INDICATORS OF PHASE TRANSITIONS

Recall that, at zeros of the partition function, various limits and derivatives do not commute, and these noncommutativities indicate that there is a phase transition. Mathematically, that lack of commutativity means that certain functions may exhibit nonanalyticity or discontinuity. Physically, the partition function measures the volume in phase space occupied by a thermodynamic system at fixed temperature, that is, the number of the various possible configurations of its components that contribute to that macroscopic state. (See chapter 3 on the Darwin-Fowler derivation, which is essentially a computation of the partition function in these terms.) And at a phase-transition point, the degrees of freedom are constrained (pressure is independent of temperature, or at critical points several variables are simultaneously fixed). The volume in phase space is not well defined at that point, for some of the dimensions of phase space have collapsed; the dimension of phase space has dropped by one at least. What was once a smooth manifold now develops creases and kinks and discontinuities. If we examine the Ising partition function near the critical point, we see that it does have such a (second-order) kink (figure 5.4).

Technically, the entropy, S, is also a measure of phase-space volume. Actually, it is k_B times the logarithm of the number of macroscopically equivalent microscopic states, n, in the microcanonical ensemble ($S = k_B \ln n$). The entropy also appears directly in a Legendre transform of the energy, U: $U(S,V) \rightarrow U - S\, \partial U/\partial S = F(T,V)$, the free energy. (And conversely, $F \rightarrow F - T\, \partial F/\partial T = U$; $F = U - TS$.) The entropy is, within a sign, the slope of the free energy, as a function of temperature. The free energy is smooth, here, for the two-dimensional Ising model, but the slope does have a rapid change at the critical point (that is, the second derivative of the free energy is large, as we might

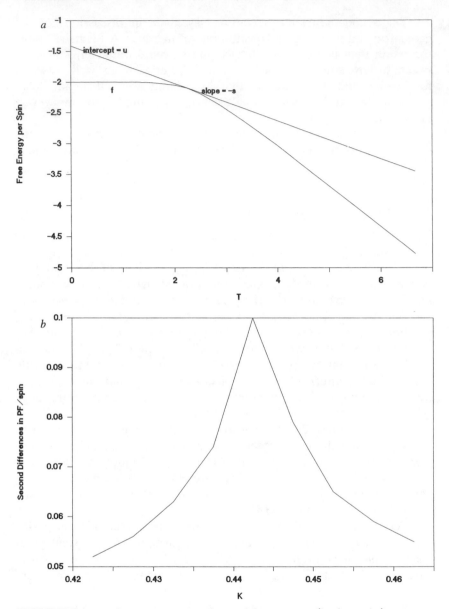

FIGURE 5.4 *a*, the entropy as a slope of free energy (in k_B units) curve ($= -\partial f/\partial T$)—the tangent line, its intercept being the internal energy. *b*, second differences in the partition function per spin, indicating a kink at the critical point.

expect for a second-order phase transition). Hence we expect a kink in the slope of the free energy (and again of the partition function) at the critical point.[54] Note as well that, since the internal energy is the intercept of that slope line, we would expect a discontinuity in the internal energy or its derivatives at that point.

We might ask, more generally, why we should find ourselves dealing with an object, such as $\partial F/\partial T$ $(= \partial/\partial T(-k_B T \ln Q) = -k_B[\ln Q + T\partial/\partial T \ln Q])$, that might be characterized as the partial derivative of the logarithm of the partition function (the latter being the generating function for the combinatorics of interest here): formally, $\partial/\partial\alpha \ln Q$. (In the microcanonical ensemble, $\ln Q$ would be replaced by S, and then $\partial S/\partial T = c_v/T$.) We might say that the system lives in phase space, more precisely, in a volume in phase space: these are all the possible microscopic arrangements of its elements that compose it macroscopically. $\partial/\partial\alpha \ln$ is an operator that measures the normalized marginal change in that volume, Q (or $\exp S/k_B$): $(\Delta Q/\Delta\alpha)/Q \approx \partial/\partial\alpha \ln Q$. It is not unreasonable to then suspect that what is crucial physically is the magnitude of that normalized marginal change (in effect, the change in the phase-space volume in which the system resides).

By the way, if we think of the partition function as a combinatorial object, then the $\partial/\partial\alpha \ln$ operator defines a generating function for the changes in the combinatorics due to a change in the parameter α: if $PF = \Sigma_n g(\alpha, n)z^n$, then

$$\partial/\partial\alpha \ln PF = \left[\sum_n \partial g(\alpha, n)/\partial\alpha \, z^n \right]\bigg/ PF.$$

(In the case of the historical numerical partition functions, which I mentioned in chapter 3, the operator leads to an interesting form: if $PF = \Pi_n 1/(1 - q_n \zeta)$, then $\partial/\partial\zeta \ln PF = \Sigma_n q^n/(1 - q^n\zeta)$.)

More generally, again, a distinctive mathematical feature is connected to a distinctive phenomenological feature. Zeros of a function, changes in symmetry, and fixed points—all indicative of sharp phase-transition points—are closely allied with a phenomenon: the behavior of the specific heat, the spontaneous magnetization, and correlation functions. Each mathematical feature is given a physical justification, while it provides a more precise meaning to that phenomenological feature. Speculatively, we might hope that we could show that all these features are the same, that zeros of the partition function, changes in symmetry, and fixed points are formally or mathematically equivalent.[55]

FORMALLY DEFINING A BULK QUANTITY: THE SPONTANEOUS MAGNETIZATION

That thermodynamics is a limit science means that the crucial features of its accounts of matter—extensive versus intensive properties, temperature, phase transitions—depend on taking infinite volume limits. In contrast, in the finite-N case, presumably intensive properties might well depend on N so that a phase is not well defined, energy and so temperature fluctuations within a phase dominate, and phase transitions are not so sharp at all.[56] Actually, as I have earlier indicated, everyday life with an Avogadro's number of molecules is almost always effectively the infinite volume limit, and that means that any theory that tries to account for ordinary experiments might well go to the abstract infinite volume limit if it is to correspond to concrete experiments.[57]

The recurrent analytical (or philosophical) task for mathematical physics is to find technical definitions of ordinary phenomena that are true to those phenomena, that can be said to correspond to experimental practice, and that are computationally tractable as well. The definition of the spontaneous magnetization of a model ferromagnet (an Ising model in two dimensions) provides another poignant example of this task, much as we found for phase transitions.[58] If the temperature is above 1043 kelvin, a permanent magnet will lose its magnetization; if it falls below that temperature, it can acquire a permanent magnetization.

Technically, given the definition of the spontaneous magnetization as $-\partial F/\partial B_{B=0} = M(0)$, we are looking for a term in the free energy linear in the magnetic field B.[59] In an actual calculation, this is not at all trivial to set up, and it requires a good deal of device so that such a linear term may be discerned. The task of definition and of computation for the two-dimensional Ising model has turned out to be one of the tours de force of mathematical physics, with something of a history: Onsager announcing an exact solution (1949); Yang actually providing a public demonstration (1952); others coming in with perhaps more straightforward solutions; then deep philosophic commentary on how one might define the spontaneous magnetization so it meets the task I outlined in the previous paragraph; and, finally, the problem disappears under even more perspicuous methods of calculation.[60]

Here is a comment by Schultz, Mattis, and Lieb (1964) in the middle of one of the attempts at clarifying the solutions.

> In contrast to the free energy, the spontaneous magnetization of the Ising model on a square lattice, correctly defined, has never been solved with complete mathematical rigor. Starting from the only sensible definition of the spontaneous magnetization $[\lim_{B \to 0+} \lim_{V \to \infty}, M_V(B)]$, the

methods of Yang and of Montroll, Potts, and Ward are each forced to make an assumption that has not been rigorously justified. The assumptions appear to be quite different; however, from the similarities between the difficulties encountered in trying to justify them, and the identity of the results obtained, one might conclude that they are closely related.

They then continue with the quote that begins this book, "the kind of mathematical questions we wish to examine *are completely ignored*" in most practice.[61] In fact, these questions are physical as well. By the way, things will turn out happily after all.

To compute the spontaneous magnetization of a crystal, namely, the magnetization that the crystal might maintain without an applied magnetic field (as in a permanent magnet) were the temperature low enough, one must take both the infinite volume limit and the limit to zero of the applied magnetic field—a field (however small it be) that presumably set the direction of alignment of the crystal initially (in effect, carving off much of phase space). To reiterate, the order in which we might take limits, and just how each is taken, may matter enormously.

That order of taking limits is not only technically crucial, it is also physically crucial. I have discussed this earlier. Here is a rather more technical yet still physical account of the problem.

> Physically, this definition [of an infinite volume limit] corresponds to the experimental situation, in which B [the applied magnetic field] is allowed to become small on a macroscopic scale and the spontaneous magnetization is the extrapolation of the $M(B)$ curve [to $B = 0$]. The values of B determining the extrapolated value are all much greater than $1/\sqrt{N}$ (in units of $k_B T/\mu_{\text{Bohr}}$ [where μ_{Bohr} is a Bohr magneton, roughly the magnetic moment of an electron, and the interaction with the field is $-\sigma B$]), so that one must pass to the limit $N \to \infty$ *before* letting B tend to zero.[62]

Namely, the magnetic field that determines the direction of the spontaneous magnetization must "always," on its way to zero, have an interaction energy much greater than the thermal energy, and this will be so if the field is much larger than the inverse size. (Assuming the spins add up randomly, the magnetic energy is about $\sqrt{N}\mu B$; the thermal energy of the spin that determines the direction of magnetization is $k_B T$.) The field itself can go to zero when that inverse size goes to zero, that is, in the infinite volume limit.

More technically, in general the magnetization in the limit will be greater if we go to the bulk before we turn off the field

($\lim_{B \to 0+} \lim_{N \to \infty} \langle M \rangle / N \geq \lim_{N \to \infty} \lim_{B \to 0+} \langle M \rangle / N$), and, as we expect, for N finite, $\lim_{B \to 0+} \langle M \rangle = 0$.[63] For finite N (and so by the definition, a nonmacroscopic regime), there can be no spontaneous magnetization.

However, while a particular technical device or definition might readily model basic physical processes, in actual calculations we might find ourselves forced to use a (slightly) different technical device or definition to make the calculation go through. And then we wonder if that new device is still good physics, whether and how it makes sense as a model of what we mean ordinarily by a permanent magnet and the spontaneous magnetization. That it produces what we take to be a correct answer is testimony to its goodness, but still we might demand an account of why it is so good.

Everyone takes as canonical the definition of the spontaneous magnetization: first measure the magnetization at nonzero field and finite size; then go to the infinite volume limit; and then, from a finite positive value of B, go to $B = 0$. Following Griffiths, we call the canonical measure m_0. However, in actual calculations this is not enough. There are two formal approaches at this point to defining the spontaneous magnetization. One may either compute moments of the magnetization, or calculate the correlation function at short range and then extend it to large distances. Of course, it is in just how we do each of these moves that the physics lies.

Technically, one might define the magnetization in terms of moments of the magnetization calculated for $B = 0$; namely, $m_j = \lim_{N \to \infty} N^{-1}[\langle |M(B)|^j \rangle]^{1/j}$, which looks at the probability distribution of the magnetization; and hopefully at $B = 0$ in one's model, it is sharply doubly peaked at $\pm m_{\text{spontaneous}}$, the spontaneous magnetization. Now, unlike the case for m_0, it is not clear that this limit exists, but at least it would seem there is just one limiting process rather than two. Also, this definition says, intuitively, that, if one were to measure the average magnetization for a large system, one would most likely find $\pm m_{\text{spontaneous}}$. One can show that $m_j \leq m_0$; that is, the probability distribution definition provides a lower bound to the canonical thermodynamic definition (the distribution might peak at zero rather than at $\pm m_{\text{spontaneous}}$).[64] And, for low temperatures, the topological argument of Peierls and Griffiths says that $m_1 > 0$ for the Ising model in two dimensions.

In a third definition, rather more in accord with calculations actually done, one looks for long-range order or correlation among spins, call it m_c, and it is much like m_2 above, since what one is concerned with are averages of the sort $\langle \sigma_i \sigma_j \rangle$ ($\approx m_c^2$), where $|j - i|$ is large.[65] The issue that is open is just how we are to take the limits. Surely, first, one goes to an infinite volume limit for some $|j - i|$ (which might well be a function of N), and "then" has $|j - i|$ go to infinity to get the long-range order. These $|j - i|$ techniques present nontrivial issues, both philosophically and technically, since ideally $|j - i|$ would be proportional to N, what is called the "long long-range order," so that as the infinite volume limit is pressed, the distance of the correlation is proportional to size. Yet, calculationally, one finds that one can take $|j - i|$ to infinity only after the infinite volume limit (the "short long-range order"). One can show formally that the long and short long-range orders are the same, and the connection with m_2 grounds the study in much experimental practice (correlations are routinely measured by neutron diffraction, for example).[66]

Let me summarize this fairly technical argument. The canonical obvious definition of the spontaneous magnetization does not lead to readily calculable results. So one develops presumably equivalent definitions, but these may not possess rigorously provable limits. Yet these definitions may well be closer to actual experimental practice and doable theoretical calculation. Along the way, in the name of actually performing a calculation, say, a particular definition is invoked so that a calculation will be doable. It is not trivial to show that such a definition is equivalent to the others—although that they would come to the same calculated results is taken as an important indicator. If the results are different (and perhaps one result corresponds with experiment, which experiment might well have been done under the influence of another definition), then one has to find out just how the physics of the two definitions do differ; if they are the same (and correspond to experiment) one might avoid engaging in theoretical reflection.

Often, the equivalence of various definitions depends very directly on the actual physics of a situation. It is not a mathematical formal identity. If there is to be a spontaneous magnetization, then the free energy, F, must be linear in the magnetic field at zero field (since, again, $-\partial F/\partial B = M$—and if it were, say, only quadratic in B, then $M = 0$ at $B = 0$). "Mathematically, although the [limit] sequence of functions $F_N(B)$ are quadratic in B, for small B, the limit function, F_∞ [$= \lim_{N \to \infty} F_N(B)$] may vary as $|B|$ for small B, and just such a behavior would lead to spontaneous magnetization. Unfortunately, no one has succeeded in calculating the function $F(B)$ for $B \neq 0$."[67] This linearity does not appear in one dimension. In a one-dimensional lattice, spin flip pairs and spin orientations are not closely connected, and this leads to only quadratic terms in B. But in higher dimensions, flips and orientations are sufficiently connected for there to be linear terms.[68] Technically, in the first case we have a cosh B term ($\approx B^2$), in the second we have an $\exp|B|$ term ($\approx |B|$).

In Yang's computation of the spontaneous magnetization of the two-dimensional Ising lattice, his magnetic field goes to zero as $1/\sqrt{N}$ (that is, $B\sqrt{N} \to 0$), an inverse linear dimension of the lattice, apparently violating the quote above about the magnetic energy and the thermal energy. Yet by a physical coincidence, it would seem, his calculation does work: "The important reason why Yang's definition appears to agree with the fundamental one is that the quadratic dependence of $F_N(B)$ persists as B increases only so long as $B = 0(1/N)$ [of order $1/N$]; for larger B, $F_N(B)$ appears to be linear in B, and presumably the slope is the same for $B = 0(1/\sqrt{N})$ and for $B = 0(1)$."[69] There is always physics behind the formalism, and that physics, how in detail the physical world actually works, is a contingent fact.

Let me briefly sum up the considerations in this chapter. Formal definitions of physical phenomena are technical, defined in terms of the realm of calculation; philosophic, saying what is necessary, crucial, and significant about a phenomenon; experimental, making possible the conversion of actual observations to an ideal form that they imperfectly imitate; and physical, dependent on particular features of how the physical world actually works. Crucially, such formal definitions are never merely a matter of greater mathematical rigor, for it seems they lead to new physical insights as they are employed.

6

Formalism

TECHNICAL DEVICES DOING THE WORK OF PHYSICS

Mathematical formalism: the physicist's toolkits, a technical preview •
What are the particles?, the right particles, phase transitions in Ising matter
• *What is the topology? The right matrix*

The principal alternative line to these algebraic methods [employing Lie algebras to solve the two-dimensional Ising model] has been to reduce the problem to one of counting polygons on a lattice—the so-called combinatorial method. . . . The combinatorial method, despite its reliance on certain ingenious topological tricks and on relatively unfamiliar entities called Pfaffians, certainly seems simple in contrast to the earlier algebraic methods and may have seemed destined to displace the former as the classical method of solution of this famous problem.

The present paper tries to restore the balance by presenting the algebraic approach in a way that is both very simple and intimately connected with the problem of a soluble many-fermion system. Except for one or two crucial steps, the approach is straightforward and requires no more than a knowledge of the elementary properties of spin 1/2 and the second quantization formalism for fermions. . . . It becomes apparent that the two-dimensional Ising model, rather than being entirely different from the trivially soluble many-body problems, reduces in some ways to one of them, being just the diagonalization of a quadratic form [as in the normal modes of a system of coupled oscillators].

T. D. Schultz, D. C. Mattis, and E. Lieb, "Two-Dimensional Ising Model as a Soluble Problem of Many Fermions"

In the summer of 1966 I spent a long time at Aspen. While there I carried out a promise I had made to myself while a graduate student, namely I worked through Onsager's solution of the two-dimensional Ising model. I read it in translation, studying the field-theoretic form given in Schultz, Mattis, and Lieb.

Kenneth Wilson, "The Renormalization Group and Critical Phenomena"

MATHEMATICAL FORMALISM

Technical devices do the work of modeling nature by the invisible hand of mathematical formalism and the more visible hand of mathematical insight. Of course, that mathematical formalism and insight also draws from observations of the natural world (through a different phenomenology, to be sure, than does physics).[1] Having adopted a model and formulated it precisely, one might well be driven to insights about that model, and about modes of solution of problems expressed in the model, through mathematical intuitions and operations rather than physical ones. Moreover, technical definitions are not merely mathematical; they are the physics, analyzing what is essential about a certain phenomenon.[2]

Those technical devices and definitions, and those mathematical formalisms and the mathematics itself, are alive with meaning—they are not abstract, at least for practicing scientists. Surely, they have mathematical meaning, referring to mathematical objects and problems. But they also are often filled with physical meaning—accreted over the years, referring to physical objects and problems, meanings that might well suggest interesting mathematical theorems that then provide suggestions along the way for how the physical world might actually work, its mechanisms.[3] Dialectically, mathematics as abstracting away from the physical world—yet sharing in the same everyday world—makes a discourse on its own with itself. Here, I have been concerned with how one might interpret formal technical devices, keeping the physics in mind.

Moreover, greater precision and rigor in analysis is not merely a cleaning up of the mathematical physics, but in general it leads to deeper insights about how the physical world works. Much of these insights may be about counterexamples and curious cases that would never occur to the practicing physicist, for they are of no interest in our experimental natural world, at least initially. But some of these "no interest" insights are in fact filled with physics, in that they lead to a rather more perspicuous account of the physical world, in saying more precisely what is essential about it.

Of course, much of what I have tried to draw here from the early original papers is much more manifest in later work—which might make the complexities of the original papers trivial by showing most clearly what is essential in them.[4] (My own experience is rather more mixed, many of the complexities now seeming even more profound, even if the work of the last fifteen years in field theory and its analogs to Ising matter are quite illuminating.) In part, this gain in perspicuousness derives from the motive of finding a cleaner technical analysis; in part, this gain comes out of a better or deeper understanding of the physics involved. And, in part, the gain comes from a sharper understanding of the motives that animated the moves in the original paper. "Although the original solution by Onsager [1944] could be easily followed from step to step, the motivations and over-all plan were obscure. In Sec. 4.2, we shall attempt to give some of the key steps in this original formulation and to describe, as we see it, the motivations that lead from one step to the next."[5]

To read Onsager now is to realize he understood perhaps better than most of his successors the meaning of the formalisms he employed. But one needs all the developments of the subsequent forty or fifty years to appreciate the meaning of what Onsager left inchoate or merely as a hint (perhaps for a promised future paper that, it seems, did not appear)—what Onsager referred to as "plantings by the wayside."[6] Perhaps this is reading more into Onsager's 1944 paper than is there, Whig historiography gone wild, but I think not. For his passing phrases can be seen, through trying to work through his solution to the two-dimensional Ising model and through the developments of other solutions, and his subsequent testimony, to almost surely reflect investigations that he performed but that he did not discuss in detail in the published paper.[7]

Although I dropped hints earlier, I have stopped short of giving the most general mathematical and physical formulations, formulations that may well show how obvious is some of the technology we employ, or that quickly identify analogies and the like. My suspicion is that these most general formulations distance one rather too quickly from the physical mechanism, at least as that mechanism was understood by the early and germinal scientists who worked on a problem. It is that play of mechanism and mathematics that concerns me here, not the most formal of ultimate justifications that show that "it is all obvious." Yet I have been guided to some extent by further developments, and in that sense my presentation is haunted by its Whiggishness. Still, I want to concentrate on the path to the most general perspicuous formulations, how the mathematical technology came to be employed.

The Physicist's Toolkits

Scientists possess toolkits of particular skills and crafts, having along the way in their training and practice specialized their knowledge.[8] If there is a happy coincidence of problem and craft skill, enormous progress can be made. Onsager could employ his knowledge of "quaternion algebra," his facility in mastering mathematical methods, and his knowledge of elliptic functions to solve the Ising model.[9] Onsager's student

Bruria Kaufman redid Onsager's solution using the rather more familiar (at least to theoretical physicists) spinors, anticommuting matrices that model, for example, rotations in space for spin-1/2 particles. And so Kaufman suggested something about how Onsager's original solution actually worked. We gather that Kaufman knew a good deal about spinors and group representations ahead of time.[10]

Again, my claim here has no direct relevance to claims about the unexpected power of mathematics in the physical realm.[11] I take the latter to be in effect theological claims, about the relationship of a sacred realm of mathematics to a mundane realm of natural science and physics. My own view of mathematics is that it, too, is a mundane realm, and so there is no miracle to be explained; rather, what is required is an account of the kinds of coordination that have occurred between two modes of production.

There are lots of physicists with lots of specializations. That some techniques and devices can be made to do physical work, and that many of them are eventually employed, is no more (and no less) interesting than is Adam Smith's description of an economy in which there is a division of labor and which is quite productive, or Joseph Schumpeter's description of the entrepreneur in capitalism.[12] The marvels of the division of labor in Smith's proverbial pin factory reflect an enormous amount of experimentation and speculation. Many enterprises fail, some succeed; the successful ones are imitated until their formula no longer succeeds, and then some; losers are forgotten; and winners are quite productive. Outsiders enter with new schemes, and they are subject to the rigors of competition. There is much mutual adjustment of production and consumption; namely, there is a market and marketing. The demand for pins affects the value of searching for new means of producing them. And so forth. Technology and science, science and mathematics, develop through such a model in such a context.

What is interesting is that, when such a happy coincidence of mathematical technology and insight and physical problem does take place, we learn new physics; we may learn more about what is essential about the physical world through our models of it. It would seem that we have developed within the context of our practical needs a formal analytical language that empowers us.[13] To solve a particular problem, we develop an instrumental calculus, which then enables us to do more than we could have imagined.[14] My working guess is that mathematics and physics are about the same everyday world, and the analytical language should be empowering insofar as that language is instrumental, about how we take hold of the world and how it responds to our grasp.[15]

In any case, here I examine how some of the mathematical apparatuses that are employed in the various ways of solving the two-dimensional Ising model, as models of the constitution of matter, appear to do
the same physical work—and how this similarity might point to the
deepest physics, say, the symmetries within such a model of nature.
Arithmetic duality, in its various technical guises, suggests that the
crucial feature of the Ising model is the bonds or pairs of interacting
neighboring spins (rather than the individual spins), as excitations from
order or deexcitations from disorder. And the star-triangle relationship
points to a canonical measure of fluctuation and order, k, which, rather
than being the average energy per degree of freedom as is the temperature, is a measure of the correlation length. k is perhaps the simplest
way of expressing duality and the star-triangle relationships: $k \rightarrow 1/k$
and k constant, respectively. Now, the topological character of duality is
marked by a dynamical variable, k, given partition function invariances.[16]
More generally, for physicists it is a truism that symmetry and conserved
dynamical quantities go together, namely, Noether's theorem.

Much of what I have written here was instigated by questions that
arise as one rereads the original papers in their original historical
context, questions like, In these Ising model solutions, why spinors? why
Clifford algebras? why elliptic functions? why fermions and Pfaffians?
These papers were known for their inventive use of unexpected mathematical techniques. Why these particular technical devices in this context—besides the fact that they lead to solutions? We want a way of
answering these questions that acknowledges that some of the why is a
matter of one's unfamiliarity with a technique (you are not the specialist
who is employed in this factory), some of the why is appreciating the
insight of the scientist who often presents work in a way that hides its
motives or its originating intuition, and some of the why is discovering
the deeper meaning, the physical symmetries, hidden from even specialist scientists because their facility with a powerful technique would
seem to preclude their seeing the deeper physics.

Nowadays, for the many physicists who practice statistical and
quantum field theory as a major component in their practical toolkit, it
has perhaps become apparent that field theory interpretations of the
Ising model (the lattice and its spins being a field) provide a powerful
answer to these why questions, in terms of the commonplaces of their
practice, those commonplaces in part historically influenced by the
various conventional solutions to the Ising model. The Ising model is a
perfect test case, exactly solvable and physically specific. Each of the
mechanisms or mathematical techniques employed in the solutions of
the two-dimensional, zero-magnetic-field, planar, Ising model—for ex-

ample, elliptic functions and anticommuting operators—is seen as a way of expressing the behavior of energy- and momentum-conserving antisymmetric or fermion fields.

For example, consider the elliptic functions taken as modular functions (functions invariant to a rational linear transformation, except for multiplication by a complex number or modulus—as is the partition function here, $k^{-N/2}Q(1/k) = Q(k)$):

> Why is it so important to get expressions in terms of modular functions? In the "low temperature" region. . . a difference in energy levels is enhanced in the Boltzmann weights. The computation in low temperature is consequently accurate enough if we sum up the contributions from the ground state and those close to it. But, however further you add, the behavior near the criticality. . . is never obtained. The modular invariance [the scaling by k] connects the behavior at criticality with behavior at low temperature. This is the crucial role of the modular functions in their work.[17]

In other words, we can scale up the low-temperature (high β) solutions to the critical temperature if they are expressed in terms of modular functions, functions that build in the deepest physics, duality, of the Ising lattice.

I shall review these more recent perspectives in chapter 8. In this chapter, I begin with a technical preview that many readers might want to skip for now. Then I describe what are the right particles for Ising matter, and why they are taken to be right. In particular, I indicate how a system of such particles can explain phase transitions in such matter in terms of excitations or creation of these particles. I do the same for the right matrix and topology.

In chapter 7, I look more closely at the symmetries of Ising matter, at duality, and at the star-triangle relation, and connect them with rather more general phenomena of physical systems and particularly perspicuous solutions of the Ising model. I close chapter 7 with a consideration of the pervasive appearance of elliptic functions in these problems.

A Technical Preview

The deeper physics is epitomized by the mathematical technology we employ: Antisymmetric functions ($f(x, y) = -f(y, x)$) and anticommuting variables ($\eta^+ \eta + \eta\eta^+ = 0$) and fermion operators ($a^+a + aa^+ = 1$)—such as those operators that represent the beginnings and ends of bonds or the creation and annihilation of fermions, and determinant-like objects or multilinear forms such as Pfaffians (defined on antisymmetric

matrices)—appear pervasively in the arithmetic and topological constitutions of Ising matter. Those operators point to the particlelike objects
in this matter—namely, quasiparticles, which are row arrangements
composed of regular periodic configurations of bonds (those pairs of
opposite spins), a regular configuration of like bonds punctuated by
unlike (characteristic of low temperature) being dual to a similarly
regular configuration of random bonds punctuated by like (high temperature). And, given that K_1 and K_2 are the horizontal and vertical
couplings, the quasiparticles' energies, ϵ_q (in $k_B T$ units, so reflecting
how excited a degree of freedom is)—

$$\cosh \epsilon_q = \cosh 2K_2 \cosh 2K_1^* - \sinh 2K_2 \sinh 2K_1^* \cos q,$$

where $Q = (2 \sinh 2K_1)^{1/2} \exp 1/4\pi \int_{-\pi}^{\pi} \epsilon_q \, dq$—are invariant to a duality
transformation $(\{K_1, K_2\} \rightarrow \{K_2^*, K_1^*\},\ k \rightarrow 1/k)$.[18]

The features of those quasiparticle energies that prove crucial in
the solution to the Ising model can be seen straightforwardly when the
quasiparticle energies are expressed in terms of hyperbolic geometry.
The formula one ends up with (after much calculation) for the energies
of the quasiparticles turns out to be the same as that for a hyperbolic
triangle (figure 6.1), again,

$$\cosh \epsilon_q = \cosh 2K_2 \cosh 2K_1^* - \sinh 2K_2 \sinh 2K_1^* \cos q.[19]$$

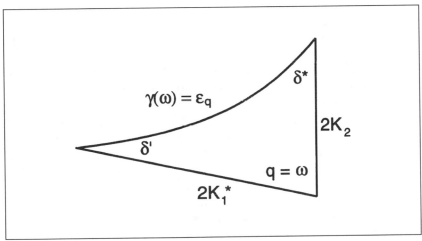

FIGURE 6.1 Hyperbolic triangle connecting ϵ_q, K_1^*, and K_2. Adapted from
Onsager, "Crystal Statistics," fig. 4.

Many of the properties of the Ising model can be discerned by studying the behavior of this triangle for different angles and sides—for example, if $q = 0$, and if K_1^* and K_2 are equal, then the quasiparticle energy is zero.

Those antisymmetric matrices and their Pfaffians point to other objects, not rows of spins but peculiar polygons of bonds. Those Pfaffians are also generating functions for a combinatorial problem (concerning random walks, or the ways of covering a decorated lattice by dominoes or chemical dimers that cover two sites at a time), a problem that may be nicely modeled on a bipartite or planar lattice—such planarity indicating how locally connected such Ising matter can be (no interactions cross over each other, for example) if it is to be tractable to these methods and models.[20]

Coupling-constant relationships that go along with topological transformations—such as duality and star-triangle transformations—characterize the geometric and holographic constitutions. Those relationships point to the crucial field in this matter, namely, k, the coupling-constant-invariant measure of fluctuation or correlation length (where in general the correlation length goes as $1/|\ln k|$, or $1/|t|$, where, near the critical point, $(T - T_c)/T_c \approx Kt \approx (k - 1)$).[21]

Those same coupling-constant relationships also define mutually consistent sets of transfer matrix operators—the matrices that build up a lattice (through calculating its partition function)—in that a lattice built up of members of one k set will have much the same symmetries and orderliness throughout and in that sense can be said to be a single phase.[22] (Not only is the correlation length a function of k, but so is the spontaneous magnetization, $\approx (1 - k^2)^{1/8}$.) Algebraically, the members of the k set commute among themselves.

Mathematically, elliptic functions, which are generalizations of the conventional trigonometric functions, are the way of highlighting the crucial character of k and of another variable u (a measure of the inequality or the asymmetry between the horizontal and vertical coupling constants). By way of analogy, k and u correspond roughly to the maximum angle and to the time, respectively, in the motion of a simple pendulum. k is the modulus of the elliptic functions ($= \sin \theta_{max}/2$); and u ($= \omega t$, where t is the time) is the argument. $\text{sn}(u)$ is a measure of the position of the pendulum bob: $k\,\text{sn}(u) = \sin \theta/2$, where θ is the angle of the pendulum. For small oscillations, this becomes $\theta = \theta_{max} \sin \omega t$.

The star-triangle relationship also has a geometric heritage, as do the ϵ_q, but here we examine the trigonometry of a right-angle hyperbolic hexagon or two polar hyperbolic triangles,[23] where $\sinh 2L_i/$

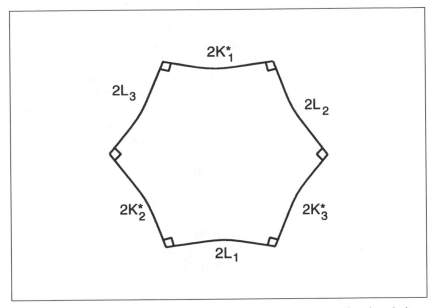

FIGURE 6.2   Hyperbolic right hexagon representing the star-triangle relation. Adapted from Stephen and Mittag, "New Representation," fig. 6a.

$\sinh 2K_i^*$ is a constant (the "sine proportion" for spherical triangles; see figure 6.2).

Finally, physically once we note and then press the analogy of statistical mechanics on a spatial lattice of dimension d (as in the Ising model) to quantum field theory on a spatial lattice of dimension $d - 1$ (the other dimension becomes time), and to symmetries of the scattering matrix, then many of the various points I have been making come together rather more naturally—albeit in part because that quantum field theory has been influenced by developments in understanding the nature of Ising matter. Hence, quantum field theory is an epitome not only logically but also historically of work on the constitution of Ising matter. Of course, statistical mechanics, as we now understand it, is much influenced by how we have come to understand quantum field theory.

WHAT ARE THE PARTICLES?

In a complex many-body problem (most of physics!), such as a lattice of atoms interacting with each other, it is often not clear just what are the particles: those objects that add up (or whose properties add up), that interact comparatively weakly among each other so that they can be seen as individuals, and that in their being created represent the excitations of the system from some highly symmetrical ground state—those excitations, when they become too numerous because the temperature is high enough, leading perhaps to phase transitions or the appearance of new phenomena and new particles.[24] Often, in condensed or bulk matter, the obvious individuals, such as the atoms or their electrons, interact so much with each other that none of them seem sufficiently individuated (in the above sense) to be particles for this system.

Put differently, how are we to model a complex system so that its behavior can be conceived of as a system of particles: the ideal gas of classical kinetic theory; or independent and identically distributed random variables (what the stability-of-matter problem reduced to in the end);[25] or a gas of harmonic oscillators, as is a blackbody, namely, its normal modes of oscillation; or a gas of quasiparticles as in what is called a Fermi liquid, a system of interacting fermions such as electrons in a solid.

As we have seen, for bulk matter the particles are those bulkettes (cubes or balls) that in their addition make up bulk matter[26] (whose interactions among each other are weak surface energies, and for which in this model the excitations are disequilibrium states, or they are volume terms at the critical point). Getting the particles right led to being able to prove the stability of matter and the existence of the thermodynamic limit.[27] Similarly, in condensed-matter physics, one finds as the right particles either collective modes of excitation, such as sound waves or crystal-lattice vibrations or phonons, which are bosons, described by symmetric functions; or elementary excitations (quasiparticles), such as electrons in a lattice (each electron mutually beclouded by it interactions with other electrons in the lattice, and with the lattice itself), which are fermions, described by antisymmetric functions— which, like the archetypal electrons, obey the Pauli exclusion principle (only one particle in each quantum state).[28] Getting the particles right allows us to account for the properties of condensed matter in terms of weak interactions of those particles. In superconductivity, for example, the right particles are the Cooper pairs of electrons with opposed momenta and spins (much as the Ising particles will be pairs with opposite momenta). For harmonic systems, such as systems of springs, one looks for the resonant pure tones, the normal modes of oscillation, and they are usually the right particles.

I have already shown how Ising matter may be constituted of putative particles: those block spins as bulkettes, those added-on sites, those momentum or Fourier components, and those polygons and polymers. Getting the particles right leads to solutions of the Ising model. Yet these particles do not seem to lead to a perspicuous understanding of what is going on physically. As Steven Weinberg has put it, "You may use any degrees of freedom you like to describe a physical system, but if you use the wrong ones, you'll be sorry."[29] Surely we are not so sorry, since we have solutions to the model; but their physics is perhaps not so manifest.

Again, duality suggests that the bonds, that is, the interactions between nearest-neighbor pairs of spins—rather than the spins themselves—are close to the crucial particles (an unlike bond [↑↓], separating sequences of like bonds [. . . ↑↑↑↑. . . ↑↓. . . ↓↓↓↓. . .] will be our focus, at least for low temperatures).

Note that changing or flipping a single spin will change perhaps more than one bond, and vice versa. Various formal schemes have been developed to allow bonds to act independently of each other, as we would hope for good particles—the major devices being either going to the dual lattice (on which the spins are the bonds on the original lattice) or decorating the original lattice with extra structure so that it is now a "counting lattice," one in which each bond can be separately accounted for. (It should be noted that a change of one "spin" on the dual lattice will have consequences for many spins on the original lattice; but there is a configuration of the underlying spins that will support any dual, pace boundary conditions. See figures 6.3 and 6.4.)

The spatial symmetry (and duality) of the lattice will lead us to row particles characterized by regular arrangements of spin flips. In order to find the good (quasi)particles, the ones that exhibit a simple Hamiltonian or partition function or transfer matrix, we make a series of transformations: from a lattice of spins, to row particles that are then made to act like fermions (by appending to them a sign indicating whether they have an even or an odd number of spin flips), to spatially regular arrangements of such flips. Finally we decouple resonant arrangements at wave numbers q and $-q$, into particles or arrangements that have the right mixture of order and disorder for the temperature and wave number.[30] This strategy parallels Onsager's as I described it in chapter 4. More generally, antisymmetry is a pervasive theme in the arithmetic and topological solutions.

After these several transformations, one ends up with particles much like the Cooper pairs of electrons in superconductivity. It turns out that this is a precise analogy. For example, in general a finite-size energy or mass gap is involved in exciting or forming these particles (in creating regular arrangements of, say, spin flips—or dually, of spin pairs

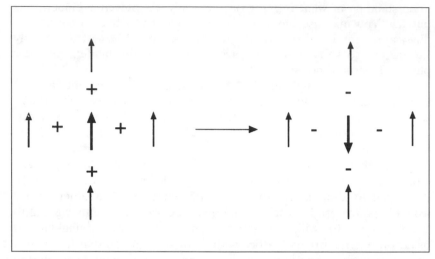

FIGURE 6.3   Changing one spin may alter many bonds; +, like spins; −, unlike spins.

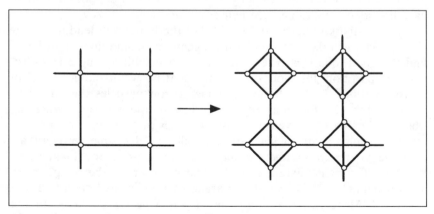

FIGURE 6.4   A lattice decorated with additional structure so that bonds are decoupled from each other. Adapted from Kastelyn, "Dimer Statistics and Phase Transitions," fig. 6.

above the critical point), except at the critical point.[31] That energy for formation is $\approx k_B T \epsilon_q$. For $q = 0$, it is called the mass gap, and equals $2(K^* - K)k_B T$. For large K (and low temperatures), the gap is negative, reflecting the asymptotic degeneracy of the even and odd ground states. For small K, the gap is positive. Its inverse in each case is a measure of the correlation length.

We note that at any bond site—representing an interaction between two spins, either like or unlike, both of which possibilities are to be added into the partition function—there can be zero or one unlike bonds. Such zero-one particles are readily represented or described by antisymmetric functions and operators: for physicists, they are fermionic particles. The trick in the above transformations was to preserve this property.

Recall that the topological constitution represents interactions as sides of a polygon that does not share sides with another polygon. Hurst and Green employ separate fermion operators to start and complete a bond or side of such a polygon—again using fermion operators, since a side is started or not, and cannot be started (or completed) twice. This is, by the way, equivalent to decorating the lattice with the structure it would need to decouple the bonds (figure 6.5).[32] "The reader familiar with the quantum theory of fields will notice that these [operators that emit a particle (actually a side) at one point and absorb it at another] are similar to the relations satisfied by fermion emission and absorption matrices, as one would expect from the fact that not more than one side of a polygon may be drawn between any pair of neighboring lattice points."[33]

Using the topological constitution, Samuel shows how integrals of anticommuting bond variables of a certain sort quite readily solve the Ising model. Along the way, he makes a crucial point about the preference for bonds over spins.

> All Ising model spin correlations may be easily calculated using the above method. The reason they result in such cumbersome expressions is the following: The variables which solve the Ising model are the η's [the anticommuting variables]. They might be called the mathematical variables because they represent it as a free field theory. Correlation functions of anticommuting variables are simple to compute. Contrast this with the spin variables. They are the physical variables. They are, however, complicated functions of the mathematical variables, the η's, which means that spin variable computations result in cumbersome expressions. In conclusion, there are two types of variables, spin variables which have a simple physical interpretation but are mathematically awkward to work

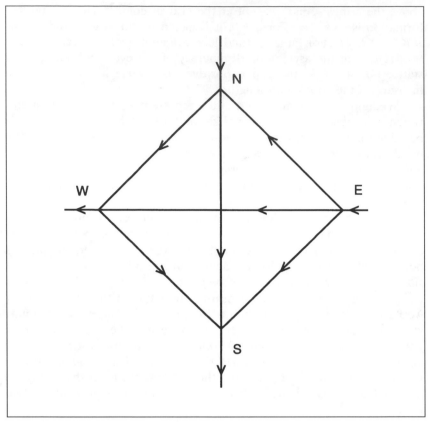

FIGURE 6.5   Decorating a vertex. Redrawn from Hurst and Green, "New Solution of the Ising Problem," fig. 1.

with and η variables which do not have as simple a physical interpretation but are easy to work with mathematically.[34]

One of my arguments here is that it pays to understand the physical interpretation of mathematical variables such as the ηs, seemingly right variables related to particles, as well as attending to the mathematical interpretation of the physical variables.

Again, antisymmetric objects appear recurrently and it seems crucially in the arithmetic and topological constitutions and solutions. When the transfer function is expressed by Onsager in his algebra—essentially his s_i and C_i, spin and spin-flip (or complementation) operators—the distinctive feature is their anticommuting character: $s_i C_i =$

$-C_i s_i$ (leading to a quaternion algebra). Kaufman employs spinors, again suitably anticommuting objects.[35] Nor, perhaps, is it surprising that multilinear forms on antisymmetric matrices (Pfaffians) appear in the topological accounts: we have to count things just once or not at all, and given that, in expanding Pfaffians one "crosses out" both rows and columns for any single index number, each site is counted only once. Again, we are counting up unlike bonds, excitations of the lattice from the orderly state, by means of enumerating the various length polygons.

A claim that the right particles are fermions is to be understood in the light provided by a set of exact solutions to the Ising model. The claim is without meaning independent of the actual models and solution. I should note that none of this is obvious or obviously true ahead of time, in the sense that only when we find solutions that work do we begin to press such an abstract analogy. Of course, along the way we may gain enormous insight or at least valuable clues from even more simplified toy models, as did Onsager.

I should also note that for actual electrons, considered as fermions, I do not feel that need to talk about models. I just talk about the particles. When we begin to do experiments on Ising row particles, or when we explain phenomena in their terms (as I shall a bit later in this chapter), then I imagine that they, too (as do the elementary excitations in crystal lattices), will have a life of their own.[36]

The Right Particles

Say bulk matter were constituted by more-or-less independent fermions. Then states of Ising matter might be known by the numbers of fermions ($N_q = 0$ or 1) present and their topology (q), which numbers tell us the energy of the system ($E = \Sigma N_q E_q +$ constant, where $E_q = \epsilon_q / \beta$). We might ask, what is the probability of each of the spin or bond arrangements in terms of their energies ($\approx \exp - \beta E_q$)?[37] The lattice is assumed to be effectively infinite in extent, at least in this infinite volume limit. There are to be no edge effects. And so we might study a single row (or column), and assume all other rows have the same probability distribution for their spin arrangements.

Ideally, the energy of the system would be the sum of the energies of its unlike bonds or its orderly arrangements of such (taken as the particles): each bond $2J_1$ or $2J_2$ above the ground state, horizontally or vertically, respectively. An unlike bond adds $2J$ units of energy to the Hamiltonian of the perhaps more-or-less independent row particles that form any such arrangement. (Namely, $H = -J\sigma\sigma'$, σ going, say, from

−1 to +1.) If we have a periodic arrangement, then we might speak of the energy per spin and so have as our unit $2J/\text{period}$.

In fact, the row particles are not independent, since there is a vertical interaction (Onsager's B operator, dependent on K_2, in effect a disorder operator whose importance is greater as the temperature is greater).[38] So, as we saw earlier for Onsager, who had to diagonalize his A and B simultaneously, we would expect that the energies of the row particles would shift when we took into account the B operator (perhaps as a perturbation). Yet the topology, q, might be unchanged.

Moreover, a thermal field, a temperature, effectively links the energies of those otherwise independent particles. Say we could find individuals, still-independent row particles under this thermal field, for each temperature, ideally with the vertical or column interactions acting as perturbations, merely shifting the energies of each row particle—their energies denominated in $k_B T$ terms. It would seem that the row particles' spatial periodicity properties should not change when we take into account the thermal field, even if their energies and their relative probabilities were to change.[39]

In effect, we would have created out of a two-dimensional lattice of spins a set of one-dimensional objects, the row particles. The row particles have well-defined masses or energies because one of the dimensions has effectively become the time dimension, and since it does not appear explicitly in the Hamiltonian, the energies of the eigenstates of the one-spatial-dimension problem are well defined. Note that we are here concerned with the partition function, a sum of an exponentiation of the Hamiltonian, and with the transfer matrix; the perturbation operates in that realm, as we shall see.[40]

The states of the system, the eigenvectors of the transfer matrix, can be expressed as just such a sum of independent row particles and their energies. This is implicit in Onsager's original expression for the eigenvalues of the transfer matrix, namely, for a $n \times m$ lattice:

$$\ln \Lambda_{\max} = 1/2\, n \ln(2 \sinh 2K_1)$$

$$+ \sum_{\substack{m \\ j \text{ odd or } j \text{ even, all combinations of sign}}} \pm \gamma(j\pi/n),$$

where the γs are for Onsager related to the excitation of order, or of disorder, in the lattice rather than the energies of particles (the $\pm$ form reflects the way the Hamiltonian is written: $+$ is the excited state of a pair of particles, $-$ is the ground state).[41] Schultz, Mattis, and Lieb show that these $\gamma(\omega)$ are just the quasiparticle fermion energies, which they call ϵ_q ($= \gamma(\omega)$, where $q = \omega$)[42] and the energies are expressed in $k_B T$ units.[43] Now we have some rather nicely defined particles. As we shall see, their energy spectrum will be quite helpful in understanding the behavior of Ising matter.

These are a different representation of the eigenvectors than Montroll's or Kramers and Wannier's of 1941, where the eigenvector was a sum of row configurations, each configuration known by its binary number. The row configurations are not independent. At zero temperature, only two configurations are possible, all up or all down. At infinite temperature, all configurations are equally excited. Montroll suggests that the critical eigenvector might be seen as a resonance of almost-ordered and almost-disordered states.

In sum, one takes an interacting set of row particles and effectively converts them to a "gas" of similar, noninteracting particles with somewhat altered properties, in this case shifted energies. The sleight of hand here is rather greater, since we go from a system of interacting spins (vertical and horizontal bonds) that are governed by a temperature, to a system of particles (row arrangements) in which the columnar interaction is absorbed into their energies. And then we take those row particles and treat them thermally, as if they were a system of particles governed or characterized by a temperature. So we have absorbed the columnar degrees of freedom. We can do so, it would seem, because of the duality symmetry of the lattice (Onsager's A and B are interconvertible). Formally, and technically, what I have described here is a paraphrase of the Onsager-Kaufman and, more particularly, the Schultz, Mattis, and Lieb solutions, in the end being the diagonalizing of exponentials of quadratic forms. Although it is fully justified mathematically and formally, one might physically justify this sleight of hand by its effectiveness in illuminating what goes on in a phase transition in Ising matter.

Phase Transitions in Ising Matter
For a square lattice, in which the couplings are the same in each direction,

$$\epsilon_q = \cosh^{-1}[\cosh 2K \coth 2K - \cos q],$$

q going from $-\pi$ to $+\pi$. q is an inverse wavelength (actually the wavelength is $2\pi/q$), reflecting how far apart unlike bonds are in this state. $q = 0$ corresponds to the lowest-energy particle (actually the odd combination of all up and all down), and to almost zero energy modes if we are very close to the critical point, where $\cosh 2K \coth 2K \approx 2^+$ and

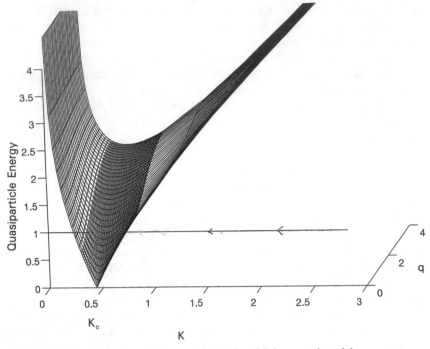

FIGURE 6.6   Energies of the two-dimensional Ising quasiparticles, ϵ_q.

so we have $\cosh^{-1}(2^+ - 1 \approx 1) \approx 0$,[44] all of this being apparent from figure 6.6.

Again, these quasiparticle energies, ϵ_q, correspond to Onsager's $\gamma(\omega)$ (except at $q = \omega = 0$), and as he says about the row configurations that go with them, "A detailed inspection of the operators Z_r [corresponding to creating quasiparticles, $q \approx r$,] can be made to show that those for which r is small are the most effective [see figure 7.12 for the amplitudes] in introducing pairs of alternations [↑ ↓] far apart, which is just what it takes to convert order into disorder."[45] In chapter 4, I described how changes in the temperature altered the amplitudes of the Z_r, and so account for the rise in disorder as the temperature increases, and for the especially dramatic behavior near the critical point (when all r, including small r, begin to count). However, in order to account for the specific heat, the following account in terms of particles is rather more perspicuous.

If the phenomena of Ising matter and its phase transition can be explained in terms of the energy level spectrum of the quasiparticles, we have further evidence of how this constitution of matter works. We are answering the question, "What distribution of energy levels for a large system gives rise to a phase transition as the temperature is varied?"[46] Figure 6.6 is crucial, and it may be seen to be derived from figure 6.1.[47] Note that the opening up of the triangle of figure 6.1, that is, q going from 0 to π, is the change along with y-axis in figure 6.6. An increasing T, with fixed coupling constant J ($K = J/k_B T$), is represented by going from right to left on the x-axis.

Literally, these quasiparticles are arrangements of unlike bonds in a row. In general, as the temperature rises from zero, more and more quasiparticles are likely to be excited, the probability of excitation being the Boltzmann factor, $\exp - \beta E$ ($= \exp - \epsilon_q$, since ϵ_q is in $k_B T$ units). But, at first, as we can see from the diagram, they are not much excited since their energy is much greater than $k_B T$. Still, as the temperature rises, the energies (in $k_B T$ units) go down, and so the excited q values rise from zero, and so the system is characterized by greater spatial inhomogeneity: smaller distances between unlike bonds, or put differently, smaller distances between changes in direction of the spins. (Again, the quasiparticle energies at each temperature are a function of that temperature. This is a curious kind of energy spectrum, to be carefully employed when analyzing phenomena dependent on the change in temperature.)

Note that the $q = 0$ mode is fully excited and dominant at low energies. It reflects the possibility of an almost degenerate odd parity eigenstate at low temperature. More precisely, at low temperatures we have two sets of almost degenerate states, all up or all down, each perhaps with some small number of spin flips. Neither is a parity eigenstate; parity converts one to the other. But their sum and difference are such eigenstates, and their difference is of odd parity. The even parity state is the conventional ground state. The odd parity particle or excitation ($q = 0$) is not stable at high temperatures, and so its probability decreases with increasing temperature.

As we come closer to the critical point from below, more and more particles are excited, in part because $k_B T$ is larger, in part because their energies (in $k_B T$ units) are even smaller and hence they are even easier to excite (the effect of the greater importance of the disorder operator, B). And so even greater inhomogeneity is introduced into the lattice. At the critical point, and only at the critical point, the lowest quasiparticle energies approach zero, and in fact the lowest energy is zero for $q = 0$. There is a continuum of states from zero energy, and so we have in effect an infinity of particles created or excited rather highly, in the infinite volume limit (but not all the particles). There is an effective infinite degeneracy of such low-energy modes. We would expect the specific heat to be much greater here, than above or below T_c, since it is proportional to the number of active degrees of freedom, and only here are there an "infinite" number of them ($\exp - \beta E \approx \exp 0 = 1$). Moreover, they dominate the grand partition function. As we saw in chapter 5, the grand partition function for such a system may well have real zeros in the infinite volume limit—a phase-transition point. What Onsager called a "conspiracy," the infinite value of the specific heat as a consequence of the behavior of three integrals ("The three integrals are infinite at the critical point, otherwise finite. The singularity results from a conspiracy.")[48] is seen here to be a physically meaningful and reasonable situation: at the critical point many scales of disorder are present, their associated quasiparticle energies are small (there is no energy gap, as there is away from the critical point), and a very large number of degrees of freedom are excited. At this point, in the infinite volume limit, there is a dense infinity of almost zero energy states, with long wavelengths and not-so-long wavelengths introducing lots of disorder. So, it would seem, not only do we have an infinite specific heat; we have the onset of pervasive disorder. (Here we might recall the problem of showing why the various phenomena happen at the same temperature.)

As the temperature rises above the critical point, one begins to deexcite the higher q states; by duality, on the other side of the critical point, they are states that represent orderliness. Eventually the lower q states as well are deexcited, so that when the temperature is large the system is fully disordered and is characterized by the disorderly ground state of the dual.

Our description, so far, started out with an initial state of the system at zero temperature, one in which all of the spins are aligned, the subsequent excitations due to heating disturbing that alignment. If we started out from infinite temperature, the ground state is the equal probability of all spin configurations, since $\exp - \beta E = 1$ if $\beta = 0$. As the temperature falls, we excite anti-quasiparticles, possessing regular degrees of orderliness, and more and more of them appear as we go toward the critical point (for much the same reason that more and more quasiparticles appeared when going up from zero temperature).

In $k_B T$ units, the energies of the quasiparticles and the anti-quasiparticles are the same ($k \to 1/k$ invariance, duality), going up or going down. A highly ordered particle at low temperature corresponds to a highly disordered particle at high temperature: both have the same energy (in $k_B T$ units). The correlation length, ξ, is a measure of the wavelength of the crucial quasiparticles. The correlation function, $g(r)$, goes as $\exp - \alpha r |\ln k|$. There is, as we might expect, a symmetry above and below the critical point (absolute value of the natural logarithm) due to the symmetry between the quasiparticles and the anti-quasiparticles.[49] At the critical point, $g(r) \approx 1/r$, in effect all scales are present, the mass gap is zero, and there is an infinite correlation length.

The free energy of a set of isolated or independent systems is the sum of their free energies, and so we might expect that the free energy of this Ising matter expressed as a system of independent q particles is just the sum of the free energies of these quasiparticles. Schultz, Mattis, and Lieb show that this is the case, the famous integral in the Ising model needed to estimate the largest eigenvalue of the transfer matrix, and so the partition function and the free energy, being merely an integral approximation to the sum of the energies of these quasiparticles:

$$Q = (2 \sinh 2 K_1)^{1/2} \exp 1/4\pi \int_{-\pi}^{\pi} \epsilon_q \, dq.$$

Note that when we go to a form with K_1 and K_2 interchanged, merely the lattice rotated 90 degrees, and we expect the partition function to be identical, the ϵ_q are different, but the first factor is now $\sinh 2 K_2$ and the resulting partition function is the same.

More generally, there are a variety of expressions for the integral that defines the partition function for the infinite lattice. Onsager's original form was identical to the form above, but the interpretation of ϵ_q was not an energy, but a side of the triangle in figure 6.1, $\gamma(\omega)$, the angles of which determine the amount of disorder in the state. Onsager also showed how to make K_1 and K_2 appear symmetrically in the integral, and that form has a natural relationship to the finite-size ($m \times n$ sites) partition function calculated by Kaufman. (It may be shown to be composed of four terms, which take into account the various combinations of $\pm \gamma(j\pi/n)$, j odd or even: $\prod \sinh$ [or cosh] $\gamma_{\text{even or odd}}$, corresponding to four different Pfaffians, and to four different boundary conditions [polygons that encircle the torus, in horizontal and vertical directions, an even and odd number of times], and to the four theta functions.) Wannier gives another form that is manifestly invariant to the $k \to 1/k$ transformation by redefining the relevant modulus as $\kappa = 2 \sinh 2 K/(\cosh 2 K)^2$, whose value is the same for k and $1/k$—and it is a restatement of Kramers and Wannier's original observation about duality. Stephen and Mittag develop a form in terms of their own (plane) triangle, whose crucial angle again refers to the amplitude of a disorder operator. Finally, there are forms of the partition function written in terms of their zeros, Baxter's $F(\theta)$ ($Q = \prod_j \exp F(\theta_j)$, θ_j being a zero, which becomes a real zero as $N \to \infty$), being a case in point (as we saw in chapter 5, note 34). In each case, the form of the integral builds in a particular physical interpretation—yet each yields the same partition function.[50]

WHAT IS THE TOPOLOGY?

As I indicated earlier, another interpretation of the solutions for Ising matter and their mathematical formulations is to say that here we have a combinatorial problem—that is, counting the number of different ways of putting together a collection of similar components given a peculiar situation: how many ways can you put m balls in n boxes given a constraint? And the (grand) partition function is the generating function for the combinatorial problem, the polynomial function whose coefficients are the combinatorics of interest: say, the number of polygons of length l, given certain constraints.[51] (The partition function for the Ising model in the high-temperature expansion is $\Sigma w^n g_n$, where $w = \tanh K$ and g_n is the combinatorics of interest; and this is formally like a grand partition function, $GPF(z)$, $= \Sigma z^n Q_n(T, P)$, for an associated lattice gas.) The various constitutions of Ising matter may then be seen as technical means to solving this combinatorial problem, counting things just right. By measuring the behavior of the specific heat of a system, we might solve a mathematical combinatorial problem. It is a remarkable fact, to be sure, that there is an arithmetic solution that does the same work as does the topological, and it is perhaps even more remarkable that the geometric and holographic solutions do it, too. It is a worthwhile task to understand how and why, as Kac and Ward suggested.[52]

The elements of the eigenvectors of the transfer matrix (in the basis in which each configuration is one dimension; hence there are 2^N dimensions) can be shown to be the probabilities for each row configuration of spins—actually, it is the square of each element (while the logarithm of the eigenvalue is effectively the free energy associated with that set of probabilities). If we randomly generated a lattice, the configuration of each row of the lattice being chosen according to the probabilities given by the vector associated with the largest eigenvalue of the transfer matrix, and we suitably marked up the resulting lattice into like and unlike bonds, we would have the polygon system that is enumerated by the van der Waerden–Kac–Ward method (or by Onsager's Lie algebra). The two-dimensional lattice of bonds or polygon links is generated by one-dimensional slices. In effect, this would be a way of generating a particular two-dimensional random walk, a Brownian motion. Now, it is a characteristic feature of such a Brownian motion that it exhibits scale invariance—it looks the same at all scales. So, for example, when we watch the Brownian motion in a microscope, we are seeing the cumulation of many many molecular steps or collisions for each observed motion of the dust particle. Yet what we see is a Brownian motion, just like the molecular-level collisions.[53]

Hence, insofar as the geometric and holographic constitutions measure the scaling and similarity effects, they are studying the same phenomena as do the arithmetic and topological constitutions. The crucial point here is that a random walk looks the same at various scales, and that walk in two dimensions might be generated by one-dimensional considerations.

The crucial feature of the two-dimensional lattice that defines this particular combinatorial problem is that the lattice is planar or bipartite: there is no crossing of bonds in the plane; or one might divide the lattice into two interleaved halves. This topological fact allows for the existence of a dual and for the exact arithmetic solution, and it is just as central for the geometric and holographic solutions. It may be seen to define a certain class of random walks in the plane, as well.

If a solution is to be doable for a lattice, exactly and practically, planarity is crucial. Say that we imbed a lattice on a surface: namely, draw the lattice on that surface without having lines of interaction cross. One can define a property of that surface, its genus, which measures its connectivity. The number of determinants or Pfaffians needed to solve the Ising model goes as 4^{genus}.[54] Now, the genus of a plane is zero, and that of a torus or doughnut is one (a torus is a finite two-dimensional lattice wrapped around to meet itself in both the x and the y directions), representing its hole. As Kastclyn puts it, "It can easily be seen that the lattice graphs for which the Ising problem could not be solved [so far] are all of infinite genus"[55]—since it can be shown that an infinite lattice with crossed bonds, or one subject to a magnetic field, cannot be imbedded in a finite-genus object. Practical issues can become in-principle issues very quickly under such a regime.[56]

THE RIGHT MATRIX

The Ising lattice itself is not directly useful if we are considering bonds, since there is a problem of multiple counting and overlap, namely, one spin may be involved with several bonds; and so one develops a counting lattice, in which each bond is independent of the others (by decorating each site with extra structure or terminals, so that one original bond does not affect another that shares a site with it). In effect, given planarity, one could cover the decorated lattice with

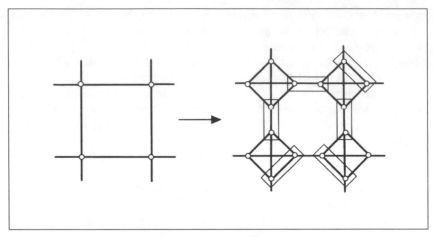

FIGURE 6.7   Dimers covering a decorated lattice.

dominoes (or chemical dimers), each domino representing a possible bond (like or unlike). In this sense, the lattice is manifestly bipartite: all one need do on the decorated lattice is to color the domino halves red and blue and treat the lattice as a checkerboard of sorts. It can be shown that the task of counting polygons is equivalent to counting the number of ways one can cover a decorated lattice with those dominoes or dimers (figure 6.7).

Whether counting polygons or dimers, we need a procedure of enumeration that is systematic and complete. If we number all the sites ($= N$), we might form an $N \times N$ matrix to indicate which site is linked or bonded to another, in the sense of its contribution to the partition function. Recall Kac and Ward's determinant counted polygons, because each matrix element represented a possible bond (and most elements were zero since bonds were only between adjacent sites); a polygon is a sequence of bonds; and terms in the determinant are sequences of multiplications of matrix elements, hence a sequence of bonds. The task was to invent a matrix—its exact arrangement and its exact elements—from which, through the determinant, the polygons were enumerated correctly. Kac and Ward's account of their motives and devices for constructing an appropriate matrix should be required reading for all apprentice theoretical physicists.[57]

Various test cases, namely peculiar polygons, are invoked to indicate why a tentative matrix would work or not, and how that tentative matrix might be altered so that the revised matrix would take into account more and more peculiar situations. (In fact, Kac and Ward end up with an $n \times 2m$ lattice.) Their great advantage was, of course, that they knew what the correct solution had to be, from Bruria Kaufman's exact solution for a finite lattice. Their motive, as they explicitly say, is to show how the combinatorics are built into the algebraic solutions of Onsager and Kaufman. So, once they can sufficiently motivate a matrix, and it gives the right answer a la Kaufman, they are through. For they have shown how a particular combinatorial problem is solved by the Ising model.

Still, this wondrous account of finding the right answer is not very satisfying, if we wanted to think physically. Here we are not talking about the right particles, but about the right matrix, in effect, the right set of interaction rules between sites. Feynman and Vdovichenko made physical what Kac and Ward performed in a rather more jerry-built fashion.[58] They first show how, rather than counting only certain polygons, one might count up all closed loops by suitably weighting each link so that the undesired polygons cancel out. Having so marked up the lattice, they then treat the problem as one about random walks on a peculiar plane. Such random walks are in effect molecular motions, and so we might end up with the physical problem of diffusion.

But in what sense does such a strategy point to the right matrix? The weights turn the problem into one that is quite familiar to practicing particle physicists and condensed-matter theorists who employ techniques from quantum field theory. In effect, one is summing up a series of Feynman graphs, now a set of random walks on a peculiar surface. The right matrix is in effect the scattering matrix for this problem— and that is, for these physicists, as physical as is anything, fully motivated in this context. The analogy with field theory proves rather more pervasive, especially since the antisymmetric character of bonds or links would suggest a different matrix and kind of determinant, the Pfaffian.[59]

Now, if one wants to count dimer coverings or configurations (on the decorated lattice, not the original Ising lattice), then the mode of enumeration has to never count any site more than once, yet to be sure to cover all the sites. Consider an $N \times N$ matrix, A_{ij} being the link of site i to site j. One needs a different sort of determinant, one that eliminates both the row and the column in its Cramer's rule expansion (since once a site is taken up it cannot be employed again, unlike in the Kac and Ward case where it has to be employed again if it is to be part of a polygon), and so formally guarantee no site will be used more than once. Once a particular site is used up, it is not coupled to another. The multilinear form that does just this work is the Pfaffian, which operates on an antisymmetric matrix (since connecting site i to site j is the same as connecting j to i with a reversal of sign).[60] Again, the problem is setting up a suitable matrix so that its Pfaffian will do the required work.

Just as in the particle formulation, where the character of bonds leads to fermions, here that same feature leads to Pfaffians (of antisymmetric matrices). And, it should be noted, Pfaffian forms appear in linearized expansions of exponentiated anticommuting operators (exp $-\beta H$; H is composed of such operators)—what is eventually the Wick theorem of quantum field theory. Of course, in each case, finding the right fermions or the right matrix is crucial, or "you'll be sorry." Even if you are able to compute the partition function, you may have little idea what is going on physically, little insight about where to go next. The random walk and the Pfaffian solutions are illuminating because you can see what is going on in terms of a problem you do understand.[61] (Computationally, it should be noted that for an antisymmetric matrix, Pf $A = (\det A)^2$, and so in fact and in practice, in the end one takes the determinant of the appropriate matrix.)

What makes fermions and Pfaffians (and the renormalization group) work practically, and it would seem physically, is in the end a topological fact—that the lattice is planar, bipartite, or orientable. Physically, bonds are in general local and do not cross, and there is no external field. Put differently, calculation rapidly becomes intractable when these features are not present. (Technically, quartic rather than quadratic terms appear in the analogous quantum field theory Hamiltonian, for example, so we no longer have a system of, say, independent harmonic oscillators, but they are now interacting since the system is anharmonic.)[62] Such a contingent fact becomes a desideratum if the calculation becomes sufficiently intractable sufficiently fast. Unless, of course, and this is historically always the big catch, new technologies break the logjam. These may include formal approximative schemes (such as Samuel's use of anticommuting variables), calculational methods and devices (numerical simulation, or formal devices such as the corner transfer matrices of Baxter), or theories that transcend previous limitations or specify very different degrees of freedom as the right ones.[63] Those technologies are as often adapted from other realms as invented anew.

Perhaps these models of matter are too parochial to be serious constitutions, for they prove systematically difficult to calculate when one adds some apparently reasonable extensions to the model (three dimensions, magnetic field, or next-nearest-neighbor interactions).[64] It would be quite surprising, but not shocking, were these models eventually to be seen to be of antiquarian interest, only by chance being at all illuminating about real matter.[65] Yet they do capture many of our intuitions and facts about actual matter, and these are constitutions that do possess infinite volume limits. Perhaps Wilson is right, that what appear as parochial tricks eventually are domesticated and become the mathematical methods of a subsequent generation. What gives me some hope is that the various methods for the solution of the Ising model have been generalized in quantum field theory, and that their particular realizations have been shown to have more general mathematical foundations. Still, it is humbling to realize we do not know how to make progress on these reasonable complications in any exactly solvable way —although we do possess powerful formal and numerical techniques for working on them, techniques that suggest why it is so difficult to solve these problems exactly, and that are related to other intractable problems.

But what does the topology mean physically? It would seem to be a way of insisting on certain symmetries, duality and the star-triangle relation and the renormalization group, which then lead to solutions— for we are now focused on the right variables, which point to or express those symmetries. These various technical apparatuses of fermions and Pfaffians are ways of insisting on the fundamental symmetries of the lattice, and in return they lead to perspicuous solutions to the Ising model.

7

Physics and Mathematics

FINDING THE RIGHT MECHANISM AND CHOOSING THE RIGHT FUNCTIONS

Star-triangle relations and duality: arithmetic duality, geometric duality, the star-triangle relation and modular (or k) invariance, fluctuation-dissipation relations, the mechanism linking the star to the triangle, tricks of the trade and terms of art, the star-triangle relation in action, the meanings of the star-triangle relation, the formalism and the physics • Elliptic functions and hyperbolic geometry: elliptic functions, elliptic functions in solving the Ising model

STAR-TRIANGLE RELATIONS AND DUALITY

In the most elegant of the accounts of the constitution of matter as an Ising model, certain symmetries would seem to lead to all one needs to know about the model. Early on, Kramers and Wannier (1941) located the critical point by means of what we now call a duality transformation. Much later, Baxter and Enting (1978) showed how the star-triangle relation leads fairly directly to the full solution. So we take duality and the star-triangle relation as symmetries of the Ising lattice, in that they represent topological symmetries of the lattice; they translate into partition function invariances; they lead to comparatively more efficient and perspicuous solutions of the Ising model; and the good particles of the system are indexed by these symmetries, in effect leading to the right degrees of freedom. As we shall see, from the perspective of quantum field theory, (arithmetic) duality points to the natural particles and their energies, while the star-triangle relation points to the natural relations among the interaction constants, in effect the geometry of coupling-constant space.[1]

That there are these very immediate consequences of symmetry suggests just what is most fundamental about the constitution of Ising matter. We shall examine each fundamental feature and see how it appears in actual accounts. My claim here is that we might well insist that some kinds of features of Ising matter are fundamental, that these technical solutions do lead to philosophic analyses, much as we saw for the stability of matter in the first three chapters.

It is a truism for physicists that the fundamental features of physical systems are expressed either by their symmetries or in the conserved quantities that characterize those symmetries: a translationally symmetric system means that momentum is conserved; if there is rotational symmetry, angular momentum is conserved; for a system in which time does not explicitly appear in the Hamiltonian, energy is conserved. Namely, a system that is invariant to a transformation conserves something in that transformation. Practically, ever since the conservation of energy became a working tool of physicists and conserved quantities came to be interpreted in terms of symmetries of the equations of motion, or of the Lagrangian ("ignorable coordinates"), it was understood that a symmetry implied a conservation law (Noether's theorem, 1918).

That the quasiparticle energies in Ising matter, ϵ_q, are indexed by q, the inverse wavelength of the row particles, is usually taken to reflect the translational invariance of the lattice (in the infinite volume limit, or with circular boundary conditions).[2] That the energies (when expressed in $k_B T$ units) are invariant to a $1/k$ or duality transformation reflects how order and disorder may be shown to be "interconvertible," that they represent flip sides of the same particles.[3] As for the star-triangle transformation, it points to k sets of what might be called thermodynamically equivalent systems, in that they have the same degrees of fluctuation and orderliness (and in this sense might be said to be a single phase).

Moreover, if we have a complete set of conserved quantities for a system, a complete description of its invariances, then it is straightforward to compute its dynamics. (In terms of its conserved quantities, it does not change at all.) The system is said to be integrable, reflecting our capacity to solve its differential equations exactly. Exactly solved problems, here in statistical mechanics, are in this sense integrable, in that they are analogous to some mechanical problem that is integrable. Still, it may not be obvious from their exact solution (here I am thinking of Onsager's solution) just which are the conserved quantities, just which are the invariances of the system, and just which is the mechanical problem.

Let us see how these symmetries play themselves out. I first examine arithmetic and geometric duality, and then the star-triangle relation, which may be seen to be intimately connected to geometric duality. In an attempt to understand the physics behind the star-triangle relation, I examine a pervasive theme within physics, fluctuation-dissipation relations that connect two different processes mediated by a shared mechanism. This leads to a suggestion for the physical mechanism behind the star-triangle relation. And then I show how that mechanism plays a role in some of the solutions to the Ising model.

Arithmetic Duality

In general, duality represents the discovery that, for two dimensions (and not for one), there exists a nonzero critical temperature and so a critical value of the coupling constants, and this is epitomized by saying that $k = 1$ is the critical point. (Note that, for one-dimensional systems with local interactions, the critical temperature is $T = 0$.)[4] Some physical phenomena are emergent when $k \leq 1$, such as the development of a nonzero "order parameter" such as the spontaneous magnetization, one that measures the degree of orderliness of a system. Other phenomena are symmetric around $k = 1$. That is, since $(k - 1) \approx Kt \approx (T - T_c)/T_c$, they are the same for positive and negative t (at least if t is small). I take duality to be a radical fact about Ising matter in two dimensions, saying that k is as important a variable as is the temperature T. Just as $T = 0$ is entrenched in the third law of thermodynamics, $k = 1$ will be entrenched in accounts of critical phenomena and duality.

In chapter 4, we discerned two kinds of duality, arithmetic and geometric. The latter showed itself in the block-spin model and in decimation, iterative averaging procedures that send a lattice to either zero or infinite temperature, except at the critical point, which is a fixed point of geometric duality. First, let us look at arithmetic duality, in this section "duality" for short.

Duality symmetry connects or converts a disorderly configuration of the lattice into an orderly one, and so a set of disorderly configurations into a set of orderly ones, a high-temperature set of dominant configurations into a low-temperature set of dominant configurations, a high-temperature partition function into a low-temperature partition function. There is a topological prescription, one that defines a dual lattice, and a rule for transforming the configuration of spins, one that takes a mostly unaligned system and converts it into a mostly aligned system. But this topological prescription and spin rule defines as well a dynamical prescription (that is, a transformation of the partition function, $Q \approx k^{-N/2}Q_{\text{dual}}$), one that Kramers and Wannier discerned in the patterns of the transfer matrix itself, and that Onsager noted in his operators that built up the transfer matrix.[5] Now, physically, duality works because it really is true, for the Ising lattice, that a small excitation in $k_B T$ units, from the ground state (at almost zero temperature, an almost completely ordered system), is of the same form and effect on the energy and the partition function as a small deexcitation from the most disorderly state.

Duality is both topological, rotating the lattice and its spin configuration into its topological dual,[6] and dynamical, inverting k, which is a measure of the temperature in terms of the coupling between spins: in each case, the spin configuration goes from orderly to disorderly, or vice versa. As we have seen, there are many technical devices for implementing duality: in the transfer matrix itself (Kramers and Wannier), or as a function of the alternation of the spins in a row versus the average value of the spins in a column (Onsager), or topologically, or in the partition function and its high- and low-temperature expansions, or in the $k \rightarrow 1/k$ replacement—but, in the end, they are lattice rotating and k inverting.

The recurrent task is to link physical insight and mathematical technology, concrete and particular physical situations to workable mathematical techniques. Let us look at Onsager's technical formulation.[7] Onsager noted that, for a rectangular lattice, one could build the lattice adding horizontal bonds first or vertical bonds first. Namely, there should be a symmetry between computing the partition function for the interactions that occur in a horizontal row and computing the contribution to the partition function for that row's (vertical) interaction with the previous row.

In Onsager's terms, his matrices B and A, representing column and row interactions respectively, are "interconvertible,"[8] And in performing such a transformation, one also goes from an orderly configuration to a disorderly one—as Kramers and Wannier originally noted by direct examination and manipulation of the transfer matrix itself (including multiplication by a symmetric matrix, then a matrix transpose, and a change from K to K^*, its dual temperature, where $\exp - 2K = \tanh K^*$). I should note, as does Kramers's biographer Max Dresden, that it is not at all clear how Kramers and Wannier found the suitable transformation matrix or the physical symmetries involved. Presumably, there would be an explanation much as Kac and Ward give for their derivation—suitable trials and errors with test cases until the answer seems inevitable. Their mode of classification of states, in terms of a binary numerical representation of a configuration of spins (say, up = 1, down = 0), does naturally lead to the dual state being something like a two's complement.

Onsager, rather more physically, observed: "For a simple try to associate order in a cold state with disorder in a hot one, it seemed reasonable to make the alternation of spins along a column [indicated by his C_i, complementation or spin-flip operators—the sum of which is his B] correspond to an average product of two adjacent spins in a row [indicated by a product of his spin operators, his $s_i s_{i+1}$, their sum being his A]."[9]

This too can be readily formalized by an operator of Onsager's invention,

$$L_{\mu\mu'} = (1/2)^{(n+1)/2} \prod \left[\delta(\mu_j - \mu_{j+1}) + \mu'_j \delta(\mu_j + \mu_{j+1}) \right],$$

operating on a state defined by sums of products of the μ'_j. So, for example, the high-temperature state is a constant—all configurations are equally likely (and the state vector has no μ dependence). $\Phi_{\text{high } T}(\mu'_j) = 1/2^{n/2}$ and $\Sigma' L_{\mu\mu'} \Phi_{\text{high } T}(\mu'_j) \approx \prod \delta(\mu_j - \mu_{j+1})$, namely, the state where all spins point in the same direction, the low-temperature eigenvector.[10]

Although the various technical realizations or formal representations of duality might be shown to be equivalent, the moves that would seem to follow in employing each of the various techniques prove to be rather different. And it is those moves, the practices employed in a particular technology, that lead to different constitutions of matter (namely, different solutions to the Ising model). In each case, however, the particular technical realization must be made to do the work of solving the model in its own context, presumably that particular realization being modified so that the work actually can be done.

It may be useful to provide another toy prescription for performing a duality transformation on a lattice, in addition to the one in chapter 4, to give a feeling for how technical the abstract formulations are, how distant they are from the actual physical objects—albeit a toy prescription that, too, is somewhat distant from the excited states themselves. In very concrete terms, if we mark up the bonds of a lattice with, say, + for like and − for unlike (figure 7.1a); we create a dual lattice (rotated π/m), its sides now marked + and − (b); if we then exchange + for − (like for unlike) (c); and go back to the original lattice and its spins (d), we get the dual configuration that Kramers and Wannier find.[11] (Doing the inversion of bonds on the dual lattice automatically solves the problem of making sure the new spins can support the inverted bonds.)

Again, my point in providing a good deal of detail is to indicate that the technical implementation of the insights about the effects of duality symmetry is not at all obvious. (And, these insights were suggested by formal attempts to solve the Ising model.) Similarly, topologically drawing the dual lattice, showing how closed polygons (the high-temperature expansion) correspond to configurations of unlike bonds on the dual (the low-temperature expansion, excitation from a perfectly ordered

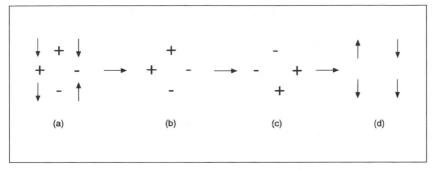

FIGURE 7.1 A duality transformation.

lattice), does not say how we are to technically implement this insight and push through to an exact solution. Connecting insight about symmetry to effective technical formalism is the very great achievement of theoretical and mathematical physics.

Geometric Duality

As indicated in chapter 4, the geometric dual averages out details and sends the lattice to zero or infinite temperature, except at T_c. In effect, block-spin averaging or decimation takes a system at k and converts it into a system at k^α, $\alpha \geq 1$.[12] There is an intimate connection with the star-triangle relation (which is, however, k preserving), which also eliminates spins by integrating out their contribution to the partition function. But in the block-spin transformation, or in decimation, one then rescales the coupling constant geometrically (perhaps having employed the star-triangle relation to eliminate some spins).

We might note that for the transformation $k \to k^\alpha$, if $\alpha = 1$, we have the star-triangle relation; if $\alpha = -1$, we have arithmetic duality; and if $\alpha > 1$, we have some form of geometric duality. In this sense, k and α may be said to index the symmetries of the lattice. From the point of view of a $k \to k^\alpha$ transformation, geometric and arithmetic duality and the star-triangle relation are generic objects, suggesting that something deeper is going on here. Topological lattice symmetries and dynamics $(k \to k^\alpha)$ are related, as we shall see.

There is a perhaps analogous situation in probability. If we take the insight that Gaussians "grow" as $N^{1/2}$, generalize it to distributions that grow as $N^{1/\alpha}$, we discover that if $\alpha = 1$ we have the Cauchy distribution, $1/(\pi(1 + x^2))$, if $\alpha = 1/2$ we have a random walk or diffusion, and so forth. And so we discover the Lévy stable distributions and their infinite divisibility (for example that a Gaussian can be divided into a sum of Gaussians).[13]

More to the point are the modular transformations of the elliptic functions. Anticipating discussion further on in this chapter, when $k \to 1/k$ (which for the elliptic function parameter τ is $\tau \to \tau/(1 - \tau)$), $sn(u, k) \to 1/k \, sn(ku, 1/k)$—recalling $Q(k) = k^{-N/2}Q_{dual}(1/k)$. ku, actually $k(I' - u)$, picks out the dual coupling, K^*, where the modulus is now $1/k$. (I' is the "imaginary" quarter-period magnitude of the elliptic functions, here a real number.) The star-triangle relation sends $k \to k$, but $u \to I' - u$: $\sinh 2K_i \sinh 2L_i = 1/k$. Geometric duality is probabil- ity best represented by Gauss-Landen modular transformations, $\tau \to \tau/2$ or 2τ, which for small k are $k \to 2\sqrt{k}$ or $k^2/4$.

(As for the star-triangle relation, since $sn(iu - iI') = 1/(k \, sn \, iu)$, and $\sinh u = -i \sin iu$, and by definition $sn \, iu_K = \sin 2iK$, then

$$\frac{1}{k \, sn(-iu_L)} = \frac{1}{-k \, sn(iu_L)} = \frac{i}{-ik \sin 2iL} = \frac{i}{k \sinh 2L}$$

$$= i \sinh 2K = \sin 2iK = sn \, iu_K = \frac{1}{k \, sn(iu_K - iI')}$$

—then $iu_L = -(iu_K - iI' + n2iI')$. To accord with other demands, $n = 0$, and so $u_L = -(u_K - I')$. And so we have recovered the definition of the transformed couplings under the star-triangle relation, since here in effect K and L [or K_1 and K_2] are interchanged for the rectangular lattice, and in that case $u_K + u_L = I'$.)

Now, we learn from the renormalization group and scaling symme- try that geometric duality reflects the persistence of pattern in a lattice, at various scales; at the critical point the lattice looks the same at an infinite range of scales; and that range of similarity diminishes as we get farther from the critical point—except for the trivially self-similar configurations at $T = 0$ or ∞.

Geometric scaling-up (or topological transformation) and dynamics are to be related, albeit differently than for arithmetic duality. Changes in geometric scale in the averaging process are also a $k \to k^\alpha$ transfor- mation. Block-spin renormalization and decimation are means of taking the physical insight about scaling, connecting geometry to dynamics, and converting it into an algorithm that leads to good quantitative numerical results.

To achieve a practical formal and exact solution, as we have for arithmetic duality, another averaging and scaling process is helpful, another sort of geometric or holographic constitution, one that scales up infinitesimally.[14] Namely, one converts a lattice of size l into a lattice of size $l - \delta$ (in effect a very modest averaging), in figure 7.2, going from the lattice of dark circles to the one of light circles, so having eliminated one site in each direction. The triangle-to-star relation is employed to transform the lattice from the dark circles to the light circles.[15] One then scales up the lattice to its original size, and thereby K (and so k) is scaled as well. In effect, the star-triangle relation (and k's being conserved in that transformation) is what might be called the deformation symmetry for this constitution of matter, one that unites a wide variety of different lattices that display similar behavior, namely, k's being a measure of macroscopic correlation length.

This constitution, like decimation, wipes out or sums over a fraction of the sites, but now the fraction is infinitesimal (rather than one-half).

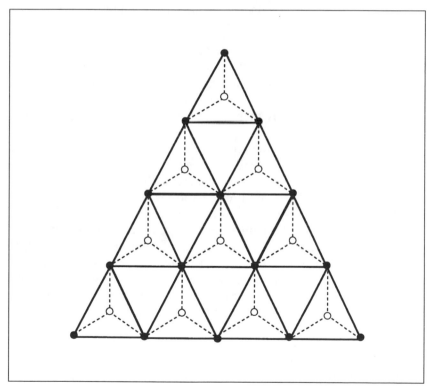

FIGURE 7.2   Infinitesimal star-triangle transformation. Redrawn from Knops and Hilhorst, "Exact Differential Renormalization," fig. 1.

This construction allows one to formally describe the spin-correlation function and to derive an exact solution for it near the critical point. Again, the actual technical implementation of the basic insight about the consequences of geometric duality symmetry is not at all obvious, and must be invented anew each time.

The Star-Triangle Relation and Modular (or k) Invariance

Onsager notes:

> A general form of the dual transformation invented by Kramers and Wannier together with a rather obvious "star-triangle" transformation are used in this connection [in computing the partition function of the hexagonal lattice], and the transition points can be computed from the transformations alone. While these were a great help to the discovery of the present results, the logical development of the latter is best relieved of such extraneous topics, and the subject of transformations will be reserved for later communication.[16]

That is, one might relate the low-temperature hexagon to a high-temperature triangle (a duality transformation, the hexagonal and triangular lattice being duals of each other), and then relate the high-temperature triangle to a high-temperature hexagon (the star-triangle transformation). It is then said that this is a side issue for the moment.[17]

It turns out that the symmetry implied by the star-triangle relation is just about sufficient for solving the Ising model. In a series of papers, Baxter and collaborators show how central is the star-triangle relation (and the elliptic functions that are used to instantiate and parametrize them) for exact solutions of two-dimensional lattice models.[18] More recent work by a cadre of scientists confirms this centrality, for example, the star-triangle relation coming to be seen as equivalent to the factor-izability of an analogous scattering matrix into two-body incoming and outgoing factors and so the conservative or symmetric character of the system; and the elliptic functions coming to be seen not merely as devices but more generally as modular functions; k is the crucial variable, conserved in the star-triangle transformation, the modulus of those modular functions. In effect, just as the star-triangle relation allowed the decimation formulation to exactly transform the lattice infinitesimally, here the star-triangle relation and modular functions describe the symmetry of the lattice as it is topologically transformed (all under the regime of the partition function).

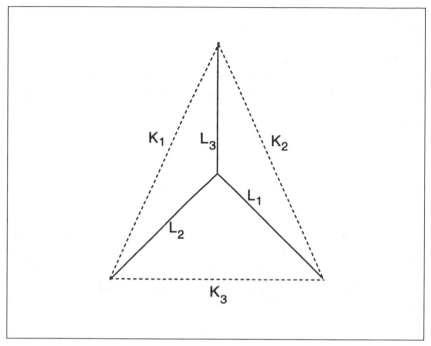

FIGURE 7.3   A triangle on a star.

Technically, by dividing the Ising lattice into alternating rows and summing the partition function over one kind of row (as in decimation), one is able to turn the partition function of a hexagon or honeycomb lattice (the "stars") into that of a triangular lattice at the same temperature: that is,

$$Q_{\text{hexagon},2N}(k,\{L_i\}) = R^N Q_{\text{triangle},N}(k,\{K_i\}),$$

where $R = 2/(k^2 \prod \sinh 2K_i)$ and $1/k = \sinh 2K_i \sinh 2L_i$ (figure 7.3).[19] (See also figure 4.7.)

It should be noted that the temperature, as such, disappears almost immediately into the $\{K_i\}$ and $\{L_i\}$, the interaction constants measured in terms of $k_B T$. In order to actually get at the experimental physics of the model, one has to bring it back in by recalling that $K_i = J_i/k_B T$.[20]

This scheme—the sum over the star leads to the partition function for a triangle, something that might seem to make topological sense—only works if there is a relationship between the interaction constants of the star (hexagon), $\{L_i\}$, and the triangle $\{K_i\}$:

$\sinh 2K_i \sinh 2L_i = \sinh 2K_j \sinh 2L_j$, $= 1/k$, $i, j = 1, 2, 3$, and $k = 1$ is the critical temperature.[21] Again, if duality inverts k, the star-triangle relation conserves it.

Note that k can be defined in terms of just the $\{K_i\}$ (or just the $\{L_i\}$), where k characterizes the correlations within a lattice. Moreover, given a $\{K_i\}$ that defines k, there are many other sets of constants with the same value of k, but only one set, here $\{L_i\}$ by definition, will be related to the $\{K_i\}$ by the star-triangle relation. As we shall see, however, the set of k-invariant $\{K_i\}$ is of more general interest, marking the set of "commuting transfer matrices.":

$$V_a(k, u_a)V_b(k, u_b) = V_b(k, u_b)V_a(k, u_a),$$

the us designating the different $\{K, L\}$ combinations that have the same k. Note, as well, that if we allow, say, $K_3 = 0$, we have just a rectangular lattice—we have wiped out the hypoteneuses of the triangles (figure 7.4)—and then it is not hard to see that L_1 now corresponds to what we earlier called the vertical interaction, K_2, and hence we have $\sinh 2K_1 \sinh 2L_1 = \sinh 2K_1 \sinh 2K_2 = 1/k$, our earlier definition of $1/k$.

Moreover, consider a lattice at one temperature, T, which is constituted by various transfer matrices ($\{J_i\}$ different) all of which commute; namely, they have the same value of k, but different us. They will have the same set of eigenvectors. But, of course, the transfer matrices will have different eigenvalues, and hence different free energies per spin. It is not obvious that the maximal eigenvalue in each case goes with the same eigenvector. In fact, it does not. But, at least if one $V(k, u)$ is most frequent and dominant, and N is large, $V^N(k, u)$ will act as a projection operator, so that there will be just one maximal eigenvector that survives.[22] The free energies and specific heats associated with each matrix may differ, yet the orderliness of the system is the same. We would seem to have a single phase. Presumably, the free energy per site would be the average of the free energies ($Q = \mathrm{Tr} \prod V_i \approx \prod \Lambda_{max, i}$, and the logarithm leads to the average). Even more curious would be a lattice built out of different k transfer matrices, all at the same temperature. In effect, the lattice then has defects.[23]

If we want to compute the correlation function for a lattice, it turns out that we might introduce "defects" (they are k-transfer matrices, but now $K \rightarrow K \pm i\pi/2$) along a path between the two spins whose correlation we want to study, and then we compute the partition function for this system.[24]

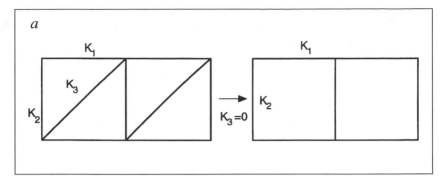

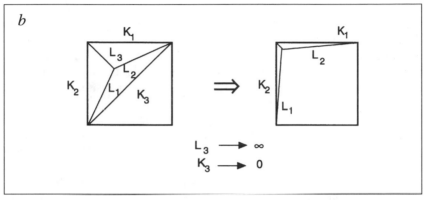

FIGURE 7.4   The star-triangle relation on a rectangular lattice: a, converting a triangular lattice into a rectangular lattice; b, letting L_3 go to infinity, so that $L_1 = K_2$.

Again, k is a very interesting variable, indeed. Early on, Kramers and Wannier defined another variable, $\kappa = 2 \sinh 2K/\cosh^2 2K$ $(= 2\sqrt{k}/(1+k))$, which was symmetric around $k = 1$ $(\kappa(k) = \kappa(1/k))$.[25] They noted that a variable, $\exp 2K$, was crucial. But it was not immediately realized that k was the modulus of an elliptic function or that κ was a well-known modular transformation of that modulus to another, or that it conveyed the crucial information about the order of the system. Onsager pointed out all of this and noted that "[i]t was interesting enough to discover that all correlation along the diagonal [of a rectangular lattice] are determined by the single parameter k, which was already known to equal unity at the critical point."[26] Even later, Wannier still suggested: "It is curious that the quantity [sinh $2K = 1/\sqrt{k}$]

entering into [arithmetic duality] is also the one entering into [the star-triangle transformation]. At this time [1945], this is not obvious from any previous reasoning."[27]

Now, thermal equilibrium is indicated by a constant temperature, and the fluctuations of a system are determined by its temperature, T. But k, while it is not that temperature scale, is the appropriate one for systems near the critical point or near infinite or zero temperature, and unlike the reduced temperature, t, k arises out of dynamical considerations. k does point directly to the critical temperature ($k = 1$), and it determines the correlations of the lattice.

Fluctuation-Dissipation Relations

There is in the physical sciences a tradition of relating seemingly rather different processes—such as stars to triangles—by connecting the variables describing one process or state with those of the other, perhaps parametrized by an external variable. For example, the wye-delta transformation in electrical-circuit theory connects the resistances R_i to r_i, so that the circuits in figure 7.5 are equivalent.

Just as Einstein (1905–1906) showed there was a connection between viscosity and diffusion, between dissipative and fluctuation phenomena, parametrized by a temperature, namely, Diffusion ≈ Temperature/Viscosity (or, in Onsager, a linear connection between the correlation length and the interfacial surface tension near the critical point);[28] and as Einstein (1916) connected the light absorption and induced emission rates for atoms, parametrized by the photon energy;

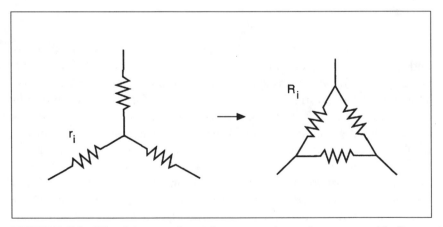

FIGURE 7.5   Wye-delta transformation connecting resistances r_i with R_i.

here, for the star-triangle relation, it turns out that we are concerned with connections between absorption and scattering of particles, in general.

Technically, the fluctuation-dissipation theorem relates diffusion to viscosity (or noise to resistance), parametrized by the temperature. The Kramers-Kronig relations (and dispersion relations in general) connect light scattering to absorption, relating the real and the imaginary parts of the index of refraction to each other, parametrized by the wavelength of light, the photon energy, or the frequency. And, in particle physics, there is a principle of duality, which I shall here denote as duality′ (since it is not quite what Kramers and Wannier meant by duality),[29] which connects particle exchange to particle annhilation/creation processes (figure 7.6).

Technically, in scattering processes $(1, 2 \rightarrow 3, 4)$ the sum of the amplitudes for direct channel resonances (absorption) is equal to the sum of the amplitudes for cross-channel exchanges (scattering), so connecting the masses and couplings for each, parametrized by the center of mass energy and the momentum transfer.[30] Mathematically, the poles in the Fourier transform of a function lead to a falloff of that function. Physically, the masses of the particles exchanged in an interaction set the range of the force. (Hence, when the mass gap is zero, the correlation length is infinite.) And phenomenologically, fluctuations lead to resistance, expressed in a linear response function (thermal fluctuations leading to a resistive drop in voltage, Ir, where I is the current and r is the resistance).

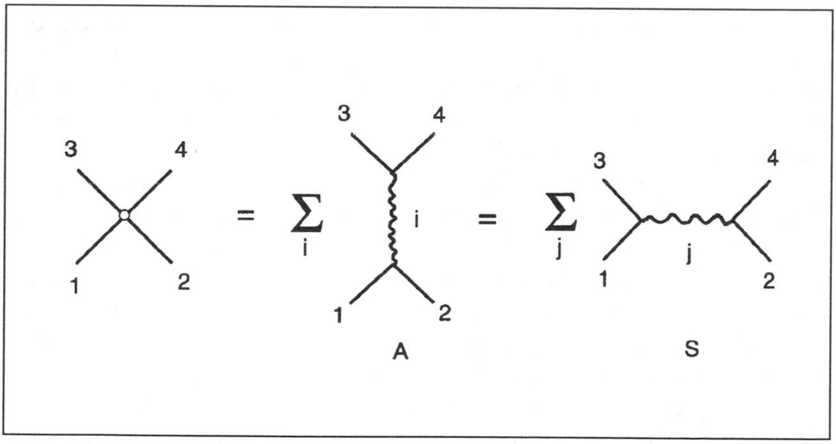

FIGURE 7.6   Duality′ in particle scattering: A, absorption; S, scattering.

In general, such processes are connected and for good reason—they both depend on the same mechanisms or intermediate particles. The fluctuating or random molecular collisions that are the mechanism of diffusion—the size of those fluctuations measured by the temperature —are also the source of the resistance that is viscosity. A spontaneous fluctuation and an externally driven disturbance (as in viscosity) are indistinguishable, and they relax to equilibrium in the same way—that is the way Onsager put it (in an entirely different context, namely, the Onsager regression hypothesis).[31] Scattering can take place by the direct creation and annihilation of intermediate particles by two incoming particles, or by the emission and absorption of exchanged particles between the incoming particles, operating through the same set of intermediate or virtual particles. And the electromagnetic dipoles of the atoms that compose matter, whose bulk effect is to endow that matter with a complex index of refraction, both reemit absorbed radiation in the forward direction (namely, coherent or elastic scattering) and in other directions (absorption).

The Mechanism Linking the Star to the Triangle
The star-triangle relation constrains the interaction constants so that only some are arbitrary, the rest being derived from the relation. The stars constrain the triangles, let us say. The question that concerns me here is, what is the physical mechanism that links the star to the triangle? I believe the answer is thermal equilibrium and the diffusive processes that achieve it.

Recall that, for duality, the crucial physical fact is that excitations of a low-temperature lattice are equivalent, as processes and in terms of the partition function, to deexcitations of the high-temperature dual lattice; and that an unlike bond affects the low-temperature lattice much as a like bond affects the high-temperature one, where T and its dual T^* (respectively, say, low and high temperature) are defined by K and K^*, as $J/k_B T$ or $J/k_B T^*$.

For the star-triangle relation, the relevant physical mechanism is that nearest-neighbor interactions mediate the induced next-nearest-neighbor interactions, through a temperature (figure 7.7). Namely, thermal fluctuations between nearest neighbors determine thermal fluctuations between next-nearest neighbors. (If there is thermal disequilibrium, diffusive processes will operate through those nearest-neighbor interactions to equalize the temperature.)

How does this physical mechanism appear in the formalism and in the physics? It appears topologically and dynamically, in that next-nearest-neighbor interactions are in a thermal bath created by nearest-

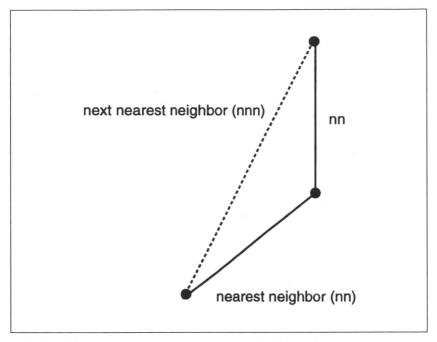

FIGURE 7.7 Relationship between nearest-neighbor and next-nearest-neighbor correlations.

neighbor interactions. It appears thermodynamically, in the fluctuation-dissipation relations. And it appears formally, in a relationship between the sum of Boltzmann factors $(\exp - \beta E)$ for the hexagon and an imputed set for the triangle.

We can make a star-triangle transformation only if the interaction constants of the star are suitably related to those of the triangle. Say we have a hexagonal lattice. We are free to draw in the triangles; any features we impute to the triangular lattice, such as triangular interactions (actually, between the original next-nearest hexagon neighbors), are parasitic on those of the hexagon (nearest neighbors), since we have added no new interactions—this is a thought experiment. Put differently, the triangle is at the same temperature (and k) as the hexagon; in effect, it is not only sitting in a thermal bath set up by the hexagon, it is also sitting in an orientation or correlation bath set up by it.[32] (Now, it is possible to conceive of couplings among the triangle that do not have an oriented state at low energy ["frustrated"], when $K_i < 0$ [actually if the strongest interaction $K_{max} < 0$, and $K_2 = K_3$].[33] But the triangle-to-star relation then leads to imaginary values for the hexagonal couplings, in effect a system that is not in equilibrium. In other words, this is not a triangle that could be embedded into an actual hexagonal lattice at a real temperature.)

Merely by following figure 4.7b, we would expect to find that the drawn-in triangles' interaction constants, $\{K_i\}$, are given by the stars' constants, $\{L_i\}$: namely, next-nearest-neighbor and nearest-neighbor interactions on the hexagon are intimately related.[34] Technically, as I have indicated, one displays this connection by summing over a star (summing nearest-neighbor interactions, "fixing" the spins on the open circles), that is, computing the partition function, and so inferring a set of triangle interaction constants (which are the hexagon's next-nearest-neighbor interactions)—recalling the effective Hamiltonian in our account of decimation—if the inferred triangle partition function were to work.

The power of the star-triangle transformation is perhaps indicated by a strategy followed by Baxter and Enting, which I have referred to already, in which the hexagonal lattice is transformed as follows (see figure 7.8):

(i) Perform a star-to-triangle transformation by summing over the centre spins of all up-pointing three-edge stars above R, and of all down-pointing three-edge stars below R. . . .

(ii) Perform a triangle-to-star transformation on down-pointing triangles above R, and up-pointing triangles below R.[35]

Thereby, one has converted a hexagonal lattice into a rectangular one, and along the way one connects nearest-neighbor correlations with next-nearest-neighbor correlations (which suggests that perhaps one can connect second-nearest neighbor to third-nearest-neighbor, and so forth). Note that this is a reversible transformation (no degrees of freedom need be lost): $\{L_i\} \to k \to \{K_i\} \to \{k, K_3, L_3\}$.

By a different but seemingly related strategy of using the star-triangle relation to shrink the lattice infinitesimally (see figure 7.2), we can exactly solve the holographic constitution of Ising matter by means of the renormalization group. In any case, the analogies do not escape the practitioners: "The points of view of Baxter and Enting's paper on the one hand, and of our RG [renormalization-group] approach on the other, are strongly divergent. The important common feature is that both demonstrate that the star-triangle transformation contains all that is needed to determine the free energy."[36]

Figure 7.9, drawn from Baxter, representing the fact that the process of summing over a star can be changed to summing over a triangle, is deliberately suggestive.[37]

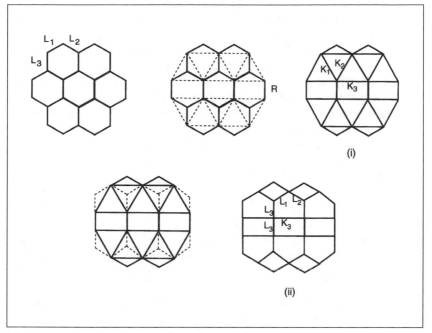

FIGURE 7.8   Baxter and Enting's topological transformation. Adapted from Baxter and Enting, "399th Solution of the Ising Model," fig. 2.

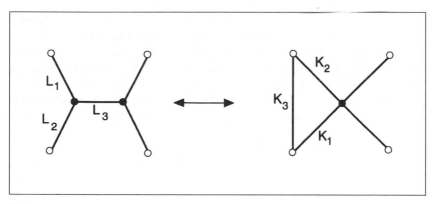

FIGURE 7.9   The star-triangle transformation seen as a transformation between scattering processes. Adapted from Baxter, *Exactly Solved Models*, fig. 7.2.

The graphical similarity to duality' in elementary particle theory—in which exchange ($\succ\!\!\prec$) and annihilation/creation ($\left.\begin{smallmatrix} Y \\ \Lambda \end{smallmatrix}\right.$) processes are intimately related—is striking.[38] Note that if $K_3 \to 0$, so that we are eliminating an "exchange," then $L_3 \to \infty$, so that we are eliminating an intermediate "particle." We might ask, in what sense is this a physical mechanism for the star-triangle relation? It is another restatement of the relationship of nearest-neighbor and next-nearest-neighbor interactions. But, more to the point here, a mechanism that is physically obvious in another context, quantum field theory and particle physics, is invoked in this analogous abstracted context.

There is, I suspect, a further analogy between duality' and the star-triangle relation, an analogy between causality and thermal equilibrium and their respective implications.

Causality and thermal equilibrium are claims that one can label objects (space-time points, systems) so that processes go in only one direction (to greater times, to lower temperatures); and, as we have just seen, duality' and the star-triangle relation each connect direct and indirect processes. Now, one can show that causality leads to duality'. Can one perhaps show that thermal equilibrium leads to the star-triangle relation?

The principle of causality says that actual causes happen before actual effects; more technically, until a signal arrives, a signal that cannot travel faster than the speed of light, there cannot be a consequence of that signal.[39] In effect, we can label events with a time variable so that events with larger times do not influence events with smaller times.

It can be shown that causality implies certain smoothness or analyticity properties of the function that describes a particle's scattering in a medium (the "scattering amplitude"), as a function of the frequency or energy taken as a complex number (imaginary frequencies being equivalent to particles that are absorbed or decay). Those analyticity properties then allow one to relate annihilation and absorption to particle exchange and elastic scattering (duality'): the smoothness requirement is so strong a constraint. Namely, we have dispersion relations that connect the imaginary and real indices of refraction to each other (as in the Kramers-Kronig relations).

Correspondingly, thermodynamic equilibrium means that we may assign a temperature to each system, so that, if those systems are thermally linked, heat does not spontaneously flow from systems of lower temperature to those of higher temperature.[40] And, in effect much like a fluctuation-dissipation relation, we would want to say something like, thermodynamic equilibrium implies analyticity or smoothness properties of the free energy (which properties are not at all otherwise automatic, and which are not the case at phase transition), which then allow us to connect nearest neighbors' to next-nearest neighbors' fluctuations, and this is just the star-triangle relation.[41] What can surely be said is that the existence of a temperature signifies a relation between nearest-neighbor and next-nearest-neighbor interactions, that the star and triangle processes are connected.[42]

Let us take our drawn-in-triangles thought experiment even more seriously. Topologically, the star-triangle relation means there is an interaction lattice upon which we can draw two different sets of figures, yet, physically, it is a lattice that is dynamically whole (say, having a uniform temperature). Put differently, decimation within the holographic constitution of matter and the star-triangle transformation both depend on our being able to take the lattice as bipartite (accommodating alternate figures, having those open and closed circles), yet its fitting together into one thermodynamic whole. Technically, just as decimation changes the mutual interaction that the unsummed spins experience (as examined in chapter 4), so the star-triangle relation converts a star or hexagonal system $\{L_i\}$ into a triangular system of unsummed spins now interacting through different couplings, $\{K_i\}$.

Technically, note that, in each case, the unsummed spins would seem to be interacting linearly with an external "magnetic" field, σL, and no longer bilinearly with each other, $\sigma_i \sigma_j L$. Namely,

$$\sum_{\substack{\text{unfixed spins, star}}} \exp(L_k \sigma_i \sigma_j) \approx \sum_{\substack{\text{fixed spins}}} \cosh\left(\sum L_i \sigma_j\right)$$

$$\approx \sum_{\substack{\text{fixed spins, triangle}}} \exp(K_k \tau_i \tau_j),$$

where $\{L_i\}$ and $\{K_i\}$ are related by the star-triangle relation.

The device that allows the star-triangle relation to go through is that one can bilinearize that linear relation in the decimated spins, which then reproduces the form of the original interaction, now on a different scale. (A similar device is employed in the block-spin transformation, where $\sum \cosh \sigma K \to \exp \sum (A + B\tau_i \tau_j + C\tau_l \tau_m \tau_n \tau_p)$.) As I indicated earlier, this is a curious device, indeed, and it would seem to depend on the two-state or up-down character of the Ising spins, and the parity invariance of the states. It would be good to have a better sense of the physics behind it.[43]

The physical process linking the star to the triangle—thermal equilibrium—appears both in the formalism and in the physics through a connection of nearest-neighbor correlations to next-nearest-neighbor correlations: geometrically, thermally, and formally (by that curious device of rewriting linears into bilinears). In sum, a lattice that possesses a temperature possesses consistent star-triangle interaction constants.

Technically, such a consistent set of interaction constants, k being conserved or constant throughout the lattice, is equivalent to the commutativity of the respective transfer matrices. Here I think of the transfer matrix as a propagator of thermodynamic equilibrium throughout the lattice, connecting up each new set of spins not only to the already accounted-for spins but to a thermal bath, respectively the $J\sigma$ and the $1/k_BT$ factors in the Boltzmann weight. In effect, the transfer matrix minimizes the free energy; it finds the largest eigenvalue and its associated eigenvector, no matter how far from the equilibrium configuration the system starts out.[44] This suggests that the physical mechanism behind the star-triangle relation is the diffusive process and heat flows that ordinarily lead to thermodynamic equilibrium. If next-nearest-neighbor interactions were not consistent with nearest-neighbor ones (say by some artifice we so set things up), then the system would relax so that they were consistent, perhaps changing its temperature or separating out into two phases.[45]

Again, although both duality and the star-triangle relation have geometric and topological interpretations, they are intimately thermodynamic, about $\exp - \beta E$ and the partition function, and dynamical, about the effective Hamiltonian. Duality and the star-triangle relation refer to systems defined by their average energy per degree of freedom, a temperature. Onsager notes this in a nice and ironically didactic way in presenting the apparently complicated technical formulas for the internal energy and the specific heat for the two-dimensional Ising model. He pulls out the thermal k_BT factor, and says: "The two terms of [the equation for the internal energy] are separately the mean energies of interaction in the two perpendicular directions in the crystal (remember the interpretation of temperature as a statistical parameter)."[46] This is a temperature that "balances" interactions, so that each degree of freedom has the same average energy. So that, for the star-triangle relation, nearest-neighbor interactions determine next-nearest-neighbor interactions. The physical mechanism that makes the star-triangle relation possible is just that diffusive one that leads to thermal equilibrium: next-nearest-neighbor interactions are mediated thermally as well as directly (or mechanically) through nearest-neighbor interactions.

Tricks of the Trade and Terms of Art
That the star-triangle relation is a statement about what is most fundamental in these statistical processes of constituting matter—

namely, the detailed consequences of possessing a temperature and of being in thermodynamic equilibrium—is physically interesting.[47] Nonetheless, its historical origin lies in a formal trick, noticed by Onsager, that makes a derivation go through—a scheme, as it seems in reading Onsager's 1944 paper, for connecting partition functions of triangular and hexagonal lattices—at first neither mathematically interesting nor physically rich.[48] The star-triangle relation plays no direct role in Onsager's derivation of the solution to the Ising model for a rectangular lattice, as he presents it in his original paper. Topologically the star-triangle relation appears at most to be about fitting triangles into hexagons, again nothing of great interest. (In later recollections of that work, Onsager does discuss the diagonal-to-diagonal transfer matrix—building up a lattice on the diagonals rather than by rows or columns, using that matrix to build up the hexagonal and triangular lattices. Here the star-triangle relation does play a larger explicit role.)[49]

By a much later device, one does see the star-triangle relation's connection with dynamics, namely, the renormalization group, redoing the geometric constitution of matter in an exactly solvable fashion. Again, topological bipartiteness plays an immediate role in duality, in writing down a star-triangle relation, and in each of the constitutions of matter represented by exact solutions to the Ising model. Still, that the star-triangle relation is a dynamical object, being about thermodynamic equilibrium and about a parameter, k, connecting two very different processes or systems, is not at all obvious from early work.

In sum, the most fundamental features of the constitutions of Ising matter may be embedded within a set of facts about k (or about the reduced temperature, t): the partition function for a lattice at k is simply related to the partition function for a dual lattice at $1/k$; the partition function of a lattice at k is related to the partition function of a lattice k^α; and the partition function for a lattice at k has a simple relation to the partition function for another lattice, a different dual, also at k. Moreover, $k = 0$, 1, and ∞ are not only mathematically interesting limits, but also physically significant—totally ordered, scale symmetric, and totally disordered lattices. When, in solving the Ising model, one employs more directly the symmetries that allow for these facts, the solution follows rather more directly—as we shall see in the next section.

The Star-Triangle Relation in Action

We might get a better feel for what the star-triangle relation means were we to examine its use in quite direct solutions to the Ising model. As mentioned earlier, Baxter and his collaborators have shown how the star-triangle relation is in fact almost all one needs to solve the model, and here I review of two of his solutions, aspects of which I have already discussed: one solution is arithmetic in its constitution, employing symmetry properties of the transfer matrix (as discussed in chapter 4), the method of commuting transfer matrices; another is holographic and is concerned with the correlations between spins.[50]

A bit more technically, the first solution concerns an evolution operator (an operator that in effect propagates an interaction through space or time), an arithmetic constitution; the second concerns the spin-spin correlation function, what in field theory is the vacuum expectation value of pairs of field operators, which suggests a constitution of matter based on correlation, what we see in the geometric and holographic constitutions. (Again, the free energy follows naturally: in the first, from computing the maximum eigenvalue of the transfer matrix; in the second, from realizing that a derivative of the free energy with respect to the coupling constant is just the correlation function.)[51]

Recall that transfer matrices build up a (two-dimensional) lattice by adding on one more row of sites/particles (and their associated bonds, for example, horizontal and vertical for a rectangular lattice). The diagonal-to-diagonal transfer matrix adds on diagonal rows.[52] Baxter finds the eigenvalues of that diagonal transfer matrix, $V(K_1, K_2)$, directly, K_1 and K_2 being the horizontal and vertical couplings. Here I want to recall the desiderata he needs to make the solution go through: essentially V's smoothness or analyticity, the star-triangle relation, and duality. (See figure 4.9.)

> The basic idea is to regard the diagonal-to-diagonal transfer matrix as a function of the two interaction coefficients K_1 and K_2. It is easily established that two such matrices commute if they have the same value of k, and for any such matrix another one can be found which is effectively its inverse. These properties are basically sufficient to obtain the eigenvalues of the transfer matrix. . . . Although indirect, this presentation has the advantage of showing that it is basically only local properties of the lattice model that are being used.[53]

Technically, four features—analyticity of V in K_1 and K_2 if $k \neq 1$, commutativity (equal to the star-triangle relation, $\sinh 2K_1 \sinh 2K_2 = a$ constant $= 1/k$), "invertibility," and "symmetry"—lead Baxter directly to the free energy.[54] For many readers of his argument, until they appreciate how systematic Baxter is in following this strategy for other models, the steps may be clear but the rhythms, motives, and general structure are not. Even if the outline is clear, the proof techniques are distinctive. (These observations might well be applied to Onsager's 1944 paper, when its logic is appreciated, although I am not sure it appeared at all direct at the time.) Baxter uses this same mode of solution for more and more complex lattices, and so it eventually becomes its own sort of commonplace.

The four features are invoked in the derivation of a functional equation for the eigenvalues of the transfer matrix. Baxter's functional equations is

$$\Lambda(u) \times \Lambda(u + I') = (-2/(k \operatorname{sn} iu))^n + (-2 \operatorname{sn} iu)^n r,$$

where n is the number of columns, r is either $+1$ or -1, Λ is an eigenvalue of the transfer matrix, and the logarithm of its maximum value is proportional to the free energy. Again, k's centrality is built into the solution by a parametrization (or elliptic substitution) of the coupling constants, K_1 and K_2, in terms of those elliptic functions in k (the modulus) and in another variable u, the argument (which measures the asymmetry, that K_1 and K_2 may not be equal): $1/k = \sinh 2K_1 \sinh 2K_2$, and $\sin 2iK_1 = \operatorname{sn} iu_1$. (Recall that a trigonometric substitution goes something like $x = \tan u$; here the elliptic substitution is, roughly, $x = \operatorname{sn} u$.) Thereby, given the explicit employment of the right variables or symmetries of the transfer function (k and u), we automatically keep in mind and take account of the fundamental symmetries of this constitution of matter. To those not used to solving equations involving functions rather than variables, especially if one is employing elliptic functions (elliptic substitutions) to solve the functional equations, Baxter's method from here on in will appear fraught with legerdemain.[55]

The lesson here is perhaps clear by now: the physics of Ising matter, in all of its constitutions, would seem to be epitomized by a modulus, k, by a duality of high- and low-temperature excitations, and by the star-triangle relation. Of course, even if we are given these facts, a solution to the Ising model depends on our realizing their exact meaning in a particular context—the physical mechanisms of Ising matter that found these facts and their formal realization within a derivation. Moreover, it is only through such derivations that we discover these facts.

For example, as mentioned earlier, Baxter and Enting have provided another solution,[56] one that even more explicitly employs the star-triangle relation: it studies the correlations of spins with each other, denoted by $\langle \sigma_i \sigma_j \rangle$, namely, how much a property at location i depends on the value of that property of location j. We might even want to designate this as an additional constitution of matter, a subset of the holographic: that is, bulk matter is rigid in that it can exhibit strong spatial correlation. In this constitution, matter is built up of those spatial correlations, where the correlation length is greater than an interatomic size.

First and topologically, as I have indicated, Baxter and Enting repeatedly use the star-triangle and triangle-star relationship on a honeycomb or hexagonal lattice, $\{L_i\}$, to create a square lattice. Second, as a consequence of this construction, they can show that the correlations among spins depend on two rather than three coupling constants $\{L_i\}$, K_3 and L_3 (k being implicit, and $\{K_i\}$ derived from the $\{L_i\}$ by the star-triangle relation); and that nearest and next-nearest neighbors' correlation functions on a honeycomb are simply related: if $g(L_3, K_3)$ is the nearest-neighbor correlation function, then $g(K_3, L_3)$ is the next-nearest (that is, the same function with coupling constants interchanged). Third, and algebraically, they do a decimation by employing the star-triangle relation, and obtain another relationship of nearest- and next-nearest-neighbor correlations.[57] They then invoke some seemingly trivial symmetries of the lattice (which Kramers and Wannier also noted),[58] go to variables $\{K_i\}$ and k (eliminating $\{L_i\}$), so that $g(K_i, L_i)$ becomes $\coth 2K_i \times h(K_i, k)$, and then solve a functional equation:

$$\sum h(K_i, k) - 1 = b(k)k^{-1} \prod \operatorname{sech} 2K_i;$$

namely, the sum of the "normalized" nearest-neighbor correlations around a triangle (next-nearest neighbors for the hexagon) is related to a simple function of the coupling constants. Along the way, one has to assume sufficient analyticity to push through the solution.

The Meanings of the Star-Triangle Relation

What are the lessons to be drawn from these two, more direct solutions? Are they cute idiosyncrasies, more a matter of tricks and devices than of physics? It would seem not. First, they do point to or incorporate some of what is physically most essential about this matter, in all of its constitutions: duality and the star-triangle relation (even for a rectangular lattice). Second, the styles of mathematics and technology they employ, as well as the physical constitutions they represent, might well be generalized to more complex situations, which is exactly what Baxter does—presumably the deepest constitutions being the most readily generalizable since they build in the physics most readily. I say "presumably" because in the end it is not often clear ahead of time what will actually work.[59] Subsequent re-provings of a difficult problem such as the two-dimensional Ising model are in effect deeper analyses, showing more perspicuously just how we might think of a constitution of matter, just how its symmetries are to operate.

I have referred to a solution of the Ising model (by Samuel) that employs anticommuting variables, in effect representing bonds, to write out the partition function in terms of polygons, as a (Grassmann) integral over all configurations of the lattice. The method is remarkably direct and cute, and it also allows for powerful generalization. As Samuel puts it: "[T]he statistical mechanics point of view allows one physical insight whereas the field theory point of view supplies the powerful mathematical tools. . . . We will write statistical mechanics systems as fermionic field theories [via integrals over anticommuting variables, or Grassmann integrals]. This is done to systems which *a priori* have no vestige of fermionic character. What is involved is a mathematical rewriting of the degrees of freedom."[60] I would add that that mathematical rewriting points up the fermionic character. There is no mention here at all, in this very direct solution, of the star-triangle relation. Rather, it is the fermionic or antisymmetric insight that it would seem is essential, a sign of that being that Samuel states that anything you want to know about the two-dimensional Ising model can be expressed as a Pfaffian.[61]

It would seem that the star-triangle relation reflects geometric duality, while antisymmetric functions or fermions reflect arithmetic duality. Namely, there is more than one account of what is physically most essential about Ising matter, more than one set of right degrees of freedom—and so there are different kinds of essential mathematical technology, generalization, and symmetries. Yet we might hope to understand how those very different technologies (and different physically essential facts) are about the same bulk matter.

As discussed in chapter 6, the different technologies are each ways of thinking about a particular kind of random walk, in effect a feature of nice random variables: the central limit theorem. In the asymptotic or infinite volume limit, counting polygons on a lattice or dimer coverings of a decorated lattice does get at scaling symmetries. Conversely, attending to correlations, and their dependence on scale or distance, does count polygons on a lattice. Random walks exhibit scale symmetry; so arithmetic duality is related to geometric duality.

In the infinite volume limit, counting polygons provides the coefficients of a series expansion for the partition function, one that asymptotically defines the exact generating function or partition function that builds in the scaling symmetry best seen near the critical point (and it seems that this series approximation works so well because we are implicitly dealing with a modular symmetry, where the modulus k of the elliptic functions allows us to scale up or down in k, from the easy k small to the difficult $k = 1$ cases). In the infinite volume limit, the strength of correlations of arbitrary not-so-nearest neighbors in effect counts the weighted number of ways they can be connected to each other (the higher the weighted number, the higher the correlation, with consideration for distance). In computing these correlations directly as a function of $\{K_i\}$ and k, we are by the way providing the asymptotic or infinite volume limit of the partition function, and hence the generating function for the polygons.

Again, random walks exhibit scaling symmetry, in effect the central limit theorem. And partition functions are in general generating functions for combinatorial problems, here a particular kind of random walk. The historical origins of statistical mechanics within combinatorial problems are here brought back to life, the recurrent elliptic theta functions in the solutions, once the way of counting partitions of the integers, being another representation of that history.

As Onsager presciently appreciated, k (and u) again turns out to be a crucial variable rather than merely a convenient constant, a discovery founded upon a series of solutions to the Ising model, each providing a different constitution of Ising matter. The star-triangle transformation is not merely a convenient trick that also happens to be k preserving; it would seem to lead to some of the most direct solutions to the Ising model and to be laden with physics.[62] It points to a set of equivalent states, where the conventional temperature does not in fact so readily point. These k-equivalent states have similar behavior in terms of their spatial orderliness and correlation, although their values for K_1 (that is, u) may well differ. The original hint that this might be the case was the discovery that the correlation length near the critical point is independent of the rectangular lattice's asymmetry, that is, $K_1 \neq K_2$.

Historically, the relevant experimental discovery was that the properties of matter near the critical point were expressible in terms of t^α: they "scaled" and were independent of the exact composition of the material (α is material insensitive, hence the term "universality"), for example, the correlation length $\xi \approx |t|^{-1}$. These facts led to the renormalization-group approaches to critical phenomena, in which scaling and universality arise naturally.

Now, $1/\xi \approx \ln k \approx k - 1 \approx t$. Hence we see how the experimental t observation and the theoretical k observation meet, the phenomenology and the dynamics are connected: both are the right variables for measuring correlation, and they point to the critical temperature. Put more curiously, universality, a t fact, would seem to be related to the fact that the eigenvectors of a k set of transfer matrices are the same (since the transfer matrices commute). Moreover, we might say that duality, arithmetic and geometric ($k \to 1/k$, or $k \to 0$, 1, or ∞), and the star-triangle relation point to topologically equivalent systems (under the demand that the partition functions be invariant).

The Formalism and the Physics

I return to a recurrent theme, one that Schrödinger and Wilson put so much faith in, a faith that is recurrently redeemed in practice. Or, better, good practice is identified with its redemption. Formalisms and sets of tricks that actually work in solving problems are likely to be filled with physics: in part, because it would seem that we have touched upon (if not explicitly realized) the symmetries in our models of nature; in part, because we are getting at the crucial physics through the kinds of quantities we use (such as antisymmetric objects, those fermions and Pfaffians), again whether or not we quite understand their implications initially; and, in part, because the formal analogies we may employ are not only formal but can be shown, perhaps with some effort, to lie at the level of mechanism as well as symbol. Here is James Clerk Maxwell:

> In our daily work we are led up to questions the same in kind with those of metaphysics; and we approach them, not trusting to the native penetrating power of our own minds, but trained by a long-continued adjustment of our modes of thought to the facts of external nature.
>
> As mathematicians, we perform certain mental operations on the symbols of number or of quantity, and, by proceeding step by step from more simple to more complex operations, we are enabled to express the same thing in many different forms. The equivalence of these different forms, though a necessary consequence of self-evident axioms, is not always, to our minds, self-evident; but the mathematician, who by long practice has acquired a familiarity with many of these forms, and has become expert in the processes which lead from one to another, can often transform a perplexing expression into another which explains its meaning in more intelligible language. . . . In this process, the feature which presents itself most forcibly to the untrained inquirer may not be that which is considered most fundamental by the experienced man of science; for the success of any physical investigation depends on the judicious selection of what is to be observed as of primary importance, combined with a voluntary abstraction of the mind from those features which, however attractive they appear, we are not yet sufficiently advanced in science to investigate with profit.[63]

Maxwell, in noting the importance of mechanical and electrical analogies, argued (in 1870) that, to know the kinds of (or "the classification of") quantities—what we would call the symmetries—one is working with, is to know a great deal of the physics of a situation.[64] Maxwell had vectors and quaternions in mind, as kinds of quantities, as well as analogies between the heat equation and the potential equation (Poisson) of electrostatics.[65]

For our purposes, one crucial kind of quantities is variances, namely, the temperature as a measure of thermal fluctuation. Another is the modulus and argument of elliptic functions, k and u. (Recall that, were we dealing with ordinary trigonometric functions, it would not be surprising to refer to amplitude, frequency, and period as such quantities, corresponding to k, u, and I and I', the latter being the quarter-period magnitudes.) And a third crucial kind of quantities would seem to be antisymmetric objects. These three kinds of quantities reflect thermodynamics, the star-triangle relation, and duality, respectively.

The deep analogy here is between statistical mechanics and quantum mechanics and quantum field theory: between the diffusion equation ($\nabla^2\phi = -D\,\partial\phi/\partial t$, where t is the time), and the Schrödinger equation ($\nabla^2\phi = -i/\hbar D\,\partial\phi/\partial t$); between the partition function as a sum over states' energies ($Q = \Sigma \exp - \beta E$) and the quantum mechanical amplitude expressed as a sum of the histories' actions, A, ($a = \Sigma \exp iA/\hbar$, $A = \int_{\text{path}} L\,dt$, where L is the Lagrangian); between correlation functions ($\langle \sigma_i \sigma_j \rangle$) and vacuum expectation values of field operators ($\langle 0|\phi_i\phi_j|0\rangle$); between a spatial lattice and a space-time lattice; and between a lattice and a continuum.

These various considerations of duality and the star-triangle relation suggest that we might think of duality as pointing to the spatial symmetry of the right particles for readily taking hold of the constitution of Ising matter, through the topological symmetries of a lattice; while the star-triangle relation points to thermodynamically equivalent systems, sets of physically equivalent coupling constants, and hence the masses and energies of the right particles: all under the regime of partition function similarities.

Or, as Baxter puts it, "[F]rom a mathematical point of view the temperature is rather an uninteresting divisor of the Hamiltonian, while the correlation length $[1/\xi \approx |t| \approx |\ln k|]$ gives valuable information on the near-critical behavior of the system."[66] It is a curious and wondrous moment when one of the greatest achievements of the nineteenth century, temperature in thermodynamics, has become an uninteresting divisor, a physical quantity has become a formal arithmetic object.

ELLIPTIC FUNCTIONS AND HYPERBOLIC GEOMETRY

Elliptic functions are the natural extensions of the trigonometric functions. Developed in the nineteenth century, they were not unfamiliar to mathematically inclined physicists of the last third of the nineteenth and early part of the twentieth centuries. The same is true for hyperbolic geometry, as an extension of plane geometry and trigonometry. While elliptic functions and hyperbolic geometry might appear a bit esoteric to many current practitioners (although not to those involved in current research in this and ancillary fields, where theta functions are everywhere present), they must be taken to be part of the toolkit of a certain era. What we have seen already is how these tools are employed to make obvious the physics of the constitution of Ising matter—the constraints, the calculations, and the results. And so we might ask, what is it about Ising matter that makes these tools so suited to it?—a question about the deepest physics of matter. (The answer will be modular or scaling symmetry.)

More generally, a symmetric formalism or a spatially symmetric diagram can sometimes make obvious what is going on in a derivation, or at least function as a mnemonic and algorithmic device, those devices themselves then becoming laden with physics. But the task is to translate what we initially know into such a formalism or diagram, and to appreciate (and, conversely, to read off) the content of that representation—although the symmetries are often apparent as symmetries to the naive, even if their meaning is not. So, for example, we shall be depending on two such diagrams, figures 6.1 and 6.2, for much of the rest of this section: one, a hyperbolic triangle that represents the relationship of the quasiparticle energies to the coupling constants,

$$\cosh \epsilon_q = \cosh 2K_2 \cosh 2K_1^* - \sinh 2K_2 \sinh 2K_1^* \cos q;$$

the other, a hyperbolic hexagon that represents the star-triangle relation, such that $\sinh 2L_i / \sinh 2K_i^*$ is a constant.

As for the hyperbolic hexagon, Onsager puts it most succinctly: "The relations between these [interaction constants for the honeycomb, the result of both the star-triangle and dual transformations] and the original interaction constants involve just the hyperbolic trigonometry of a totally improper triangle: The quantities of the hexagon whose corners are all right angles."[67] It is clear that L_i and K_i^* play symmetric roles, but what does that mean? (Recall that K_i^* is the dual of K_i, appearing as small expansion parameters respectively in the low- and high-temperature expansions of the Ising model, $\tanh K_i = \exp - 2K_i^*$.)

From the hyperbolic triangle, we find we have a lovely representation of all the relevant variables; but, again, what does that mean?

These technical devices and diagrams highlight how k, duality, and the star-triangle relation are significant, by providing them with prominent roles in a canonical and seemingly manifest description, as we saw earlier in the discussion of the right particles. At the same time, such devices may distance us so far from the physical mechanisms that we lose touch with what is actually happening physically—sometimes a good thing, since the formalism and devices may point us automatically to a solution by following through with the mathematics. Yet, as we shall also see, the physical insights as a consequence of becoming formal can be quite suggestive about the meaning of the formalism.[68]

Elliptic Functions

From a mathematical point of view, all these [Ising-like] models have two common features. One is that their solution leads sooner or later to the introduction of elliptic functions. . . . The other common feature is that they can all be solved by some appropriately generalized Bethe ansatz.[69]

Before going further, I offer a quick introduction to elliptic functions, recalling some of what I stated earlier. Elliptic functions are the doubly periodic functions—

$$-\operatorname{sn}(u + 2I) = \operatorname{sn}(u + 4I) = \operatorname{sn}(u + 2iI') = \operatorname{sn}(u),$$

with quarter-period magnitudes I and I'—much as trigonometric functions are the singly periodic ones (namely, $\sin(x + 2\pi) = \sin x$), where the period is 2π.[70] They are the natural functions to describe the motion of a pendulum, one that makes large oscillations and that could even go all the way around (figure 7.10).[71]

The modulus, k, of such functions measures the maximum size of those oscillations ($k = \sin \theta_{\max}/2$, for $\theta_{\max} \leq \pi$), $0 \leq k < 1$ being an ordinary pendulum, $k > 1$ (suitably defined) being a pendulum that goes all the way around, and $k = 1$ being a pendulum that just goes to the top (with an infinite period, by the way, much as the Ising model has an infinite correlation length at the critical point). By convention, let us call the period of the pendulum $4I$. One can describe another such pendulum, gravity now pointed in a reversed direction, and that pendulum has another period, $4I'$, which proves to be the second period of the elliptic functions (times i). Finally, a measure of the amplitude as a function of the time elapsed, $u = \omega t$ (the argument), is $\operatorname{am}(u, k)$. Here ω is the angular frequency. And $\sin \operatorname{am}(u)$ abbreviated as $\operatorname{sn} u$, while

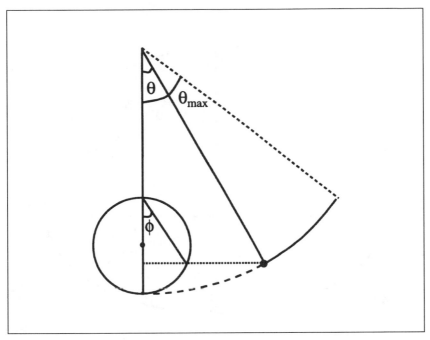

FIGURE 7.10 The simple pendulum. $k = \sin \alpha/2$, $\alpha = \theta_{max}$; $\phi = \pi/2$ when $\theta = \alpha$.

$\cos \operatorname{am}(u)$ as $\operatorname{cn} u$, are the Jacobian elliptic functions; for the pendulum, $k \operatorname{sn} u = \sin \theta/2$. (Note that $\operatorname{am}(u) = \phi$, the amplitude, is not the same as the angle, θ, the pendulum makes with its axis, but, as can be seen from the diagram, it is a differently defined angle, defined so that the amplitude always goes from 0 to $\pi/2$.) For $k \ll 1$, it can be shown that one gets the familiar small oscillations $(\operatorname{sn}(u, \text{small } k) = \sin \omega t)$, namely, sine-wave periodic behavior, while for $k \gg 1$, we have a rapidly rotating pendulum.[72] (In effect, we have the easy low- and high-temperature approximations to an Ising model.)

Alternatively, a different and eponymous model is provided by the ellipse with semiaxes a and b (figure 7.11).[73]

Let us take k as the eccentricity, $\sqrt{(1 - (b/a)^2)}$, and the amplitude, ϕ, as the "eccentric angle" of a point from the center (again going from 0 to $\pi/2$), measured from the minor axis. (Note the use of a projection onto a circle to measure the amplitude, as for the pendulum.) Then the argument u (where $\operatorname{am}(u, k) = \phi$) is an effective angle (u goes from zero to I) such that $\operatorname{sn}(u) = x/a$, $\operatorname{cn}(u) = y/b$. Were the eccentricity

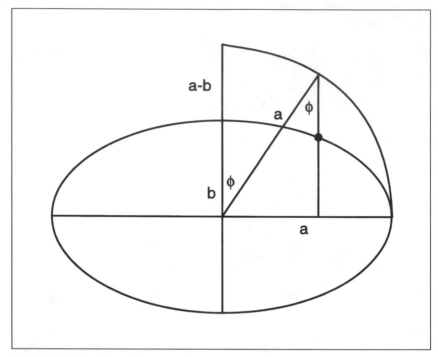

FIGURE 7.11   The ellipse.

zero, and we had a circle, u would be identical to the amplitude; but instead it is the linearized angle (much as ωt is for the pendulum), and sn u ($=\sin\phi$) gives us the right values for the oscillatory but not simple-harmonic motion of a point projected onto the x-axis. A unit of length along the ellipse, ds, is $a\sqrt{(1-k^2\sin^2\phi)}\,d\phi$, or $a(1-k^2\mathrm{sn}^2(u))\,du$. The equation for the ellipse, $(x/a)^2+(y/b)^2=1$, is a restatement of the identity that is a foundation for these definitions, $\mathrm{sn}^2(u)+\mathrm{cn}^2(u)=1$.

It should be noted that the amplitude and argument have a long history in celestial mechanics, where they are known as the eccentric anomaly and the mean anomaly (from circular motion), respectively. (These may be defined with respect to the major rather than the minor axis.) In particular, the mean anomaly is the angle from the center of the ellipse that would be made by a planet moving uniformly along the ellipse, with the same period as it would have in regular (nonuniform) orbital motion.

For the Ising lattice, k is a measure of the distance to the critical temperature (again, $k = 1$, $t = 0$, $T = T_c$; $k - 1 \approx Kt$ near the critical point, where K is the coupling constant; for a square lattice, t or $k - 1$ is a measure of $2(K^* - K)$, the mass gap, near the critical point). The imaginary amplitude, $\text{am}(iu_j, k)$, equals $2iK_j$. u is an angle, actually what is called an elliptic measure of an angle (since $\text{sn}\, iu = \sin 2iK$, the formula for such a measure), representing the strength of the effective coupling constant (since k is implicit here), as well as the degree of inequality between the horizontal and vertical coupling constants, K_1 and K_2. In effect, k and u are normalized variables, representing temperature (coupling-constant strength) and lattice asymmetry, respectively. (u/I' equals $\pi/4$ if the lattice is square, and is otherwise between 0 and $\pi/2$.)

We might summarize these considerations in terms of the model, the modulus (k), the argument (u), the amplitude (ϕ), and the quarter-periods (I, I'), respectively—pendulum: maximum angle of swing, time ($u = \omega t$, where t is the time and $\omega = 2\pi/4I$), normalized amplitude of swing (always $0 \rightarrow \pi/2$), and quarter period; ellipse: eccentricity, mean anomaly or distance along the ellipse, eccentric angle, actual quarter period of planetary motion; Ising model: orderliness or normalized temperature, effective coupling and asymmetry (here, the imaginary argument), the actual coupling, a conserved quantity among equivalent transfer matrices (the sum of the us) and in the star-triangle relation.

In thinking in terms of elliptic functions, rather different variables than might seem at first obvious are suggested—such as k as a modulus for the Ising model, or the eccentric angle as an amplitude for the ellipse.

Just as all ellipses with the same eccentricity have the same shape, all lattices with the same k have the same orderliness. And just as u/I measures a common point on every ellipse, so u/I' for the Ising model measures a common lattice asymmetry. A star-triangle transformation finds a symmetric point on an ellipse through a diagonal ($u \rightarrow I - u$), and a duality transformation finds a dual point (on an associated hyperbola).

One last example, which I discussed in chapter 3. If we want to solve the heat equation, for a two-dimensional infinite slab (in the xy direction), then the four elliptic theta functions are the solutions. (The Jacobian elliptic functions, such as $\text{sn}(u)$, may be expressed as ratios of theta functions.) The nome is a measure of the exponential time decay of each normal mode, $q = \exp - 4\alpha t$ (where the nome is conventionally q and here t is the time), and z is the vertical distance from the center plane of the slab. The different theta functions correspond to different boundary conditions for the slab, much as they correspond to different boundary conditions in the Ising model.[74] Recall, as well, that the theta functions are the natural objects for describing numerical partitions (and ζ equals $\exp 2iz$).

Elliptic Functions in Solving the Ising Model

It is at first curious that in Onsager's original 1944 paper there is a long appendix on elliptic functions and their use in working out the various integrals in this problem (in part, to set the notation), and that in Baxter's book on exactly solved models (1982), there is again a thorough appendix on elliptic functions. Both are enthusiastic about them, and Baxter encourages the reader to join in his enthusiasm. Now, in general, ours is not an era when most theoretical physicists are intimately acquainted with special functions (rather than with group theory and topology, say) and read Whittaker and Watson's *Modern Analysis* (1902; fourth edition 1927), more precisely, *A Course of Modern Analysis: An Introduction to the General Theory of Infinite Processes and of Analytic Functions, with an Account of the Principal Transcendental Functions* or its modern progeny, as a bible, as did Onsager.[75] So it makes sense that elliptic functions need be recommended to the reader since they prove so useful in describing Ising matter. And there may be physics to be found in using them. I want to examine how they appear in the analysis. But first, some lovely remarks.

Here is Baxter: "Onsager (1944) noted that this [elliptic substitution] was an obvious thing to do; . . . the K_i and L_i [the triangular and hexagonal coupling constants, respectively] satisfy [star-triangle] relations similar to those of hyperbolic trigonometry. It is well known that these can be simplified by using elliptic functions. Onsager calls this a 'uniformizing substitution.' "[76] Namely, as Onsager put it, "The trigonometric functions of the angles and sides of a spherical triangle with two fixed parts [figure 6.1 above] can always be expressed in terms of single-valued functions by a suitable elliptic substitution. The same is true for hyperbolic (and plane) triangles."[77] Onsager presumably obtained the term *uniformizing substitution*, used in appendix 2 of his paper, from the culture of Whittaker and Watson, who "employ the term *uniform* in the sense of one-valued."[78]

The star-triangle relation, duality, and what we might call k invariance, is mirrored both in hyperbolic geometry and in elliptic functions; what is derived from one point of view is obvious when seen from another; the physics, the diagrammatics, and the mathematics are made congruent among each other.[79] And so elliptic functions and hyperbolic trigonometry are the formal means by which duality and the star-triangle relationships are built into the analysis of the Ising model. The formal means may then lead to deeper physical insight.

What is perhaps most striking, and it was surely that to Onsager, was the solution for the angles, $\gamma(\omega)$ (or in later terms, the quasiparticle energies, ϵ_q) for the rectangular lattice: again,

$$\cosh \gamma(\omega) = \cosh 2K_2 \cosh 2K_1^* - \sinh 2K_2 \sinh 2K_1^* \cos \omega,$$

which is an expression in hyperbolic geometry for the triangle in figure 6.1.

Onsager is sufficiently impressed that he says, "It seems rather likely that this result [here he is referring not only to the above formula but to an even more symmetric one as well] could be derived from direct algebraic and topological considerations without recourse to the operator method used in the present work."[80]

Note that k is just the ratio of the hyperbolic sines of two of the sides of this triangle, and $am(iu, k)$ is proportional to one of the sides: more precisely,

$$1/k = \sinh 2K_1 \sinh 2K_2 = \sinh 2K_1/\sinh 2K_2^* ;$$

$$am(iu_i, k) = 2iK_i.$$

Moreover, the star-triangle relation can be expressed in term of ratios of the sides of the spherical hexagon (figure 6.2), namely, $(\sinh 2L_i)/\sinh 2K_i^*$ being a constant, a trigonometric fact for this figure.[81]

These diagrammatic and trigonometric facts make manifest the crucial formal relations for Ising matter, contingent relations discovered only by actually solving the model: here the physics precedes the diagrammatics. As I indicated earlier, such diagrams must then be interpreted if the physics and the physical mechanism is to be seen as being represented by those diagrams. This is no mean task. So, in Kramers and Wannier's 1941 paper, duality is displayed as particular symmetries of actual matrices; yet how those symmetries are physically interesting is not addressed at all. And, as described in chapters 4 and 6, Onsager provides a narrative relating the hyperbolic triangle diagram and changes in the coupling constants and in ω, to physical phenomena. In particular, he describes how disorder is introduced into the orderly ground state as the amplitudes for the disorder operators become larger as the temperature grows toward the critical temperature.[82] (The physical source and mechanism for this is the growing importance of the disorder operator, B, as the temperature rises toward the critical point.) He can read off the general behavior from the hyperbolic triangle of my figure 6.1, his figure 4. And he displays another diagram for the

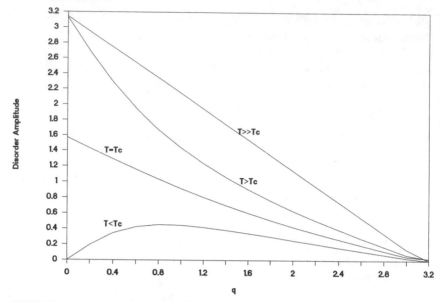

FIGURE 7.12   Amplitude, δ^*, for Onsager's disorder operator, for (from the top) $T \gg T_c$, $T > T_c$, $T = T_c$, and $T < T_c$. q is the wave number. Adapted from Onsager, "Crystal Statistics," fig. 5.

amplitude for disorder, δ^* (his figure 5, my figure 7.12). The y-axis is the amplitude for disorder; the x-axis is proportional to the inverse wavelength of the disorder.

Technically, elliptic functions are the means by which the crucial quantity k is embedded into the analysis, either in parametrizing the star-triangle relation or in doing the integrals for the partition function and the free energy. That k is a constant for the pairs of K_1 and K_2 for commuting transfer matrices means that we might eliminate, say, K_2—so making k explicit. Yet, in doing a change of variables in order to build in k invariance, one encounters square roots, and so undetermined signs. In using an elliptic substitution, the sign of the square roots will become precise. The big and surprising payoff is that k then proves to be the modulus of the elliptic functions.[83] In retrospect, this is perhaps not so surprising, given our earlier formula for the relationship between the dual's and the original's partition functions.

Specifically, recall that Baxter shows for the rectangular lattice how letting $\sinh 2K_1 = x$ and $\sinh 2K_2 = 1/kx$, thereby building in k invariance ($\sinh 2K_1 \sinh 2K_2 = 1/k$), and then letting $x = -i \operatorname{sn} iu$ and so $\sin 2iK = \operatorname{sn} iu$, "conveniently" lets the eigenvalues of the transfer matrix depend only on u (for a fixed k). (u may be defined in terms of an elliptic integral in K_1 and k: $\int_0^{2K_1} dx \, 1/\sqrt{(1 + k^2 \sinh^2 x)}$.) If $k = 1$, we can solve the problem nicely by letting $\sinh 2K_1 = \tan u$, and $\sinh 2K_2 = \cot u$; again, $\tan u \cot u = 1$.[84]

We get a hint at u's significance from the $k = 1$ case. Here u, as a function of the couplings, goes between 0 and $\pi/2$, and for $K_1 = K_2$, $u = \pi/4$. If $K_1 < K_2$, u is between 0 and $\pi/4$, otherwise u is between $\pi/4$ and $\pi/2$. Hence, as I indicated earlier, u is a normalized measure of the coupling constant (k and K_1 are involved in its definition) and of the inequality between horizontal and vertical couplings. By the way, it can be shown, as a consequence of the star-triangle relation, in general that if u_i is defined as $\mathrm{am}(iu_i, k) = 2iK_i$, where $\{K_i\}$ are the triangle couplings, then $\Sigma u_i = I'$, the imaginary quarter-period magnitude of the elliptic function.[85] (Here, at $k = 1$, for a rectangular lattice, where one u [say u_3] is zero, $u_1 + u_2 = \pi/2$.)

The possibility of a star-triangle relation and of a duality transformation, which are k preserving and k inverting, respectively, is now manifestly built into the formulation by the adoption of k and u as the variables of interest.

Again, it should be noted that these elliptic substitutions allow for two kinds of work. In Baxter, they epitomize the symmetries of the problem (the star-triangle relation). Moreover, he shows how he might introduce elliptic functions at various stages in the calculation:

> In the above working I have delayed introducing elliptic functions for as long as possible: until Section 7.8. There they were needed in order to express $\exp(2K_1), \exp(2K_2)$ as meromorphic functions of some variable u, while satisfying (7.6.1) [the usual star-triangle relation] for k independent of u [so parametrizing the star-triangle relation and allowing for (single-valued) elliptic substitutions].
>
> This equation (7.6.1) is the "commutation condition": two transfer matrices with the same value of k, but different values of u, commute. . . .
>
> Thus it would have been perfectly natural to have introduced elliptic functions as early as Section 6.4 [when the star-triangle relation is first introduced as a connection between the Boltzmann sums for the star and the triangle, how microtransfer matrices build up the lattice], so as to obtain a parametrization. . . of the full star-triangle relations.[86]

Baxter, once he is in the realm of k and u, can elegantly express the symmetries of the transfer matrix, and so work out his solution. Once Baxter achieves his solution for the free energy, he then throws away these elliptic substitutions, since the symmetries they embody are no longer needed, and he shifts to more conventional functions of K_1 and K_2.

In Onsager, the elliptic substitutions come in even later, rather than being central to the strategy for working out the crucial equations and their solutions (where it would seem duality, built into his formalism, plays that central role for Onsager). The substitutions allow for the evaluation of the crucial integrals in computing the partition function and the related thermodynamic properties, evaluating integrals of the form $\int \gamma(\omega) \, d\omega$, and understanding the behavior of those solutions (where the γs are the side of the triangle opposite the angle ω, and note that $\gamma(\omega) = \epsilon_q$ [except at $q = 0$], the energies of the particles that I discussed earlier in this chapter). Onsager repeatedly refers to the hyperbolic triangle to appreciate features of the energies and of the integrals he is evaluating. Elliptic substitutions and hyperbolic geometry illuminate each other (as would trigonometric substitutions and plane geometry) and so the physics of what is going on. Building in the star-triangle relation (and in k, duality), the substitution leads to geometric figures whose behavior models the behavior of the relevant physical quantities.

Onsager and Baxter both suspect there is more here to hyperbolic geometry and the associated elliptic functions than mathematical devices or merely a method of keeping track of relationships among coupling constants and temperatures, and more than some mathematical relations between hyperbolic geometry and elliptic functions.[87] In Onsager's words:

> The relations between these [interaction constants for the triangle, $\{L_i^*\}$, the result of both the dual ($\{K_i\} \to \{K_i^*\}$) and star-triangle ($\{K_i^*\} \to \{L_i^*\}$) transformations] and the original interaction constants [$\{K_i\}$] involve just the hyperbolic trigonometry of a totally improper triangle: The quantities of a hexagon whose corners are all right angles. The two triples of geodetics are mutually polar, and all the rules of spherical trigonometry have counterparts. For example, the ratios of hyperbolic sines are the same for all three pairs of opposite sides [constant k]. At the critical points this ratio equals unity. For equal interactions between vertices of order $m = (3, 4, 6)$ [triangles, rectangles, hexagons] the critical condition could be summarized in the form $2 \arctan(\tanh K) = \mathrm{gd}\, 2K = \pi/m$ [where gd is the Gudermannian angle, $\mathrm{gd}\, 2K = \arctan \sinh 2K$ ($\mathrm{gd}\, a = b$, if $\sinh a = \tan b$), and $\mathrm{am}(u, 1) = \mathrm{gd}\, u$], which was ultimately generalized to $\Sigma \, \mathrm{gd}\, 2K_j = \pi/2$ [which, given that $\mathrm{gd}\, 2K = u$, eventually is shown to be a $\Sigma u_i = \pi/2$ fact, which is what we expect from the star-triangle relation and $k = 1$ here].[88] The obvious is not always trivial.[89]

And in Baxter's:

> Other corollaries of the star-triangle relations. . . can be interpreted in terms of hyperbolic trigonometry. It seems likely that many of the properties of the star-triangle relation. . . could be conveniently interpreted in this way: as far as I know, this has not yet been done.[90]

"The obvious is not always trivial." I should note that we have just spent a good deal of effort unpacking Onsager's very compact summary of what can be made obvious to the eye diagrammatically, but which is not all so obvious or trivial physically.

Moreover, continuing in this vein of the diagrammatics epitomizing the physics,[91] trying to understand singularities at the critical point, Onsager notes that "the singularities of the various functions at the critical point are completely described by the degeneration of the period parallelogram [for elliptic functions, $4I \times 4iI'$, into $\infty \times 2\pi i$ at $k = 1$]."[92] He also notes that "[t]he three integrals [needed for computing the thermodynamic properties, derived from the integral of $\gamma(\omega)$ or ϵ_q above, namely, the free energy] are infinite at the critical point, otherwise finite. The singularity results from a conspiracy: In the case $K_1 = K_2^*$ we have $\gamma(0) = 0$ [$\epsilon_0 = 0$] and at the same time $\delta'(0) = \delta^*(0) = \pi/2$ [the collapse of the hyperbolic triangle]"[93] (see figure 7.13).

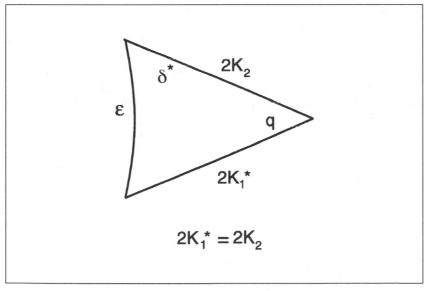

FIGURE 7.13   The collapse of Onsager's triangle.

As indicated earlier, Onsager's "conspiracy" is of course physical and not only geometric, so to speak. One has for an isosceles triangle $(2K_2 = 2K_1^*)$ an infinity of almost zero energy modes, when $q \approx 0$ and $N \approx \infty$, and hence a very sharp peak in the specific heat and a phase transition.

In sum, duality transformations and star-triangle relations are parametrized by elliptic functions and are nicely represented by hyperbolic geometry. Crucial features of the solution and of the energy spectrum of the quasiparticles are similarly parametrized and represented. Those formalisms build in enough of the symmetries of the Ising model so that they automatically do lots of the work of solving the model. Recall, however, that this automaticity is a very great achievement (Onsager did have to solve the model), not at all freely available unless one has been properly prepared: as indicated earlier, Baxter's commuting-transfer-matrix derivation is straightforward, if as he says "indirect" ("showing that it is only basically local properties of the lattice model that are being used"),[94] until he unleashes the analysis he needs to find the solution to his functional equation. This is a lush jungle of elliptic functions and their wondrous properties. Perhaps this ought not be a surprise, since according to Weierstrass, a polynomial functional equation in $f(x)$, $f(y)$, and $f(x+y)$ has as solutions elliptic functions. And it is a procedure that generalizes nicely for very much more complicated lattices. It is, however, initially not a very physical garden, given the culture of most physicists.

Onsager's procedure, when it was proposed in 1944, was also not a very physical garden, much of it being a marvelous essay in applied algebra—reflecting earlier practical insights Onsager had achieved working out explicit solutions to toy models.[95] (Onsager chose his method of exposition for purposes other than to show how his proof actually worked.) And it is true that, while Kaufman's 1949 solution placed the Onsager procedure in a rather more domesticated form of spinor algebra, the actual physical mechanism was still hidden. On the other hand, Schultz, Mattis, and Lieb's solution is nicer for the contemporary (1964) physicist in that it is much closer to the realm of particles (quasiparticles, here) and quantum field theory and standard physical technology, namely, diagonalizing quadratic forms, the undergraduate's story of the normal modes of coupled harmonic oscillators. Hence, Kenneth Wilson read the Onsager solution in this "translation," as he puts it.

Finally, much as the anticommuting character of Onsager's s and C operators, a seemingly occasional move (one that does lead to a quaternion algebra and a solution) is eventually epitomized into a crucial physical and mathematical feature of the constitution of Ising matter (a feature he might well have appreciated), so the appearance of elliptic functions in Onsager and then in Baxter is eventually epitomized into another crucial physical and mathematical feature.

More recently, the star-triangle relation has come to be seen as the main symmetry (now called the Yang-Baxter equation), and researchers want "to play Baxter's ball game."

> The rules of the game are as follows. . . . Are the exact results related to the theta functions?. . . Some are related to theta functions, others are not. . . . In [Baxter's] work, the role of the theta functions was dominant and crucial.
>
> (i) Parametrize the local Boltzmann weights in terms of elliptic theta functions so that they satisfy the star-triangle relation. . . . The star-triangle relation implies that the local state probabilities [the eigenvectors] depend only on the elliptic nome q $[q = q(k)$; q here is *not* the wave number], and not on the spectral parameter u.
>
> (ii) Reduce the 2D configuration sums expressing the local state probabilities to 1D configuration sums. The method is called the corner transfer matrix method.
>
> (iii) Evaluate the 1D configuration sums in terms of elliptic modular functions [as did Onsager].[96]

Ising matter is both fermionic and modular. It seems that these features are connected by the nature of random walks, both in how we might generate self-avoiding ones, and in their scaling symmetries.

While I have been to some extent guided by more recent developments, the exposition I have presented here has a cutoff point of about 1975–1980.[97] There are many hints of interesting congruences and coincidences, and lots of the time we seem to be slogging through peculiar details of particular derivations, notable for both their idiosyncrasy and their pregnancy. In the next chapter, I bring the story up to date, showing how these many loose ends are weaving together—yet, of course, there are new loose ends, as well as others that remain part of a frayed fringe.

One other point, as I have indicated occasionally: I might have sketched a rather more straightforward path to the insights I have garnered, by immediately noting the formal analogy of quantum field theory on a space-time lattice with statistical mechanics on an Ising lattice.[98] I have chosen not to do so. Rather, I have been trying to show how the original mathematics builds in the physics already, rather than to show how the most general formulation makes the various solutions and their particular mathematical technologies perspicuous. I wanted to stay close to the technology as once employed in its context. Also, the most general technology tends to abstract even further from the actual physical mechanisms (spins interacting with each other, and particular sets of spin configurations as the good particles in this story) than has most of the earlier formalism. Of course, the abstractions suggest new ways of thinking about those mechanisms, in time the abstractions being seen as being as concrete as what were originally taken as the actual physical mechanisms.

I have been insisting, all along, that mathematical technology is physics, and that what is interesting is not that abstract claim, as such, but rather its particular exemplifications. Interpreting that technology as physics requires very deep insight, in part because it is the function of any technology to hide the values it exemplifies and to be seen merely as automatic and autonomous.[99] Antisymmetry and k invariances, fermions and temperatures, are surely constructs of the imagination. But objects with these properties—bonds, and lattices composed of them—are palpably present in our lives.

8

The Physics and the Mathematics

The mathematics and the foundations • The physics • Some facts •
The constitutions of matter • Mathematical representation and physical
reality

We have taken a tour through about forty years of mathematical-physical model building and solving. Complementary to features displayed in proving the stability of matter, our souvenir is a sense of some of what is crucial about the constitution of (Ising) matter, as that is recurrently and more deeply revealed in the various models and solutions: the lattice's regularity superimposed on a gas of fermions or a gas of paths.[1] Those crucial features might even be necessary, otherwise hidden but for our efforts of model building and solving, which in the models' success limn what we might take as the essential features of the constitutions of matter.[2] Or we might take it that those crucial features are displayed comparatively and contrastively, among and between the various models and solutions, their being about actual matter's real features (which, for many a practicing physicist, surely do exist) being a concern we put aside for the moment—although, in any case, those features do influence how we think of and take hold of actual matter. What I hope to have displayed is "how physics is a culture of worldly theorizing and how the mathematics is hopelessly and irremediably bound up with its potentialities, realizations, and philosophical interrogations of real-worldly matters."[3]

As I have indicated, in the last two decades there have been substantial developments in our understanding of the constitutions of Ising matter. The crucial features I have deliberately teased out from earlier papers are now taken as commonplaces, fully embedded within contemporary thinking about statistical mechanics and quantum field theory.[4] Many features would seem to follow naturally as a consequence of more generic and deeper mathematics. However, my concern for the most part has been to display how those features made their earliest appearances and became more and more substantial with further physical insight and technical mathematical developments.[5]

Still, I want to bring the discussion rather more up to date, to display or at least mention how the various features are now embedded in thought and practice. While this part of the chapter will be quite technical, I have also tried to convey in general terms the tone of these more recent developments. In closing, I return to the central limit theorem as a model of this particular embeddedness.

THE MATHEMATICS AND THE FOUNDATIONS

We have slogged through extensive discussions of arithmetic and geometric duality and of the star-triangle relation. The centrality of k, in indexing these transformations, would seem to be the clue to how they are all about one story: k invariance and a deformation symmetry. But there is another story, perhaps even more illuminating.

First, some observations: (1) The diagrams of arithmetic duality and the star-triangle relation are different but curiously alike. (2) Partitions may be understood as processes of inclusion-exclusion, in effect counting up all the particular subsets of a set. This is also true of the statistical mechanical partition function, where the sum over states is a sum over the various subsets given certain constraints.[6] (3) Fourier transforms are in effect measures of inclusion and exclusion, at finer and finer partitions. That is what the plus-minus character of the sine function does at smaller and smaller wavelengths. (4) Construction of the dual, in particular construction of the dual polygons, is an inclusion-exclusion process, drawing links to create boundaries or inclusions of droplets of like spins in seas of oppositely pointed like spins. This process occurs at finer and finer resolutions, so that we can pick up islands of opposite like spins within islands of like spins, and so on.[7] (5) And theta functions (and more generally, q-series)[8] are often the devices of choice for doing the enumerations of the arithmetic partitionings.

So, perhaps, it is not too surprising that arithmetic duality and geometric duality (and the star-triangle relation) may be seen to be in effect Fourier transforms of the partition function of the lattice under different symmetry groups, as McKean has shown. Technically, "like the Jacobi transformation [of modular functions, from say T to $1/T$], the Kramers-Wannier duality [our arithmetic duality] is an instance of a Poisson summation formula [$\sum_n g(n) = \sum_n \hat{g}(n)$, where $\hat{g}$ is the Fourier transform of g]."[9] In our case, the partition function for a lattice and for its dual (in effect its inclusion-exclusion dual) are simply related.

Such a Poisson formula may be taken to say that the sum of the Fourier coefficients is equal (within a factor) to the sum of the original amplitudes of the function on the lattice. The sum of the statistical weights ($\exp - \beta E$) for the original lattice is proportional to the sum for the dual lattice at the dual temperature. Put differently, music may be appreciated either in terms of its frequencies (as when one is watching a display on a high-fidelity amplifier) or in terms of its temporal development as a piece of music.

The discussion of fluctuation-dissipation relations in chapter 7 is another form of these kinds of relationships. Scattering and absorption are special kinds of transforms of each other. Technically, dispersion relations connect the two. The Poisson sum of the original function can be seen to be the sum of the inelastic or absorptive scattering amplitudes, while the Poisson sum of the dual leads to the amplitude for forward or elastic scattering. Moreover, that β and time are related, as they would seem to be from the analogy of statistical mechanics to quantum mechanics, is of deep physical significance. What is crucial for causality (and dispersion relations) is the existence of a time that parametrizes events; what is crucial for statistical mechanics is the existence of a β (a temperature) that parametrizes thermodynamical processes (the direction of heat flow).[10] Finally, recalling the Darwin-Fowler derivation of the canonical ensemble and its parameter β in chapter 3, we might take as its quantum mechanical equivalent a derivation of the evolution operator $(\exp - iHt/\hbar)$ and its parameter, t, the time.

The Darwin-Fowler derivation presumed that we had some idea of what a thermodynamic equilibrium state was, and from that it found the relative probability of a state of energy E to be given by the canonical ensemble, $\exp - \beta E$ (in what is effectively an infinite volume limit). Modern developments of the foundations of thermodynamics and of statistical mechanics focus rather on the notion of a state and of a phase, and provide a philosophic analysis of what we might mean by these notions through the kind of mathematical-physical analysis I have described. Here is a nice summary of the strategy, as well of some of the themes throughout this book.

> The contrast between the microscopic and the macroscopic level is the starting point of Classical Statistical Mechanics as developed by Maxwell, Boltzmann, and Gibbs. . . . The microscopic complexity can be overcome by a statistical approach; the macroscopic determinism then may be regarded as a consequence of a suitable law of large numbers. . . . The above distribution [$\exp - \beta H$] is called the Gibbs distribution (or, somewhat old-fashioned, the Gibbs ensemble) relative to H. . . . According to a standard rule of mathematical thinking, the intrinsic properties of large objects can be made manifest by performing suitable limiting procedures. It is therefore a common practice in Statistical Physics to pass to the

infinite volume limit. (This limit is also referred to as the thermodynamic limit.) However, instead of performing the same kind of limit over and over it is often preferable to study directly the class of all possible limiting objects. . . . [W]e are faced with the problem of describing an equilibrium state of such a system by a suitable probability measure on an infinite product space [the phase space]. . . . [The canonical ensemble probability measure] is uniquely determined by the property that each finite subsystem, [probabilistically] conditioned on its surroundings [the temperature], has a Gibbsian distribution relative to the Hamiltonian that belongs to this subsystem [what is called the DLR equation, for Dobrushin, Lanford, and Ruelle]. . . . [T]he physical phenomenon of phase transition should be reflected in our mathematical model by the non-uniqueness of the Gibbs measure for the prescribed specification. Fortunately, and somewhat surprisingly, the model developed above is indeed realistic enough to exhibit this non-uniqueness of Gibbs measures in an overwhelming number of cases in which a phase transition is predicted by physics. This fact is one of the main reasons for the physical interest in Gibbs measures [the canonical ensemble].[11]

A state is given by a set of probabilities, one for each microscopic configuration, and an equilibrium state is such that a given region is in thermal equilibrium with its surroundings, namely the canonical ensemble (that is, $\exp - \beta E$). The task is to work out the implications of these notions for a phenomenology of states, phases, and phase transitions.[12] Many of the issues that came up in chapter 5 are dealt with in an entirely different manner, in part because the infinite volume limit issue is now finessed (by the assumption of that canonical ensemble, and its temperature bath). In effect, we have returned to Gibbs's thermodynamic concerns of 1875, with a new level of technical apparatus.[13]

THE PHYSICS

Let us begin with some quotations, albeit jargon-filled, that summarize developments of the last two decades. The various constitutions and methods are now seen as part of a mathematical whole.

A whole field seems thereby to be arising, connecting solvable models [such as the two-dimensional Ising model], integrable systems, knot theory [commutativity properties], Lie algebras [the A_i and G_k of Onsager that generate the polygons], and quantum groups [critical exponents].[14]

That mathematical whole has direct implications for quantum field theory, and there is a sweet connection between exactly and practically solvable two-dimensional models and formally integrable (or "conserved," as in conservation of energy and momentum) one-dimensional models. The star-triangle relation is just what points to the symmetries and conserved quantities in these systems.

> The reduction of the different methods of solution to a unique global general method has also had unexpected successes. . . .
>
> The logical steps lead from the dimer covering technique, through the introduction of Grassmann [anticommuting] variables, to the Pfaffian method, and from the latter, by the very general procedure of the transfer matrix, to a quantum one-dimensional problem [scattering on a line].[15]
>
> Thus the equilibrium configurations of classical two-dimensional systems with nearest neighbour interactions were shown to exhibit a strict topological analogy with the space-time trajectories of a dynamical system in one spatial dimension. . . . The integrability of the transfer matrix (and that of the dynamical system) is connected with the existence of ternary relations among micro-transfer-matrices [the star-triangle relation]. Such relations are remarkable realizations of subgroups of the permutation group (braids), and imply the existence of one-parameter [k, indexed by u] families of mutually commuting transfer matrices (whereby the integrability follows). Also Yang's method [of solving the one-dimensional chain of hard-core bosons] foreshadows the inverse scattering technique thoroughly developed by Faddeev, which links the symmetry conditions with those of periodicity [action-angle variables], beautifully setting in operatorial form Bethe's Ansatz [that multiple scattering is a sum of two-body scatterings, that bulk matter is a sum of two-body interactions]. It is therefore in a way no more surprising that Zamolodchikov's factorization equations for the S-matrix, representing the conditions which are necessary to factorize a multi-particle scattering matrix into two-particle ones, coincide with Yang's equations and Baxter's "triangle" equations.[16]

Onsager's work may be seen as prefiguring all future developments in this field (both statistically and quantum theoretically), although such prefigurement is only apparent when it is fulfilled through the ingenious work of Onsager's successors, such as Fisher, Lieb, and Baxter:

> Now, nearly fifty years later, we can appreciate Onsager's methods much better. He was the first to introduce the star-triangle equation to statistical mechanics, even though now the name Yang-Baxter equation is commonly used. He also was the first to introduce loop algebras [for generating the polygons, and the description of the orderliness of the lattice] as a solvability principle. Onsager and Kaufman have clear priority

over Wick for the Wick theorem and, together with Bethe, they scouted a new area of mathematics which is now called quantum groups.

Onsager's (only partly published) work on two-point functions in the Ising model got extended in 1966, but it was only in 1973, when Wu, McCoy, Tracy, and Barouch announced the Painlevé equation for the scaled two-point correlation, that the theory of the two-dimensional Ising model went beyond Onsager's level. Also, the first two-dimensional models solved that were more complicated than the Ising model were Lieb's ice model of 1967 and Baxter's eight-vertex model of 1973.

Onsager's loop-algebra solution method [Onsager's A_i and G_k] was generalized only in 1985 when Von Gehlen and Rittenberg solved the Dolan-Grady criterion [for exact integrability as a consequence of self-duality] within a one-dimensional generalization of the quantum Potts model. The connection with Onsager's 1944 paper was noticed by Perk, showing that the chiral Potts model is the first genuine generalization of the Ising model, with its "superintegrable" case solvable for two reasons: star-triangle integrability and loop-algebra integrability [arithmetic duality]. The chiral Potts model upgrades the fermions of Kaufman to parafermions. It also provides an infinite hierarchy of quantum groups at roots of unity, with Ising as its first entry. Thus the works of Onsager are still at the center of attention.[17]

The archetype for our contemporary understanding is ancient, the regular polyhedrons or solids, one of the Pythagoreans' achievements, developed from the Egyptians' work. Eventually, those polyhedrons were understood as geometric objects, and the five particular ones—tetrahedron, cube, octahedron, dodecahedron, and icosahedron—were accounted for by symmetry groups representing spatial rotations and reflections in three dimensions. Here we have a similar mystery, the star-triangle relation, or more generally k invariance or conservation (or transfer-matrix commutativity) as I described it earlier. Eventually, that has come to be seen as a rather more general invariance, the Yang-Baxter relation (which turns out to be the same as saying that a thermodynamically consistent lattice may be built up of a mixture of transfer matrices, or of microtransfer matrices that add in just one link, as long as they share a common k). That relation was understood as a model of the braid group of knot theory, or perhaps the other way around. And the particular lattice systems with their nice rational numbers for their critical exponents are taken as representations of what are called quantum groups (descendants of the Lie algebras Onsager employed in his solution). The lattice's symmetries, under a partition function, lead naturally to the critical exponents. Technically, much as the various atomic angular momentum states (as representations of the rotation group) have well-defined eigenvalues, $L(L + 1)$, which then also specify the radial behavior of the wave function (the leading exponents of r in the various special functions or polynomials), here the algebra that generates the configurations of the system then leads to the exponents of $|t|$ in the critical region.

We have the following analogy or ratio—Yang-Baxter relation : knot theory : quantum groups :: regular solids : geometry : symmetry groups.[18] Put differently, and recalling the quotations above—Yang-Baxter = star-triangle = commuting transfer matrices = factorizability of the S-matrix = two-body interactions = exactly solvable models = integrability. And—braid group = Temperley-Lieb algebra = commuting transfer matrices and star-triangle relation = building up the lattice out of two-body interactions or links = building up the lattice (namely, its partition function).

SOME FACTS

Some of the various connections mentioned above may be apparent from what I have said in the earlier chapters. But the connections were quite fragmentary, as of say 1980, and only of late has there been a more systematic understanding of what is going on. I want to convey some of the insight that has been earned, albeit without delivering on the technical details.[19] In order to do so, I shall review a variety of facts that are prominent in my earlier discussions of the constitutions of Ising matter, and indicate how they are keys to the various connections. Those facts are k invariance, the exact solvability of the Ising model, the analogy of quantum field theory and statistical mechanics, and the particle spectrum of Ising matter.

First of all, we have uncovered how crucial is k, the modulus of elliptic functions that appear in all the various constitutions. k appeared as a modulus early on, when it was noted that, in going to the dual, the form of the partition function remained the same except for being multiplied by $k^{-N/2}$. It is also notable that the transformation to the dual is a $1/k$ transformation, a modular transformation ($\tau \to \tau/(1 - \tau)$). Or, if $w = \tanh K$, then $w^* = \tanh K^* = (1 - w)/(1 + w)$, a conformal mapping (one that preserves angles and transforms circles into circles), a mapping in the complex plane. And the star-triangle relation, which is the statement of k invariance and the commutativity of transfer matrices, is nicely parametrized by (Jacobian) elliptic functions.

As I mentioned in chapter 3, combinatorial partition functions are in fact the model for the theta functions, and eventually for the elliptic functions. When one takes Kaufman's exact solution for the Ising model for a finite-sized lattice $(m \times n)$ and finds its asymptotic behavior at the problematic (slowly converging) critical point, it "can be reduced to elliptic theta functions of nome $q = \exp(-\pi m/n)$ [where the nome q and the modulus k are monotonic functions of each other from zero to one]."[20]

All of these various hints have borne fruit in current theorizing, in which the elliptic functions and conformal invariance have become the means of incorporating scaling, duality, and the star-triangle relation (in two-dimensional models).[21] So, for example, three 1984 papers, on theta functions, on conformal symmetry, and on infinite-dimensional Lie algebras and modular forms, respectively, are referred to by one author as "The Trinity," appearing at "first independently, and later on as closely interrelated."[22]

u as a variable becomes rather more prominent. It is called the spectral parameter, since it indexes the zeros of the partition function $(Q(u_j) = 0)$ as well as the commuting transfer matrices (and their different maximal eigenvalues, namely, their spectrum, their different free energies or ground states or masses). A duality transformation not only inverts k, it transforms $u \to ku$ (the dual u), more precisely, $u_K \to k(I' - u_K) = I'^* - ku_K = u_{K^*}^*$, where the asterisk means that the quarter-period magnitude and the argument are evaluated for the dual modulus, $1/k$. The star-triangle transformation conserves k, while $u_{K_i} \to I' - u_{K_i} = u_{L_i}$. The sum of the u_i equals suitable multiples of I', both for the $\{K_i\}$, where $\Sigma u_{K_i} = I'$, and for the $\{L_i\}$, where $\Sigma u_{L_i} = 2I'$. And, in the analogy to particle scattering, the u_i turn out to be momentum variables ("rapidities"), the star-triangle relation being a statement of the relationship between two different scattering processes and the sums of their rapidities.[23]

We know that transfer matrices commute with each other if they have the same value of k. This fact was originally realized by Onsager; Stephen and Mittag, and Baxter made elegant use of this fact to find lovely modes of solving the Ising model.[24] More generally, if matrices commute, they share the same eigenvectors, although they may have different eigenvalues. (Similar matrices share eigenvalues as well.) Those eigenvectors, in this case, measure the kinds of orderliness or correlation to be found in a lattice: the square of each element of the maximal eigenvector (for a symmetric matrix) is just the probability that a row would have the configuration of up and down spins that goes with that element. Namely, if $\psi_{\max} = (a_1, \ldots, a_n)$, and a_j refers to the row configuration j (so a configuration of spins, 1 being ↑ say, $01101010001001_{\text{base2}} = j = 6793_{\text{base 10}}$), then a_j^2 is the probability that a row of the lattice has that configuration. (Clearly, if a row is in configuration j, then the probability of an adjacent row being in configuration m is not in general a_m^2, except at infinite temperature when all are equally likely.) In this sense, commuting transfer matrices constitute matter (in the sense of the partition function), matter that has bulk properties that either are extensive (as is V^N) or are intensive (as is $V = V(k, u)$), the intensive ones to some extent fully characterized near the critical point by the correlation length, ξ, or by k (recall that $|\ln k| \approx 1/\xi$), or by $|t|$.

Put differently (see figure 8.1), one can show that if one builds up a triangular lattice, $\{K_i\}$, by adding on a horizontal (K_1), then a vertical (K_2), and then a horizontal/diagonal bond (K_3); or, if one builds up a hexagonal/star lattice, $\{L_i\}$, by adding on a vertical (L_3), then a horizontal (L_2), and then a vertical/diagonal bond (L_1); and, if the triangle's coupling constants are related to the star's by the star-triangle relation; then the contributions to the partition functions, expressed in terms of the transfer matrices, are the same: Horizontal/Vertical/Horizontal$_{\text{triangle}}$ = Vertical/Horizontal/Vertical$_{\text{star}}$.

If we have a set of transfer matrices, $\{U_m\}$, which have the same k and which are distinguished by their values of u, and if odd subscripts denote horizontal bonds and even subscripts vertical bonds, then we may rephrase the above as (say, p is even)

$$U_{p+1}(u_1)U_p(u_1 + u_3)U_{p+1}(u_3) = U_p(u_3)U_{p+1}(u_1 + u_3)U_p(u_1),$$

where u_i refers to the u defined by $\{K_i\}$; and $U_pU_j = U_jU_p$, if p and j differ by more than one.[25] (Recall that $\Sigma u_i = I'$, and hence, rather than

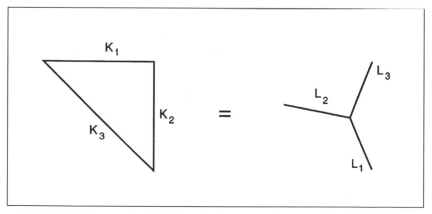

FIGURE 8.1 The star-triangle relation as a Yang-Baxter equation.

u_2, it turns out we can have $u_1 + u_3$ above.) This is a generic form of what is called the Yang-Baxter equation, and is our star-triangle relation.[26]

Again, the invariance here is thermodynamic (a partition function invariance), and it depends on the k invariance of the transfer matrices. In each case, a lattice is built up of links or bonds between two sites, and conventionally only nearest neighbors are so linked.[27] The lattices we are concerned with can be invariantly made up of two-body linkages or single bonds.

Roughly contemporaneous with the development of the solutions to the two-dimensional Ising model, there developed within mathematics a theory that could describe sets of topologically equivalent knots, in particular knots that could be conceived of as a series of links, what is called a braid. (Figure 8.2).[28]

Let b_p denote the pth string passing over the $p + 1$st string (and b_p^{-1}, if it passes under), strings being known by their location in terms of the upper bar. Topological equivalence of different representations of the same knot (rather than partition function and thermodynamic

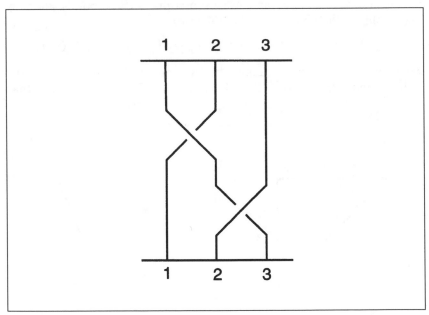

FIGURE 8.2   Braid representing $b_1 b_2^{-1}$. Adapted from Deguchi, Wadati, and Akutsu, "Knot Theory," fig. 21.

equivalence) is guaranteed by the following rules: $b_p b_{p+1} b_p = b_{p+1} b_p b_{p+1}$; and $b_p b_j = b_j b_p$, if p and j differ by more than one—rules formally identical to the commuting transfer matrix formula above, except that the $\{u_i\}$ do not play a role here. (For in effect $k = 1$ and $u = \pi/2$, so that $\tan u = \sinh 2K$ is infinite, and so $T_c = 0$).[29] Thermodynamically equivalent bulk matter or lattices (same k and different u_1 or K_1, say) and topologically equivalent knots, each of which are made up of two-body interactions, of yes-no linkages (fermions), of over or under braids, have a powerful mutual analogy (at least at the zero-temperature critical point). Just as the partition function is a polynomial (say, in $z_i = \tanh K_i$) that represents or generates that lattice and its equivalents, so there exist polynomials that can represent these equivalent knots.

More generally, Yang showed that if a multibody scattering process can be decomposed into a set of two-body interactions—or, as others showed, if an Ising lattice could be understood as built up of rows, each row defined by superpositions of regular spin flips or up-down pairs or dislocations (in effect, what is called the Bethe *Ansatz*)—then one ends up with a Yang-Baxter equation for the scattering amplitudes:[30] it does not matter in which order the scatterings take place (or in which order a knot is constructed, or in which order a lattice is built up), in the sense that a process is invariant: the physical scattering process and its scattering matrix, the actual knot and its knot polynomial, the actual thermodynamic system and its partition function, remain the same.

Much as in a melodrama, it is no coincidence that k plays such a crucial role in solutions to the Ising model. The modulus, k, is the key to all the invariances that make it possible to solve the problem exactly. As the quote on modular functions (elliptic or theta functions) at the beginning of chapter 6 indicates, k allows one to scale up from a simple ground state (at $T = 0^+$, all but a few spins aligned) to the critical point. k goes from 0^+ to 1^-—by modular transformations that take k to k^α (actually, $k \to 2\sqrt{k}/(1 + k)$, Gauss's transformation of modular functions), and they also transform the partition function suitably once it is expressed as elliptic functions. A series of these transformations in effect stair-step k from 0^+ to 1^- (Figure 8.3).

Technically, the Gauss transformation and its inverse, the Landen transformation (the one in effect employed in the block-spin transformation, going from $k = 1^-$ to 0^+) are simple transformations on the nome ($q \to q^{1/2}$ or q^2) or on the parameter ($\tau \to \tau/2$ or 2τ, where $q = \exp i\pi\tau$) of the elliptic functions. And the nome plays a central role in more complex lattice models.[31] The rich literature on q-series suggests that the nome, q, should be our preferred physical variable here.

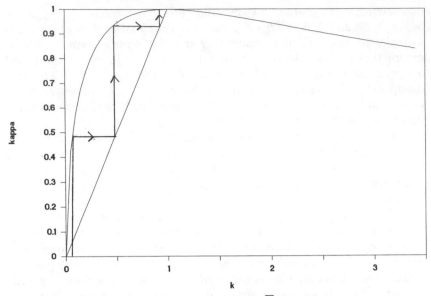

FIGURE 8.3 Modular transformation, $k \rightarrow 2\sqrt{k}/(1+k) = \kappa$, between just above zero temperature and just below the critical temperature (when $k = 1$).

This would mean that, rather than focusing on the trigonometric analogs, the Jacobian elliptic functions, sn, cn, and dn, we would focus on the theta functions.[32] Put differently, we find theta functions when we solve the heat equation, where the nome represents the time decay of a mode; we find Jacobian elliptic functions when we solve the simple pendulum, where k is related to the amplitude and period; and we find Weierstrass $\wp$ functions, as in a spinning top, when the periods are our central concern. Just which is the best analog here remains open to investigation.

Moreover, these invariances are generalizable to other physical and mathematical situations, perhaps because they reflect some of the most long-standing conventions and developments in the practice of mathematical physics—the solvable two-body problem (and, more generally, perturbation expansions in terms of sums of two-body interactions).

Even more poignantly, a formally or technically exactly-solvable two-dimensional problem in statistical mechanics is seen to be the same as an integrable one-dimensional problem in quantum mechanics, characterized by an evolution operator that is just the transfer matrix—where the k associated with a transfer matrix is one conserved quantity, and the $\{u_i\}$ (as a set, $\Sigma u_i = I'$, the quarter-period magnitude of the elliptic functions) are the others, and where an integrable problem is one in which we may describe the solution in terms of a complete set of conserved quantities. (Technically, recall that the evolution operator, $U = \exp - iH_{qm}t/\hbar$, such that $U(t)|0\rangle = |t\rangle$, the state at time t; and the transfer matrix is such that $\mathrm{Tr}\, V = (\Sigma \exp - \beta H_{sm})^{1/N}$.) The Yang-Baxter equation is the key to the exact solvability of the Ising model, as we saw in chapters 6 and 7; it is also a key to the integrability of an associated one-dimensional quantum mechanical Hamiltonian.

Again, that one-dimensional problem might be a set of hard-core bosons restricted to a line, and the Yang-Baxter relationship here allows for the wave function to be expressed as plane waves (the Bethe *Ansatz*) whose phase shifts, the ϕs in $\exp - i(kx + \phi)$, are those of two-body scatterings. The plane waves correspond to a row configuration composed not of interacting hard-core bosons but of spin flips.

Moreover, the symmetries of a row configuration of spins (a one-dimensional object) define the solution to the two-dimensional Ising problem. Those symmetries are described by Onsager's A and B operators and the Lie algebra, $\{A_i, G_k\}$, they generate. Technically, $A_0 = -B, A_1 = A$; in Fourier transforming the set of these, one discovers the natural symmetries; and the associated one-dimensional Hamiltonian is $B + k^{-1}A$. The symmetries then may be shown to lead to the nice rational numbers that define the critical exponents of Ising matter, in effect the $|\ln k|^\alpha$ dependence of its critical properties.

Rather than being merely a sweet problem allowing for all sorts of tricks on the way to its exact solution, the two-spatial-dimension Ising model is of much wider interest, in part because it as well a one-spatial-dimension integrable model (the other dimension is time). Not only is the two-dimensional Ising model a lattice model of a continuum field theory (the free field being at $k = 1$, and it is still quasi-free, the mass matrix not diagonal, for $k \neq 1$), the field values given only at the lattice points; it is as well, through the Yang-Baxter or star-triangle relation, a model of scattering processes, k and u being related to a measure of the energy and the momentum in such scattering.

Besides k and exact solvability, another fact about the constitution of Ising matter is the formal analogy between quantum mechanics, quantum field theory, and statistical mechanics in this context. When Montroll, and Kramers and Wannier in 1941 did some of the early work on the two-dimensional Ising model, it was made to look much like an eigenvalue problem typical of quantum mechanics.[33] Even the ψs of quantum mechanics are present, albeit they are row vectors rather than wave functions. Montroll spoke of a resonance at the critical point between the almost ordered (low-temperature) and almost disordered (high-temperature) states, and he got good numerical results. Ashkin and Lamb's paper two years later is even more quantum mechanical in its formalism, and when compared to Lassettre and Howe's paper on the same topic, it seems to me to be more a physicist's presentation than a chemist's.[34] (But recall that Onsager was a professor of chemistry.) In Onsager's 1944 paper, the quantum mechanical tone is rather clearer, although, again, this is a classical statistical mechanics problem. (Note that the largest eigenvalue is related to the ground-state energy of the associated quantum mechanical problem: $E_{ground} = \lim_{\beta \to \infty} - 1/\beta \ln \Lambda_{max}$.)[35] And when Yang's paper on the spontaneous magnetization is described, the language is of quantum mechanical perturbation theory, what is perhaps best suited to the expected reader, even though it is a problem in classical statistical mechanics.[36]

In any case, there is a formal analogy at the level of the differential equations (the Schrödinger equation is the heat or diffusion equation in imaginary time, both being $\nabla^2 \psi = -\alpha \, \partial \psi / \partial t$); at the level of the quantum mechanical probability amplitude ($a_{jp} = \Sigma \exp iA/\hbar$, where A is the action, an integral of the Lagrangian over each of the paths from state j to state p) or the partition function ($Q = \Sigma \exp - \beta H$); and at the level of random walks.[37]

In particular, the heat or diffusion equation may be understood as a consequence of the random walk of molecules; quantum mechanical amplitudes are a weighted sum of all histories or paths between two points; and the constitution of (Ising) matter is the partition function, a sum over all states. Ising matter is constituted by a particular kind of random walk, a particular kind of self-avoiding one, in the sense that its partition function is just such a weighted sum of paths (the polygons). (More precisely, a self-avoiding random walk is a zero-component Ising model, while the usual Ising model I have discussed is a one-component model, assigning a number rather than, say, a vector, to each spin.)[38] Put differently, Ising matter's scattering matrix is a sum of all interactions among its molecules. The anticommuting Grassmann or fermionic variables, or Onsager's quaternion algebra, or the Pfaffian forms are devices to make sure that this particular kind of walk is counted just right.

> With monotonous regularity each method [of solving the Ising model] has reproduced virtually the same results as those listed above and has added virtually no new ones on the analytic [exact] side. . . .
>
> This experience is almost unique in mathematical physics. (Nearly always a valid new treatment of a problem produces new results as well as

reproducing old ones.). . . It has gradually been realized over the years that all the workable methods so far discovered are really particular cases of one very general treatment fairly well known in field theory. The close relationship between field theory and both "lattice" and "continuum" statistical mechanics have tended to be obscured by wide differences in notation and outlook and have come to be realized only very gradually over the years.

The general theorem, usually known as Wick's Theorem, . . . relates either the trace or the "vacuum to vacuum expectation" of a product of linear sums of operators, known as Clifford [as in Onsager] or Fermi operators, to what is known mathematically as a Pfaffian.[39]

What the Pfaffian or the antisymmetric operators do is to make sure that the random walk or the polygons are properly enumerated, with no bond counted twice in any single polygon or any configuration of polygons.

Our last fact is the character of the particles that may be said to constitute Ising matter. They are fermions, whose two-state, zero-one character is sourced in the twofold character of a bond, either parallel or antiparallel, $\uparrow \uparrow$ or $\uparrow \downarrow$, or in the corresponding requirement that the polygons not share sides. But those bonds are not the independent particles. Bonds are interacting fermions, since they depend on the underlying spins (duality), there is a thermal coupling, and the lattice is translationally invariant. The fermion particles that do have well-defined energies (or masses), and are free or noninteracting, are regular arrangements of spin flips.[40]

Duality says that adding one particle to a low-temperature lattice is equivalent to taking away the complement of that particle from (or adding an antiparticle to) the very high temperature lattice. Near the critical point, from the low-temperature direction, all the particles are excited and present (while from the high-temperature direction, all the antiparticles are excited and present)—in part because their effective masses or energies are quite small (in $k_B T$ units) at that point.

Now, if there is translational invariance, the standard move is to go to momentum space, a Fourier transform. In fact, all excitations or deexcitations are pairs (except at q equals zero or π), in Fourier transform terms $\{q, -q\}$, where q is momentum or wave number. The Hamiltonian or energy is expressed as bilinear products (say, $a_q^\dagger a_{-q}$) of creation and annihilation operators in q combined with $-q$.[41] One then diagonalizes the Hamiltonian. The eigenstates are $\cos \phi_q |q\rangle + \sin \phi_q |-q\rangle^\dagger$. The mixing angle, ϕ_q, shares the behavior of Onsager's disorder amplitude δ^* (but for K^* rather than K). At the critical point, there is a continuum spectrum of zero-to-low energy particles (no mass gap), and a degeneracy between order and disorder.

I should note that I have not mentioned a variety of other lattice models that mostly confirm but sometimes disrupt my analysis.[42] My main point is, what was once esoteric is now central; what was once particular is now generic; and what was once mysterious and surprising is now a commonplace.

THE CONSTITUTIONS OF MATTER

The central limit theorem says, roughly, that when one adds up mildly independent, finite variance, mean-zero random variables, what one ends up with is a Gaussian whose variance is the sum of the variances of its components, as long as N is reasonably large. (If the variances are not finite, there still exist such similar "stable" distributions, but rather than the sum scaling with $N^{1/2}$, it scales as $N^{1/\alpha}$.) What is important here is that the details of the individual distributions, and the degree and the details of their mutual dependence, become irrelevant if N is large enough: the asymptotic Gaussian form settles in, so to speak. The system will begin to look the same at various scales, just as a derived random walk constructed out of an original random walk, in which we average pairs of the original steps, will look the same as the original (suitably scaled).

The constitutions of matter would seem to be a matter of the central limit theorem and of a particular random walk (described by fermions, a walk of actual electrons for the stability of matter, or the random walk that is provided implicitly by rows of spins or bonds for Ising matter). Adding on one more particle, like adding on one more random variable, does not cause things to blow up. And the scaling, and the irrelevance of lattice details in Ising matter, is nicely analogous to the properties of sums of random variables.

Of course, we discover that we might get away with this analogy only after a great deal of careful mathematical analysis that justifies such a simplified picture. Only in terms of that analysis might we say that the stability of Coulombic matter depends on electrons being fermions; or that the behavior of Ising matter depends on the planarity of the lattice (at least for exact solvability) and the fermionic character of bonds.

Along the way, we have seen how the careful technical definition of bulk matter, or of stability, or of a phase transition, is an enormous achievement, building on various reasonable but insufficiently analyzed previous notions. And the working out of the various techniques for solving the Ising model are sleights of hand that are only later understood as more generic procedures justified by the deepest physics.

Exact solutions are testimony to mathematical technologies, physical intuition, and the capacity to create a model that is rich enough but tractable as well.[43] The exact solutions also allow us to extract the physics from more approximative schemes. They guide us in constructing computer simulations, as in simulated annealing as a mode of optimizing in mathematical programming, and in lattice gauge calculations of elementary particle properties. But there is one avenue I have not pursued here. Computational algorithms might be taken as exact models, whether they be particle simulations, or ways of calculating perturbative expansion coefficients, or cellular automata. This would be a very different physics than I have studied here, and a very different sort of mathematics. It might well lead to new constitutions of matter and new insights about what is crucial about that matter.[44] Of course, the mathematical and physical developments I have adumbrated at the beginning of this chapter are also leading to new insights, albeit more in the line of their predecessors.

Mathematical Representation and Physical Reality

It is perhaps useful to recall an earlier preview of my argument: People make models of the world, to some extent formal and mathematical. Rather than experimenting on the world, one experiments on or plays with the model to see if some generic phenomena appear when certain features are added or not, and so one discovers if these features and phenomena are fundamental to the model (and so, presumably, to the world).

Now, a variety of technical moves are made within the model, in computation or in formalism, to solve it and to derive its consequences. The mechanism within the model is usually taken to say something about the physical world and how it works. Here I have focused on how its technical "working" (Baxter's phrase) or detailed calculation, mathematical and formal though it be, might also be about that mechanism and about the physical world. So, if one needs quantities of a certain kind to solve a model, then those kinds or symmetries may be intrinsic to the physical model and to the physical situation as well, even if the symmetries are not at all apparent in the initial formulation of the model.

Modeling here is both substantive description and philosophical analysis, varying features to find out what matters and how. A model, drawn from a more general theory, if it can exhibit physically interesting phenomena, allows us to believe in and further understand the mechanism provided by that theory. But, also, we can discover what is crucial in that mechanism by dropping particular features and seeing if the model still does the work. And the mode of mathematical solution of that model may say something about how the mechanism does its work within the model.

I believe that what gives physicists confidence in this approach is that a rather small number of generic models keep reappearing in very nontrivial fashions (what I have called "the physicist's toolkit"),[45] and their technical details do eventually prove illuminating about the physics of a situation.[46] For our purposes, concerned as they are with the constitutions of matter, the collisions of billiard balls with each other, particle motion under Newton's laws, the harmonic oscillator, rotational and translational symmetry of a crystal lattice, the random walk, the sum of random variables and the central limit theorem, and the ideal gas as a sum of random variables prove to be some of those recurrent models.

Moreover, one discovers that if one sets up even more simplified models—toy models, models that can be readily calculated, often by modestly brute force—one finds that they lead to a deeper understanding of what is going on and so eventually to a rather more generic, not-so-toy model. Results and solutions of toy models suggest hints about how they are to be derived, exactly, from more realistic ones.

However, we also know that just because we have a formalism we have not at all settled the physics. Maxwell's equations were understood by Maxwell and the Maxwellians (in the last third of the nineteenth century) in a very different fashion (in terms of fluids and media) than they are now understood (in terms of particles and fields).[47]

A characteristic feature of all of these endeavors is the recurrent re-provings and rederivations of known results ("399th Solution of the Ising Model"). Proofs become more perspicuous, and what is essential about the models and so the constitutions of matter is highlighted. We come to see just what is going on in the model, and in the earlier derivations.[48] And it seems we learn to answer the question, "What more do we know about a *true* statement when we learn that it is *provable* in a particular theory?"[49] We learn the physics behind the truth. And this is not merely a phenomenon that appears in physical science. Here is a description of mathematical work.

> [In Husserl's philosophy] the necessity of iteration [re-proving of theorems, say] in a project of fulfillment of evidence [so discovering the

physics or the mathematics] is singled out for special description. The gradual development of the project of discovery in the light of an item [a theorem] beckoning for evidence is shown to fragment into a temporality of its own, not related to physical temporality. The imperfection, the lack of certainty, the insecurity of all evidence, are described by Husserl as constitutive features of the project of evidence.

Husserl's description is confirmed by some observations of mathematicians at work. Mathematicians are frequently frustrated by the experience of repeatedly verifying the correctness of every single step of a mathematical proof, without arriving at the [self-]evidence of the statement being proved. Such frustration is the rule rather than the exception. The deepest results of mathematics take centuries to be understood after the first proof is given. A proof that shines with the light of authentic evidence usually comes late after the discovery of a mathematical item. It is a fact that mathematicians do not rest after the first proof of a new item is given. Most of the research work of a mathematician goes into figuring out simpler and more conceptual proofs of already discovered items, until an [self-]evidence-yielding argument is found. Thus, whereas mathematicians may claim to be after the truth, their own work belies this claim. Their primary concern is not truth, but evidence. The gradualness of mathematical evidence, together with the search for evidence over and above truth. . . .[50]

Our task is to take the most ordinary of phenomena and so appreciate what is crucial about them now that we have re-presented the world back to ourselves in a more illuminating albeit sometimes more technical fashion. Perhaps that has been achieved in the most mundane and the most supramundane of arenas: ordinary bulk matter, and the universe. We now have a very good feel for why matter is stable and bulky, and why it might well make transitions of phase. And we now have a very good feel for how our universe has evolved since the big bang, the big bang itself and its immediate sequelae being taken as phase transitions. These re-presentations are of cosmological importance, for our cosmos so becomes meaningful. We have a way to think about how it is put together and maintains itself. We know better what is essential about it. I take these to be very great achievements, largely of our century. We might well be proud. Of course, pride goeth before the Fall. So it will be interesting to see how these cosmologies are transformed in subsequent generations by new empirical discoveries, deeper physics, and more abstract mathematics.

Original page numbers for the appendixes are reproduced in brackets at the top inside margin.

APPENDIX A

Reprinted from *Journal of Physical Chemistry*, volume 43, pp. 189–196. Published 1939 by the American Chemical Society.

ELECTROSTATIC INTERACTION OF MOLECULES[1]

LARS ONSAGER

Department of Chemistry, Yale University, New Haven, Connecticut

Received October 12, 1938

In the electrostatic theory of crystals and liquids the existence of an additive lower bound for the electrostatic energy is generally taken for granted as a very trivial matter, and it does not appear that a rigorous proof, even for a simplified model, has ever been attempted. For an assembly of ions or even electric dipoles, a computation of the Coulomb energy by direct summation meets with considerable difficulty because the convergence is conditional at best, and the sequence of summation must be considered with some care even in the case of a perfect crystal lattice. However, it is easy to show that for no assembly of ions is there an *additive upper bound* for the energy: Consider the cations stacked in one pile, and the anions in another!

The idea of a lower bound for the electrostatic energy *per se* cannot be separated from the assumption of a closest distance of approach between the centers of charges. We shall therefore restrict our considerations to "hard billiard ball" models of molecules and ions, and attempt no more refined representation of the short-range repulsive forces. There need be no restriction as to the shape of the "billiard balls"; only the hardness is essential, and the particles may have a dielectric constant, or even more general assumptions about their polarizability are admissible.

To obtain a lower bound for the electric energy of an assembly of such particles, let us imagine that we have on hand a large quantity of a continuous conducting fluid. According to electrostatics, the energy released by the immersion of a charged body in such a fluid is finite; we shall call it the *proper energy* of the particle i (species). Without risk of confusion, this proper energy may be denoted simply by u_i, for we shall have no other occasion to assign energies to individual particles. The energy of interaction between any two particles we shall denote by u_{ik}.

Consider all the particles of our assembly immersed one at a time in our conducting fluid. When all the particles are immersed, the energy of the whole system equals

$$- \Sigma u_i$$

[1] Presented at the Symposium on Intermolecular Action, held at Brown University, Providence, Rhode Island, December 27–29, 1938, under the auspices of the Division of Physical and Inorganic Chemistry of the American Chemical Society.

regardless of the arrangement of the particles. Now the removal of the fluid cannot release any more energy; on the contrary, some energy must in general be expended to accomplish it. Thus, however the particles may be arranged, we know a lower bound for the energy:

$$U = \Sigma u_{ik} \geqq - \Sigma u_i \tag{1}$$

For an ion of charge e and radius a, the proper energy is given by the familiar formula

$$u_i = e^2/2a$$

and equals the total Maxwell energy of the external field of the ion:

$$u_i = e^2/2a = \frac{1}{8\pi} \int\limits_{r>a} \mathrm{grad}^2 \, (e/r) \, 4\pi r^2 \, dr \tag{2}$$

For a permanent dipole of strength μ possessed by a particle of radius a and polarizability

$$\alpha = a^3(\epsilon - 1)/(\epsilon + 2) \tag{3}$$

we find

$$u_i = \mu^2/2(a^3 - \alpha) = (\epsilon + 2) \, \mu^2/6a^3$$

$$= \frac{1}{8\pi} \int\limits_{r>a} \mathrm{grad}^2 \, (\mu \cos \theta/r^2) \, dV + \frac{\epsilon}{8\pi} \int\limits_{r<a} \mathrm{grad}^2 \, (\mu r \cos \theta) \, dV \tag{4}$$

Here not only the external field energy

$$\mu^2/3a^3$$

of the electric dipole, but an additional amount

$$\epsilon\mu^2/6a^3 = (a^3 + 2\alpha)/6a^3(a^3 - \alpha)$$

of *internal energy* is released by the immersion of the particle in a conductor. Thereby the electric moment increases in the ratio

$$\mu_\infty/\mu_0 = (\epsilon + 2)/3 = a^3/(a^3 - \alpha) \tag{5}$$

In general, for a multipole described by a spherical harmonic of order n in a spherical particle of effective dielectric constant ϵ, the ratio of internal and external energies will be

$$u_{\mathrm{int}}/u_{\mathrm{ext}} = n\epsilon/(n + 1) \tag{6}$$

and the strength of the multipole increases by immersion, in the ratio

$$M^{(n)}_\infty/M^{(n)}_0 = (n\epsilon + n + 1)/(2n + 1) \tag{7}$$

The lower bound given by equation 1 cannot, in general, be realized. It is of interest to see how closely it can be approached by arrangements consistent with the models.

For a given type of ionic lattice the electric energy depends on the minimum distance between anion and cation, but the proper energies of the ions involve the radius of each ion. To render the question definite, we shall compare the lattice energy with least proper energy which is consistent with a given mean diameter r_0 of the ions:

$$a_+ + a_- = r_0 \tag{8}$$

For a compound, $C_n^{(m+)} A_m^{(n-)}$, the self energy is

$$\sum u_i = e^2 \left(n \frac{m^2}{2a_+} + m \frac{n^2}{2a_-} \right) \tag{9}$$

The minimum of this quantity under the restriction 8 is obtained when

$$\frac{e_+}{a_+^2} + \frac{e_-}{a_-^2} = \frac{m}{a_+^2} - \frac{n}{a_-^2} = 0 \tag{10}$$

and equals

$$\mathrm{Min}(\Sigma u_i) = mn(\sqrt{m} + \sqrt{n})^2 / 2r_0 \tag{11}$$

We obtain for ionic compounds CA, CA_2, and C_2A_3, respectively,

$$1.1(1 + 1)^2/2 = 2$$
$$1.2(1 + \sqrt{2})^2/2 = 3 + 2\sqrt{2} = 5.82843 \tag{12}$$
$$2.3(\sqrt{2} + \sqrt{3})^2/2 = 15 + 6\sqrt{6} = 29.6969$$

In table 1 the energies of some important lattices, taken partly from Born and Goeppert-Mayer (1) and partly from Sherman (4), are expressed in fractions of the least proper energies. Lattice types marked by an asterisk have adjustable parameters which have been chosen so as to give the lowest possible electrostatic energy.

Lattices made up of identical dipole molecules involve the complications of polarizability and of possible resultant electric moments.

On account of the polarizability, the moment of a molecule in the lattice will be

$$\mu = \mu_0 + \alpha E \tag{13}$$

where E is the electric field due to the other molecules. (In the lattices which we shall consider here, the directions of μ and E are parallel.) The field E is, of course, proportional to the actual moments μ, but the energy will be

$$\Sigma u_{ik} = -\tfrac{1}{2}NE\mu_0 \tag{14}$$

The ratio E/μ is given by the geometry of the lattice. We shall assume spherical molecules of radius a, and characterize the type of lattice by the dimensionless ratio γ, defined by the equation

$$E = \gamma\mu/a^3 \tag{15}$$

Then the lattice energy will be

$$U = \Sigma u_{ik} = -\tfrac{1}{2}N\mu_0^2\gamma/(a^3 - \gamma\alpha) \tag{16}$$

$$= -\Sigma u_i \times \gamma(a^3 - \alpha)/(a^3 - \gamma\alpha)$$

Here equation 1 implies $\gamma \leqq 1$. For "electret" arrangements in which the crystal has a net electric moment per unit volume, the electric field *outside the crystal* due to this moment will in general contain an energy comparable to the total proper energy of the molecules, but dependent on the shape of

TABLE 1

Energies of some important lattices expressed in fractions of the least proper energies

LATTICE	SYSTEM	COÖRDINATION NUMBER	$-\Sigma u_{ik}/\Sigma u_i$
Cesium chloride, CsCl...............	Cubic	8	0.8813
Sodium chloride, NaCl..............	Cubic	6	0.8738
Zinc blende, ZnS....................	Cubic	4	0.8190
Wurtzite, ZnS......................	Hexagonal	4	0.8197
Calcium fluoride, CaF$_2$..............	Cubic	4–8	0.8645
Rutile, TiO$_2$.......................	Tetragonal*	3–6	0.8263
Anatase, TiO$_2$......................	Tetragonal*	3–6	0.8236
Cuprite, Cu$_2$O......................	Cubic	2–4	0.7061
Corundum, Al$_2$O$_3$....................	Hexagonal*	4–6	0.8429

*Parameters adjusted to optimum values.

the crystal and negligible for a very long needle polarized lengthwise (5). Other ways to eliminate the external field are twinning—a common habit of ferromagnetic crystals—or the immersion in a conductor. Any such device which reduces the boundary effectively to an equipotential will lead to a periodic potential (not merely a periodic field) in the interior of the crystal, and to a minimum energy which will be referred to as the energy of the electret.

For certain particularly simple cubic lattices, known as *diagonal lattices* (1), the energy can be computed very simply from a classical consideration of symmetry. In such cases the energy equals

$$U = -\frac{4\pi}{3}N\mu^2/2 \tag{17}$$

where N is the number of dipoles per unit volume. The important characteristic of a diagonal lattice is that four trigonal axes of symmetry pass through every occupied point.   In table 2 the constant γ defined by equation 15 has been computed for some important electret lattices and for simple linear, quadratic, and cubic lattices of dipoles arranged so that the orientations of nearest neighbors are reflected in the line connecting them. The energies of the "reflected" lattices and the energy differences between hexagonal and cubic electrets have been computed by an adaptation of Madelung's method (2).   Some other lattices of alternating orientations have been discussed by Van Vleck (5).   Arrangements of alternating orientations have not been listed for coördination numbers greater than 6, because none has been found of energies lower than those of the corresponding electrets.   It seems a safe conjecture that the hexagonal close-packed lattice, completely polarized in the direction of the major axis, has a lower energy than any other arrangement of spheres carrying electric

TABLE 2

Values of γ computed for some lattices

LATTICE	COÖRDINATION NUMBER	γ^*	$\gamma\dagger$
A. Electrets:			
Close-packed cubic......................	12	$\pi\sqrt{2}/6 =$	0.7405
Close-packed hexagonal.................	12	0.7413	0.7400
Simple cubic............................	6	$\pi/6 =$	0.5236
Cube-centered.........................	8	$\pi\sqrt{3}/8 =$	0.6802
Diamond...............................	4	$\pi\sqrt{3}/16 =$	0.3401
Ice.....................................	4	0.3485	0.3359
B. Reflected orientations:			
Simple cubic...........................	6	0.6692	
Simple quadratic.......................	4	0.6374	0.3307
Linear..................................	2	0.601	0.225

* Dipoles oriented parallel to major axis or plane.
† Dipoles oriented perpendicular to major axis or plane.

dipoles.   However, if for some reason the dipoles are arranged in the 6-coördinated simple cubic lattice, or in one of the 4-coördinated lattices, then lower energies can be attained by "reflected" orientations.   For the simple cubic lattice one can easily show that arrangements of reflected orientation have the lowest possible energy.   The proof depends on the observation that the entire space of the crystal can be divided by plane equipotential surfaces into cubes, each containing one dipole at the center, so that the energy of the lattice will be the same as if the dipoles, however oriented, were enclosed in individual Faraday cages.   For the 4-coördinated diamond and ice lattices, it is also possible to find arrangements of lower energies than the electrets.   The value of γ for a zigzag string with tetrahedral angles is 0.43 when the dipoles are oriented lengthwise, and the *average* value of γ for all arrangements of such strings in a given lattice is necessarily the same; the maximum of γ for these lattices is probably not far from 0.53.

The comparison between electrets and other arrangements is of interest to the theory of dielectrics, because the temperature coefficient of the dielectric constant and the possibility of a Curie point depend on the relative energies of polarized and unpolarized states. The contrast between dense and loose packings may be general.

Tables 1 and 2 show that, in favorable cases, the lower bound given by equation 1 is a remarkably close one. It is easy to understand that the proper energies of ions can be released more completely than those of dipoles. The electrostatic energy of ions is all contained in their external fields, but in the case of dipoles at least one-third of the energy is internal, and even their external energy is more closely confined.

It has been shown that the fraction of the proper energy released by the formation of an ionic lattice depends on the ratio of the diameters of the ions. For the optimum, equation 10 requires equality of the *surface field intensities* of the ions. This condition is rather imperfectly realized in nature, for cations tend to be small while anions tend to be large, and more so the greater the valence of either. That highly charged ions get along better than they should on the basis of our considerations depends on the shortcomings of the ionic model when applied to cases which really represent a transition between ionic and homopolar interaction.

Chemical evidence indicates that in spite of the oversimplification of the model, the condition expressed by equation 10 has something to do with the relative stability and solubilities of ionic compounds, although in typical stable insoluble salts the cations have generally stronger fields than the anions. This may be just a matter of competition in a world of small cations and large anions, or else indicate a measure of homopolar character inherent in all ionic interaction.

The "matching condition" for the surface fields is at least capable of characterizing those liquids which are good solvents for electrolytes, namely, those polar liquids whose molecular surface fields are as intense as the fields of the ions. For a spherical dipole the maximum intensity of the surface field equals

$$E = 2\mu/a^3$$

If this is the decisive characteristic, then the solvent power for electrolytes is by no means a function of the dipole moment μ alone, nor of the proper energy

$$u_i \sim \mu^2/a^3$$

nor of the dielectric constant,

$$D \sim u_i/kT$$

Instead, among two liquids of the same dielectric constant, the one of lower molecular volume will generally have the greater ability to dissolve electrolytes.   However, it will be obvious that localization of the surface field will have the same effect as a small molecular volume. A comparison of these conclusions with experience would amount to a reiteration of the outstanding empirical rules.

Since the surface field of a molecule depends on induction from the environment, it seems worth while to indicate a scheme of description which eliminates some of the resultant complications.   There are many possible arrangements of charges in the interior of a particle which will give rise to the same external field, but there is exactly one distribution of charges over the surface which will produce a given field.   For an ion of spherical symmetry, charge e and radius a, the required charge density is obviously

$$\sigma = e/4\pi a^2 \tag{18}$$

A spherical dipole molecule carries the fixed charge distribution

$$\sigma = 3\mu_\infty \cos\theta/4\pi a^3 = 3\mu_0 \cos\theta/4\pi(a^3 - \alpha) \tag{19}$$

in the notation of equation 5.   In addition to these fixed charges, *there are induced charges whenever the surface is not an equipotential.*   For an isolated dipole molecule, the induced surface charge equals

$$\sigma_{\text{ind}} = -3\alpha\mu_\infty \cos\theta/4\pi a^6$$

so that the total electric moment is μ_0 instead of μ_∞.   The generalization to higher multipoles is obvious.

In this scheme of description the condition for minimum energy is the best possible compensation of neighboring surface charges.

As an illustration we may construct an electrostatic model of the water molecule which will account for the known structure of the ice lattice, including the type of randomness assumed by Pauling (3).   The shape of the molecule may be spherical, and the charge distribution may consist of fairly concentrated positive charges near two corners of an inscribed tetrahedron and similar negative charges near the other two corners.   If such molecules are arranged in a lattice of tetrahedral coördination and oriented so that positive and negative charges compensate each other at all points of contact, practically the whole proper energy of the molecules will be released.   According to Pauling's estimate, there are $(3/2)^N$ ways of orienting N molecules with matching of the charges, and from the agreement with the observed residual entropy of ice, he infers that those $(3/2)^N$ configurations have all about the same energy.   This hypothesis would indeed be true on the basis of our model, for when the compensation of adjacent charges is perfect, there are no stray fields left.

Incidentally, it is possible to show that the number of Pauling's "configurations" is larger than $(3/2)^N$, but a tentative estimate of the correction indicates that it is well within the experimental error of the residual entropy.

While our purely electrostatic model of a "proton bond" may need refinements, the general scheme would seem to accomodate them readily.

REFERENCES

(1) BORN, M., AND GOEPPERT-MAYER, M.: Handbuch der Physik, 2nd edition, XXIV/2. Julius Springer, Berlin (1933).
(2) MADELUNG: Physik. Z. 19, 524 (1918).
(3) PAULING, L.: J. Am. Chem. Soc. 57, 2680 (1935).
(4) SHERMAN, J.: Chem. Rev. 11, 93 (1932).
(5) VAN VLECK: J. Chem. Phys. 5, 556 (1937).

APPENDIX B

Reprinted with permission from *Physical Review*, volume 65 (February 1, 15, 1944), pp. 117–149. Copyright 1944 The American Physical Society.

Crystal Statistics. I. A Two-Dimensional Model with an Order-Disorder Transition

Lars Onsager

Sterling Chemistry Laboratory, Yale University, New Haven, Connecticut

(Received October 4, 1943)

The partition function of a two-dimensional "ferromagnetic" with scalar "spins" (Ising model) is computed rigorously for the case of vanishing field. The eigenwert problem involved in the corresponding computation for a long strip crystal of finite width (n atoms), joined straight to itself around a cylinder, is solved by direct product decomposition; in the special case $n = \infty$ an integral replaces a sum. The choice of different interaction energies ($\pm J$, $\pm J'$) in the (0 1) and (1 0) directions does not complicate the problem. The two-way infinite crystal has an order-disorder transition at a temperature $T = T_c$ given by the condition

$$\sinh(2J/kT_c)\sinh(2J'/kT_c) = 1.$$

The energy is a continuous function of T; but the specific heat becomes infinite as $-\log |T - T_c|$. For strips of finite width, the maximum of the specific heat increases linearly with $\log n$. The order-converting dual transformation invented by Kramers and Wannier effects a simple automorphism of the basis of the quaternion algebra which is natural to the problem in hand. In addition to the thermodynamic properties of the massive crystal, the free energy of a (0 1) boundary between areas of opposite order is computed; on this basis the mean ordered length of a strip crystal is

$$(\exp(2J/kT)\tanh(2J'/kT))^n.$$

INTRODUCTION

THE statistical theory of phase changes in solids and liquids involves formidable mathematical problems.

In dealing with transitions of the first order, computation of the partition functions of both phases by successive approximation may be adequate. In such cases it is to be expected that both functions will be analytic functions of the temperature, capable of extension beyond the transition point, so that good methods of approximating the functions may be expected to yield good results for their derivatives as well, and the heat of transition can be obtained from the difference of the latter. In this case, allowing the continuation of at least one phase into its metastable range, the heat of transition, the most appropriate measure of the discontinuity, may be considered to exist over a range of temperatures.

It is quite otherwise with the more subtle transitions which take place without the release of latent heat. These transitions are usually marked by the vanishing of a physical variable, often an asymmetry, which ceases to exist beyond the transition point. By definition, the strongest possible discontinuity involves the specific heat. Experimentally, several types are known. In the $\alpha - \beta$ quartz transition,[1] the specific heat becomes infinite as $(T_c - T)^{-\frac{1}{2}}$; this may be the rule for a great many structural transformations in crystals. On the other hand, supraconductors exhibit a clear-cut finite discontinuity of the specific heat, and the normal state can be continued at will below the transition

[1] H. Moser, Physik. Zeits. **37**, 737 (1936).

point by the application of a magnetic field.[2] Ferromagnetic Curie points[1, 3] and the transition of liquid helium to its suprafluid state[4] are apparently marked by essential singularities of the specific heats; in either case it is difficult to decide whether the specific heat itself or its temperature coefficient becomes infinite. Theoretically, for an ideal Bose gas the temperature coefficient of the specific heat should have a finite discontinuity at the temperature where "Einstein condensation" begins.

In every one of these cases the transition point is marked by a discontinuity which does not exist in the same form at any other temperature; for even when a phase can be continued beyond its normal range of stability, the transition in the extended range involves a finite heat effect.

Whenever the thermodynamic functions have an essential singularity any computation by successive approximation is difficult because the convergence of approximation by analytic functions in such cases is notoriously slow.

When the existing dearth of suitable mathematical methods is considered, it becomes a matter of interest to investigate models, however far removed from natural crystals, which yield to exact computation and exhibit transition points.

It is known that no model which is infinite in one dimension only can have any transition.[5, 6]

The two-dimensional "Ising model," originally intended as a model of a ferromagnetic,[7] is known to be more properly representative of condensation phenomena in the two-dimensional systems formed by the adsorption of gases on the surfaces of crystals.[8] It is known that this model should have a transition.[9] Recently, the transition point has been located by Kramers and Wannier.[10] In the following, the partition func-

tion and derived thermodynamic functions will be computed for a rectangular lattice with different interaction constants in two perpendicular directions; this generalization is of some interest and does not add any difficulties.

The similar computation of the partition functions for hexagonal and honeycomb lattices involves but relatively simple additional considerations. A general form of the dual transformation invented by Kramers and Wannier together with a rather obvious "star-triangle" transformation are used in this connection, and the transition points can be computed from the transformations alone. While these were a great help to the discovery of the present results, the logical development of the latter is best relieved of such extraneous topics, and the subject of transformations will be reserved for later communication.

OUTLINE OF METHOD

As shown by Kramers and Wannier,[10] the computation of the partition function can be reduced to an eigenwert problem. This method will be employed in the following, but it will be convenient to emphasize the abstract properties of relatively simple operators rather than their explicit representation by unwieldy matrices.

For an introduction to the language we shall start with the well-known problem of the linear lattice and proceed from that to a rectangular lattice on a finite base. For symmetry we shall wrap the latter on a cylinder, only straight rather than in the screw arrangement preferred by Kramers and Wannier.

The special properties of the operators involved in this problem allow their expansion as linear combinations of the generating basis elements of an algebra which can be decomposed into direct products of quaternion algebras. The representation of the operators in question can be reduced accordingly to a sum of direct products of two-dimensional representations, and the roots of the secular equation for the problem in hand are obtained as products of the roots of certain quadratic equations. To find all the roots requires complete reduction, which is best performed by the explicit construction of a transforming matrix, with valuable by-products of

[2] W. H. Keesom, Zeits. f. tech. Physik 15, 515 (1934); H. G. Smith and R. O. Wilhelm, Rev. Mod. Phys. 7, 237 (1935).
[3] H. Klinkhardt, Ann. d. Physik [4], 84, 167 (1927).
[4] W. H. and A. P. Keesom, Physica 2, 557 (1935); J. Satterly, Rev. Mod. Phys. 8, 347 (1938); K. Darrow, ibid. 12, 257 (1940).
[5] E. Montroll, J. Chem. Phys. 9, 706 (1941).
[6] K. Herzfeld and M. Goeppert-Mayer, J. Chem. Phys. 2, 38 (1934).
[7] E. Ising, Zeits. f. Physik 31, 253 (1925).
[8] R. Peierls, Proc. Camb. Phil. Soc. 32, 471 (1936).
[9] R. Peierls, Proc. Camb. Phil. Soc. 32, 477 (1936).
[10] H. A. Kramers and G. H. Wannier, Phys. Rev. 60, 252, 263 (1941).

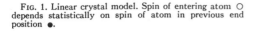

Fig. 1. Linear crystal model. Spin of entering atom ○ depends statistically on spin of atom in previous end position ●.

identities useful for the computation of averages pertaining to the crystal. This important but formidable undertaking will be reserved for a later communication. It so happens that the representations of maximal dimension, which contain the two largest roots, are identified with ease from simple general properties of the operators and their representative matrices. The largest roots whose eigenvectors satisfy certain special conditions can be found by a moderate elaboration of the procedure; these results will suffice for a qualitative investigation of the spectrum. To determine the thermo-dynamic properties of the model it suffices to compute the largest root of the secular equation as a function of the temperature.

The passage to the limiting case of an infinite base involves merely the substitution of integrals for sums. The integrals are simplified by elliptic substitutions, whereby the symmetrical parameter of Kramers and Wannier[10] appears in the modulus of the elliptic functions. The modulus equals unity at the "Curie point"; the consequent logarithmic infinity of the specific heat confirms a conjecture made by Kramers and Wannier. Their conjecture regarding the variation of the maximum specific heat with the size of the crystal base is also confirmed.

THE LINEAR ISING MODEL

As an introduction to our subsequent notation we shall consider a one-dimensional model which has been dealt with by many previous authors.[7,10] (See Fig. 1.)

Consider a chain of atoms where each atom (k) possesses an internal coordinate $\mu^{(k)}$ which may take the values ± 1. The interaction energy between each pair of successive atoms depends on whether their internal coordinates are alike or different:

$$u(1, 1) = u(-1, -1) = -u(1, -1)$$
$$= -u(-1, 1) = -J,$$

or

$$u(\mu^{(k)}, \mu^{(k-1)}) = -J\mu^{(k)}\mu^{(k-1)}. \quad (1)$$

The partition function of this one-dimensional crystal equals

$$e^{-F/kT} = Q(T) \quad (2)$$
$$= \sum_{\mu^{(1)}, \mu^{(2)}, \cdots, \mu^{(N)} = \pm 1} \exp(-\sum_k u(\mu^{(k)}, \mu^{(k-1)})/kT).$$

Following Kramers and Wannier,[10] we express this sum in terms of a matrix whose elements are

$$(\mu \mid V \mid \mu') = e^{-u(\mu, \mu')/kT}. \quad (3)$$

Then the summation over $(\mu^{(2)}, \cdots, \mu^{(N-1)})$ in (2) is equivalent to matrix multiplication, and we obtain

$$Q(T) = \sum_{\mu_1, \mu_N} (\mu^{(1)} \mid V^{N-1} \mid \mu^{(N)}) = \lambda_m{}^N O(1), \quad (4)$$

where λ_m is the greatest characteristic number of the matrix V. Since all the elements of this matrix are positive, λ_m will be a simple root of the secular equation

$$\mid V - \lambda \mid = 0,$$

and the identity

$$\lambda_m \psi_m(\mu) = (V, \psi_m(\mu)) \quad (5)$$

will be satisfied by a function (eigenvector) $\psi_m(\mu)$ which does not change sign.[11] In the present case we have

$$\psi_m(1) = \psi_m(-1)$$
$$\lambda_m = 2 \cosh(J/kT). \quad (6)$$

It will be convenient to use the abbreviation

$$J/kT = H. \quad (7)$$

The matrix V, written explicitly

$$V = \begin{pmatrix} e^H, & e^{-H} \\ e^{-H}, & e^H \end{pmatrix}, \quad (8)$$

represents an operator on functions $\psi(\mu)$, which has the effect

$$(V, \psi(\mu)) = e^H \psi(\mu) + e^{-H} \psi(-\mu). \quad (9)$$

This operator is therefore a function of the "complementary" operator C which replaces μ by $-\mu$

$$(C, \psi(\mu)) = \psi(-\mu); \quad C^2 = 1, \quad (10)$$

---

[11] S. B. Frobenius, Preuss. Akad. Wiss. pp. 514–518 (1909); R. Oldenberger, Duke Math. J. 6, 357 (1940). See also E. Montroll, reference 5.

FIG. 2. Cylindrical crystal model. Configuration of added tier—○—○—○— depends on that of tier —●—●—●— in previous marginal position.

and whose eigenfunctions are the (two) different powers of μ

$$(C, 1) = 1, \quad (C, \mu) = -\mu, \quad (11)$$

with the characteristic numbers $+1$ and -1, respectively. The linearized expression of V in terms of C is evident from (9), which is now written more compactly

$$V = e^H + e^{-H}C. \quad (12)$$

The eigenfunctions of V are of course the same as those of C, given by (11); the corresponding characteristic numbers are

$$\lambda_m = \lambda_+ = e^H + e^{-H} = 2 \cosh H$$
$$\lambda_- = e^H - e^{-H} = 2 \sinh H. \quad (13)$$

We shall deal frequently with *products* of many operators of the type (12). It is therefore a matter of great interest to linearize the *logarithm* of such an operator. Following Kramers and Wannier,[10] we introduce the function of H

$$H^* = \tfrac{1}{2} \log \coth H = \tanh^{-1} (e^{-2H}), \quad (14)$$

and obtain

$$V = e^H(1 + e^{-2H}C) = e^H(1 \perp (\tanh H^*)C),$$

which may be written

$$V = (e^H/\cosh H^*) \exp (H^*C)$$
$$= (2 \sinh 2H)^{\frac{1}{2}} \exp (H^*C). \quad (15)$$

The first, less symmetrical form serves to keep a record of the sign when H is not a positive real number. For example,

given by Kramers and Wannier.

$$\sinh 2H \sinh 2H^* = \cosh 2H \tanh 2H^*$$
$$= \tanh 2H \cosh 2H^* = 1. \quad (17)$$

Alternatively, we can express the hyperbolic functions in terms of the gudermannian angle

$$\mathrm{gd}\ u \equiv 2 \tan^{-1}(e^u) - \tfrac{1}{2}\pi = 2 \tan^{-1} (\tanh \tfrac{1}{2}u), \quad (18)$$

whereby

$$\sinh u \cot(\mathrm{gd}\ u) = \coth u \sin(\mathrm{gd}\ u)$$
$$= \cosh u \cos(\mathrm{gd}\ u) = 1. \quad (19)$$

Then the relation between H and H^* may be written in the form

$$\mathrm{gd}\ 2H + \mathrm{gd}\ 2H^* = \tfrac{1}{2}\pi. \quad (20)$$

THE RECTANGULAR ISING MODEL

We consider n parallel chains of the same type as before. We shall build up these chains simultaneously, tier by tier, adding one atom to each chain in one step (see Fig. 2). With the last atom in the jth chain is associated the variable μ_j, capable of the values ± 1. A complete set of values assigned to the n variables describes a configuration $(\mu) = (\mu_1, \cdots, \mu_n)$ of the last tier of atoms. The operator which describes the addition of a new tier with the interaction energy

$$u_1((\mu), (\mu')) = -\sum_{j=1}^{n} J\mu_j\mu_j' = -kTH\sum\mu_j\mu_j' \quad (21)$$

is now

$$V_1 = \prod_{j=1}^{n} (e^H + e^{-H}C_j)$$
$$= (2 \sinh 2H)^{n/2} \exp(H^*B), \quad (22)$$

whereby

$$B \equiv \sum_{j=1}^{n} C_j, \quad (23)$$

and the individual operators $C_1, \cdots, C_n$ have the effect

$$(C_j, \psi(\mu_1, \cdots, \mu_j, \cdots, \mu_n))$$
$$= \psi(\mu_1, \cdots, \mu_{j-1}, -\mu_j, \mu_{j+1}, \cdots, \mu_n). \quad (24)$$

Next, let us assume a similar interaction between

that the atoms in a tier are numbered modulo n,

$$\mu_{j+n} = \mu_j,$$

the total tierwise interaction energy will be

$$u_2(\mu_1, \cdots, \mu_n)$$

$$= -J' \sum_{j=1}^{n} \mu_j \mu_{j+1} = -kTH' \sum_{j=1}^{n} \mu_j \mu_{j+1}. \quad (25)$$

The effect of this interaction is simply to multiply the general term of the partition function, represented by one of the 2^n vector components $\psi(\mu_1, \cdots, \mu_n)$, by the appropriate factor $\exp(-u_2(\mu_1, \cdots, \mu_n)/kT)$; the corresponding operator V_2 has a diagonal matrix in this representation. It can be constructed from the simple operators $s_1, \cdots, s_n$ which multiply ψ by its first, $\cdots$, nth argument, as follows

$$(s_j, \psi(\mu_1, \cdots, \mu_n)) = \mu_j \psi(\mu_1, \cdots, \mu_n) \quad (26)$$

$$A = \sum_{j=1}^{n} s_j s_{j+1}, \quad (27)$$

$$V_2 = \exp(H'A). \quad (28)$$

We now imagine that we build the crystal in alternate steps. We first add a new tier of atoms (see Fig. 3), next introduce interaction among atoms in the same tier, after that we add another tier, etc. The alternate modifications of the partition function are represented by the product

$$\cdots V_2 \, V_1 \, V_2 \, V_1 \, V_2 \, V_1$$

with alternating factors; the addition of one tier of atoms with interaction both ways is represented by $(V_2 V_1) = V$. The eigenwert problem which yields the partition function of the crystal is accordingly

$$\lambda \psi = (V, \psi) = (V_2 V_1, \psi)$$

$$= (2 \sinh 2H)^{n/2} (\exp(H'A) \exp(H^*B), \psi). \quad (29)$$

Viz., in the more explicit matrix notation used by previous authors[12]

$$\lambda \psi(\mu_1 \cdots \mu_n) = \exp\left(H' \sum_{1}^{n} \mu_j \mu_{j+1} \right)$$

$$\times \sum_{\mu_1' \cdots \mu_n' = \pm 1} \exp\left(H \sum_{1}^{n} \mu_j \mu_j' \right) \psi(\mu_1' \cdots \mu_n').$$

[12] See H. A. Kramers and G. H. Wannier (reference 10). E. Montroll (reference 5); E. N. Lassettre and J. P. Howe, J. Chem. Phys. 9, 747 (1941).

This 2^n-dimensional problem we shall reduce to certain sets of 2-dimensional problems. So far, we have expressed A and B rather simply by the generating elements $s_1, C_1, \cdots, s_n, C_n$ of a certain abstract algebra, which is easily recognized as a direct product of n quaternion algebras. However, the expansion (27) of A is not linear. Following a preliminary study of the basis group, we shall therefore construct a new basis for a part of the present algebra, which allows linear expansions of both A and B in terms of its generating elements.

QUATERNION ALGEBRA

We have introduced two sets of linear operators $s_1, \cdots, s_n$ and $C_1, \cdots, C_n$ by the definitions

$$(s_j, \psi(\mu_1, \cdots, \mu_j, \cdots, \mu_n))$$
$$= \mu_j \psi(\mu_1, \cdots, \mu_j, \cdots, \mu_n),$$
$$(C_j, \psi(\mu_1, \cdots, \mu_j, \cdots, \mu_n)) \quad (30)$$
$$= \psi(\mu_1, \cdots, -\mu_j, \cdots, \mu_n).$$

We shall see that the group generated by these operators forms a complete basis for the algebra of linear operators in the 2^n-dimensional vector space of the functions $\psi(\mu_1, \cdots, \mu_n)$. Their abstract properties are

$$s_j^2 = C_j^2 = 1, \quad s_j C_j = -C_j s_j;$$
$$s_j s_k = s_k s_j, \quad C_j C_k = C_k C_j; \quad (31)$$
$$s_j C_k = C_k s_j; \quad (j \neq k).$$

The algebra based on one of the subgroups $(1, s_j, C_j, s_j C_j)$ is a simple quaternion algebra. The group of its basis elements includes $(-1, -s_j, -C_j, C_j s_j)$ as well; but this completion of the group adds no new independent element for the purpose of constructing a basis. If we disregard

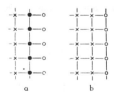

FIG. 3. Two-step extension of a two-dimensional crystal. (a) A new tier of atoms $\bigcirc$ is added (V_1); their configuration depends on that of atoms $\bullet$ in previous marginal position. (b) Interaction energy between marginal atoms $\bigcirc$ is introduced (V_2), which modifies the distribution of configurations in this tier of atoms.

the sign, each element is its own inverse. Except for the invariant element (-1) and the unit proper, each element together with its negative constitutes a class. This algebra has just one irreducible representation

$$\mathbf{D}_2(1)=\begin{pmatrix}1&0\\0&1\end{pmatrix}; \qquad \mathbf{D}_2(s_j)=\begin{pmatrix}1&0\\0&-1\end{pmatrix};$$

$$\mathbf{D}_2(C_j)=\begin{pmatrix}0&1\\1&0\end{pmatrix}; \qquad \mathbf{D}_2(s_jC_j)=\begin{pmatrix}0&1\\-1&0\end{pmatrix}.$$

Our algebra of operators in 2^n-dimensional space can be described accordingly as the direct product of n simple quaternion algebras, which is still referred to as a (general) quaternion algebra. Each one of its 4^n independent basis elements can be expressed uniquely as a product of n factors, one from each set $(1, s_j, C_j, s_jC_j)$. Still, each element furnishes its own inverse and, except for (-1) and the unit, each element with its negative constitutes a class, which immediately characterizes all possible representations of the algebra: *The characters of all basis elements with the sole exception of the unit must vanish.* This of course is a consequence of the direct product construction. Another is the recognition that the algebra has only one irreducible representation

$$\mathbf{D}_2\times\mathbf{D}_2\times\cdots\times\mathbf{D}_2=\mathbf{D}_{2^n}$$

which is of 2^n dimensions; our definitions (30) describe an explicit version of it.

With the customary restriction that the elements of the basis group are to be represented by unitary matrices, the representations of the 4^n independent basis elements must be orthogonal to each other (in the hermitian sense) on the 2^{2n} (or more) matrix elements as a vector basis. For the vector product in question is nothing but

$$\sum_{r,s=1}^{2^n} D(Q_1)_{rs}\overline{D}(Q_2)_{rs}$$
$$=\text{Tr.}\ (\mathbf{D}(Q_1Q_2^{-1}))=\text{Tr.}\ (\mathbf{D}(Q_3)),$$

and if Q_1 and Q_2 are independent elements of the basis group, then their quotient Q_3 is an element different from the unit and (-1), so that its character vanishes.

Since orthogonality guarantees linear independence, the irreducible 2^n-dimensional representation of our 4^n independent basis elements

will suffice for the construction of any $2^n\times 2^n$ matrix by linear combination; *our quaternion algebra is equivalent to the complete matrix algebra of 2^n dimensions.* The choice between the two forms is a matter of convenience; the quaternion description seems more natural to the problem in hand. One of the main results obtained by Kramers and Wannier will be derived below; the comparison with their derivation is instructive.

In the following, the entire basis group will not be needed. Since our operator A involves only products of pairs of the generating elements $s_1, \cdots, s_n$ and B does not involve any of these but only $C_1, \cdots, C_n$, we shall have to deal only with that subgroup of the basis which consists of elements containing an *even* number of factors s_j. This "even" subgroup, which comprises half the previous basis, may be characterized equally well by the observation that it consists of those basis elements which commute with the operator

$$C=C_1C_2\cdots C_n \qquad (32)$$

which reverses all spins at once

$$(C, \psi(\mu_1, \cdots, \mu_n))=\psi(-\mu_1, \cdots, -\mu_n). \quad (32a)$$

The representation (30) is now reducible because the 2^{n-1} *even* functions satisfying

$$(C, \psi)=\psi$$

are transformed among themselves as are the *odd* functions satisfying

$$(C, \psi)=-\psi.$$

The corresponding algebras are both of the same type as before, with $(n-1)$ pairs of generating elements

$$s_1s_n, s_2s_n, \cdots, s_{n-1}s_n; \quad C_1, C_2, \cdots, C_{n-1}$$

and either algebra is equivalent to the complete matrix algebra of 2^{n-1} dimensions. The previously independent operator C_n is now expressible by the others because either $C=1$ or $C=-1$, according to whether we are dealing with even or odd functions. This choice of sign determines the only difference between the two versions of the algebra in question. Moreover, the operator

$$A=\sum_{i=1}^{n}s_js_{j+1}$$

is the same in both versions, but obviously

$$B = \sum_{j=1}^{n} C_j$$

is not; here the interpretation of C_n makes a difference.

The arbitrary elimination of one of the sets of operators which enter symmetrically into the problem causes much inconvenience. The symmetry can be retained if we start with the two series of operators

$$s_1 s_2, \ s_2 s_3, \ \cdots, \ s_n s_1; \quad C_1, \ C_2, \ \cdots, \ C_n \quad (33)$$

required to satisfy the identities

$$s_1 s_2 s_2 s_3 \cdots s_n s_1 = s_1{}^2 s_2{}^2 \cdots s_n{}^2 = 1$$

$$C_1 C_2 \cdots C_n = C = \begin{cases} 1; & \text{(even functions)} \\ -1; & \text{(odd functions).} \end{cases} \quad (34)$$

The commutation rules are simply described when the two sets are arranged in the interlacing cyclic sequence

$$s_n s_1, \ C_1, \ s_1 s_2, \ C_2, \ \cdots, \ s_{n-1} s_n, \ C_n, \ (s_n s_1, \ \cdots). \quad (35)$$

Here each element anticommutes with its neighbors and commutes with all other elements in the sequence.

Dual Transformation

The fundamental algebraic reason for the dual relation found by Kramers and Wannier is now readily recognized: *In the representation which belongs to even functions*, the two sets (33) are alike because the starting point of the sequence (35) is immaterial. Neither does it matter if the sequence is reversed, so that the general rules of computation remain unchanged if each member of the sequence (35) is replaced by the corresponding member of the sequence

$$C_k, \ s_k s_{k \pm 1}, \ C_{k \pm 1}, \ s_{k \pm 1} s_{k \pm 2}, \ \cdots.$$

Accordingly, the correspondence

$$s_j s_{j+1} \to C_{k \pm j}, \quad C_j \to s_{k \pm (j-1)} s_{k \pm} \quad (36)$$

describes an automorphism of the whole algebra. Under this automorphism we have in particular

$$A \to B$$
$$B \to A$$

or, compactly

$$f(A, B) \to f(B, A),$$

all in the algebra of even functions. Similarity can be inferred from the additional observation that the characters of all representations are determined by the commutation rules, or equally well from the theorem that every automorphism of a complete matrix algebra is an internal automorphism. When we return to our original mixed algebra of even and odd functions, our result takes the form

$$(1+C)f(A, B) \sim (1+C)f(B, A). \quad (37)$$

Application to the operator V given by (29) shows that the two operators

$$(\sinh 2H)^{-n/2}(1+C)V(H', H)$$
$$= 2^{-n/2}(1+C)\exp(H'A)\exp(H^*B),$$
$$(\sinh 2H')^{n/2}(1+C)V'(H^*, H'^*)$$
$$= 2^{-n/2}(1+C)\exp(H'B)\exp(H^*A)$$

are similar, so that the part of the spectrum of $V(H^*, H'^*)$ which belongs to even functions can be obtained from the corresponding part of the spectrum of $V(H', H)$ by the relation

$$\lambda_+(H^*, H'^*)$$
$$= (\sinh 2H \sinh 2H')^{-n/2}\lambda_1(H', H). \quad (38)$$

This result applies in particular to the largest characteristic numbers of the operators in question because both can be represented by matrices with all positive elements so that, by Frobenius' theorem,[11] the corresponding eigenvectors cannot be odd.

In the "screw" construction of a crystal preferred by Kramers and Wannier,[10] the addition of one atom is accomplished by the operator ($\mathfrak{M}$ in their notation)

$$\exp(Hs_{n-1}s_n)(\sinh 2H)^{\frac{1}{2}}\exp(H^*C_n)T$$
$$= (\sinh 2H)^{\frac{1}{2}}\exp(Hs_{n-1}s_n)T\exp(H^*C_1),$$

where T is a permutation operator which turns the screw through the angle $2\pi/n$. They indicate in explicit matrix form a transformation which effects one of the automorphisms (36), namely,

$$s_j s_{j+1} \to C_{n-j}, \quad C_j \to s_{n-j+1}s_{n-j}$$

whereby naturally

$$T \to T^{-1}.$$

The equivalent of (38) follows. In consequence, if the crystal is disordered at high temperatures (small H), a transition must occur at the temperature where the equality

$$H = H^* = \tfrac{1}{2} \log \cot \pi/8 \qquad (39a)$$

is satisfied.

Unfortunately, our simplified derivation fails to exhibit one of the most important properties of the "dual" transformation: When applied to large two-dimensional (∞^2) crystal models, it *always* converts order into disorder and vice versa; this result is readily obtained by detailed inspection of the matrix representation. The consequent determination of the transition point for the rectangular crystal by the generalization of (39a)

$$H' = H^*; \quad (kT = kT_c) \qquad (39b)$$

will be verified in the following by other means.

On the other hand, the given results are still but special applications of a transformation which applies to all orientable graphs, not just rectangular nets. Except for its "order-converting" property, this transformation is best obtained and discussed in simple topological terms. The properties of the "dual" and certain other transformations will be given due attention in a separate treatise; it would be too much of a digression to carry the subject further at this point.

Fundamental Vector Systems

We shall give brief consideration to the vector basis of our quaternion algebra, as described by the operator formulation (30). The systems of eigenvector which belong to the two sets of operators $(s_1, \cdots, s_n)$ and $(C_1, \cdots, C_n)$ are easily identified. The representations (30) of the former are already diagonal, so that their eigenvectors are δ functions

$$(s_j, \delta(\mu_1 - \mu_1', \cdots, \mu_j - \mu_j', \cdots, \mu_n - \mu_n'))$$

$$= \mu_j' \delta(\mu_1 - \mu_1', \cdots, \mu_j - \mu_j', \cdots, \mu_n - \mu_n'). \qquad (40)$$

The eigenvectors of $(C_1, \cdots, C_n)$ are the different products of $\mu_1, \cdots, \mu_n$, with the normalizing factor $2^{-n/2}$

$$(C_j, (\mu_1^{\nu_1} \cdots \mu_n^{\nu_n})) = (-)^{\nu_j}(\mu_1^{\nu_1} \cdots \mu_n^{\nu_n}). \qquad (41)$$

The explicit representation of a transformation which leads from one of these systems to the other is obtained from the expansion of the identity

$$\delta(\mu_1 - \mu_1', \cdots, \mu_n - \mu_n')$$

$$= 2^{-n}(1 + \mu_1'\mu_1) \cdots (1 + \mu_n'\mu_n). \qquad (42)$$

This product description of a δ function is often more convenient. The two systems (40) and (41) yield diagonal representations of the operators A and B, respectively. The construction of the latter is a little simpler. With a slight variation of the notation we find

$$(B, (\mu_{j_1}\mu_{j_2}\cdots\mu_{j_k})) = (n - 2k)(\mu_{j_1}\mu_{j_2}\cdots\mu_{j_k}), \qquad (43a)$$

so that any homogeneous function of $(\mu_1, \cdots, \mu_n)$ is an eigenvector of B. With no repeated factors allowed, the number of independent homogeneous functions of degree k is the binomial coefficient $\binom{n}{k}$. Accordingly, the secular polynomial is

$$|B - \lambda| \equiv \prod_{k=0}^{n} (n - 2k - \lambda)^{\binom{n}{k}}. \qquad (43b)$$

Whereas B measures the degree of a function $\psi(\mu_1, \cdots, \mu_n)$, the operator A simply counts alternations of sign in the configuration $(\mu_1, \cdots, \mu_n)$. The analog of (43a) is now

$$\left(A, \left(\prod_1^{i_1} (1 + \mu_j) \prod_{i_1+1}^{i_2} (1 - \mu_j) \cdots \prod_{i_{2k}+1}^{n} (1 + \mu_j)\right)\right)$$

$$= (n - 4k)\left(\prod_1^{i_1} (1 + \mu_j) \prod_{i_1+1}^{i_2} (1 - \mu_j) \cdots \right.$$

$$\left. \times \prod_{i_{2k}+1}^{n} (1 + \mu_j)\right). \qquad (44a)$$

The alternations of sign are located arbitrarily behind $j_1, j_2, \cdots j_{2k}$; but their total number must be even $= 2k$. On the other hand, the result (44a) is equally valid when μ_j is replaced throughout by $-\mu_j$; so that the spectrum of A repeats in duplicate that part of the spectrum of B which belongs to even functions, and its secular polynomial is

$$|A - \lambda| \equiv \prod_{k=0}^{[n/2]} (n - 4k - \lambda)^{2\binom{n}{2k}}. \qquad (44b)$$

By an obvious transformation, one-half of the

paired characteristic numbers of A may be assigned to even functions and the other, identical half to odd functions. We note again that the three operators

$$(1-C)A \sim (1+C)A \sim (1+C)B$$

are similar, whereas $(1-C)B$ has a different structure.

Some Important Elements of the Basis

As a preliminary to the simultaneous reduction of the representation of the operators A and B, we shall consider certain elements of the basis group which will recur frequently in the process and introduce a special notation for them. We have already encountered

$$P_{aa} = -C_a, \quad P_{a,a+1} = s_a s_{a+1}.$$

In general we define P_{ab} as the product

$$P_{ab} = s_a C_{a+1} C_{a+2} \cdots C_{b-1} s_b;$$
$$(a, b = 1, 2, \cdots, 2n). \quad (45)$$

These operators are elements of the "even" basis and commute with C. They satisfy the recurrence formula

$$P_{a,b+1} = P_{ab} P_{bb} P_{b,b+1} = P_{ab} s_b C_b s_{b+1}. \quad (45a)$$

The period of this recurrence formula is $2n$; increasing either index by n has the effect of a multiplication by $(-C)$

$$P_{a,b+n} = P_{a+n,b} = -CP_{ab} = -P_{ab}C. \quad (46)$$

The product (45) obviously may be written in reverse order when $(b-a) \leq n$, and by (46), this result is easily extended to all (a, b). Accordingly,

$$P_{ab}^2 = s_a C_{a+1} \cdots C_{b-1} s_b s_b C_{b-1} \cdots C_{a+1} s_a = 1. \quad (47)$$

More generally, we thus obtain the "rule of the wild index"

$$P_{ac} P_{bc} = P_{ad} P_{bd}, \quad P_{ac} P_{ad} = P_{bc} P_{bd}. \quad (48)$$

The analogous result for indices congruent $(\mathrm{mod} = n)$ is obtained immediately by combination with (46)

$$P_{ac} P_{b,c+n} = P_{ad} P_{b,d+n} = P_{ad} P_{b+n,d},$$
$$P_{ac} P_{a+n,d} = P_{bc} P_{b+n,d} = P_{bc} P_{b,d+n}. \quad (49)$$

While the period of the recurrence formula (45a)

is $2n$, it is evident from (46) that the period of the commutation rules will be just n; for the factor C commutes with all the operators P_{ab}. For this reason we shall need a modified Kronecker symbol with the period n, as follows

$$D(0) = D(n) = 1, \quad D(m) \sin(m\pi/n) = 0. \quad (50)$$

Indices which differ by 0 or n $(\mathrm{mod} = 2n)$ will be called *congruent*; (in the former case we call them equal). Only the congruence of *corresponding* indices matters; no important identity results from the accident that a front index coincides with a rear index. The distinction between front and rear indices represents the distinction between the opposite directions of rotation $(1, 2, 3, \cdots, n)$ and $(1, n, n-1, \cdots, 2)$ around the base of the "crystal."

The effect of simultaneous congruence is evident from (47) together with (46); one obtains either the unit (1) or the invariant basis element $(-C)$. More generally, a product of the form

$$P_{aa'} P_{bb'} \cdots P_{kk'}$$

with any number of factors equals one of the invariant elements ± 1, $\pm C$ if both front and rear indices congruent to each of the possible values $1, 2, \cdots, n$ occur an even number $(0, 2, 4, \cdots)$ of times. This rule can be obtained from (48) by simple manipulation; compare the derivation of exchange rules below. In addition, thanks to the relation

$$P_{11} P_{22} \cdots P_{nn} = (-)^n C_1 C_2 \cdots C_n = (-)^n C, \quad (51)$$

an invariant element also results if both front and rear indices congruent to each of the numbers $1, 2, \cdots, n$ occur an *odd* number of times. A count of the group elements shows that there can be no further identities of this sort; accordingly, if a product satisfies neither of the given conditions, it must be a basis element different from ± 1 and $\pm C$.

The commutation rules for the operators P_{ab} can be broken up into rules which govern the exchange of indices between adjacent factors in a product. These in turn can be obtained by the rule of the wild index from the commutation rules which involve $P_{aa} = -C_a$. We first consider the product

$$P_{ab} P_{bb} = -P_{ab} C_b = -s_a C_{a+1} \cdots C_{b-1} s_b C_b.$$

When $a \equiv b$, the commutative character of the multiplication is trivial. Otherwise, there is only one factor s_b in P_{ab}; so that this operator anticommutes with $C_b = -P_{bb}$. Accordingly,

$$P_{ab}P_{bb} + (-)^{D(a-b)}P_{bb}P_{ab} = 0.$$

In view of (48) this may be written equally well

$$P_{ac}P_{bc} + (-)^{D(a-b)}P_{bc}P_{ac} = 0.$$

However, the result does not depend on the presence of a wild index, because

$$P_{ac}P_{bd} = P_{ac}P_{ad}P_{ad}P_{bd} = P_{bc}P_{bd}P_{ad}P_{bd}$$
$$= -(-)^{D(a-b)}P_{bc}P_{ad}P_{bd}^2.$$

The same reasoning applies to the exchange of rear indices, and we obtain the general rule

$$P_{ac}P_{bd} = -(-)^{D(a-b)}P_{bc}P_{ad} = -(-)^{D(c-d)}P_{ad}P_{bc}$$
$$= (-)^{D(a-b)+D(c-d)}P_{bd}P_{ac}. \quad (52)$$

Accordingly, the exchange of indices between adjacent factors in a product takes place with change of sign unless the trivial exchange of congruent (wild) indices is involved.

The product $P_{ac}P_{bd}$ is seen to be commutative when neither $(a-b)$ nor $(c-d)$ divides n, and in the trivial case when both do. When one and only one congruence exists, the factors anticommute.

For future reference we give here the direct operational definition of P_{ab}. In the case

$$\cot(a-\tfrac{1}{2})\pi/n > \cot(b-\tfrac{1}{2})\pi/n$$

we have

$$(P_{ab}, \psi(\mu_1, \cdots, \mu_n)) = \mu_a\mu_b\psi(\mu_1, \cdots, \mu_a,$$
$$-\mu_{a+1}, \cdots, -\mu_{b-1}, \mu_b, \cdots, \mu_n); \quad (53\text{a})$$
$$(\sin((b-a-\tfrac{1}{2})\pi/n) > 0)$$
$$(P_{ab}, \psi(\mu_1, \cdots, \mu_n)) = -\mu_a\mu_b\psi(-\mu_1, \cdots,$$
$$-\mu_a, \mu_{a+1}, \cdots, \mu_{b-1}, -\mu_b, \cdots, -\mu_n); \quad (53\text{b})$$
$$(\sin((b-a-\tfrac{1}{2})\pi/n) < 0).$$

When $\cot(a-\tfrac{1}{2})\pi/n \leqq \cot(b-\tfrac{1}{2})\pi/n$, substantially the same formulas apply, only the changes of sign are arranged $(-\mu_1, \cdots, -\mu_{b-1},$

$\mu_b, \cdots, \mu_a, -\mu_{a+1}, \cdots, -\mu_n)$ in the case (a) and

$(\mu_1, \cdots, \mu_{b-1}, -\mu_b, \cdots, -\mu_a, \mu_{a+1}, \cdots, \mu_n)$

in the case (b).

The given formulas describe the effect of P_{ab} in terms of the vector basis (40). The effect of this same operator on a vector which belongs to the fundamental system (41) can be described conveniently with the aid of functions which introduce a change of sign once in a period $(1, 2, \cdots, n)$. Individually, these functions have the period $2n$. We shall use the standard notation

$$\text{sgn } x = \begin{cases} -1; & x < 0 \\ 0; & x = 0 \\ 1; & x > 0. \end{cases}$$

Then the function

$$\text{sgn}(\sin(j\pi/n))$$

changes sign at $j = 0$ and $j = n$ and vanishes for these values of j, while the function

$$\text{sgn}(\sin(j - \tfrac{1}{2})\pi/n)$$

has everywhere the absolute value 1, with changes of sign between 0 and 1, and between n and $n+1$. With this notation we obtain

$$(P_{ab}, \mu_{j_1}\mu_{j_2}\cdots\mu_{j_k}) = \text{sgn}[\sin((b-a-\tfrac{1}{2})\pi/n)$$
$$\times \prod_{h=1}^{k}(\sin(j_h - a - \tfrac{1}{2})\pi/n \, \sin(j_h - b + \tfrac{1}{2})\pi/n)]$$
$$\times \mu_a\mu_b\mu_{j_1}\cdots\mu_{j_k}. \quad (54)$$

This formula can be verified by direct count of the sign changes caused by the operator

$$C_{a+1}C_{a+2}\cdots C_{b-1}$$

acting on each factor of the product

$$\mu_b\mu_{j_1}\mu_{j_2}\cdots\mu_{j_k}.$$

THE BASIS OF A REDUCED REPRESENTATION

Some Important Commutators

The operators

$$A = \sum_1^n s_j s_{j+1} = \sum_1^n P_{a,a+1}, \quad B = \sum_1^n C_j = -\sum_1^n P_{aa}$$

are invariant against the dihedral (rotation-reflection) group of transformations

$$\mu_j \rightarrow \mu_{k\pm j}; \quad (j = 1, 2, \cdots, n). \quad (55)$$

From the operators P_{ab} we can form $2n$ linear combinations which exhibit the same symmetry, most simply as follows:

$$A_m = \sum_{a=1}^{n} P_{a,a+m}, \qquad (56)$$

whereby

$$A_0 = -B, \quad A_1 = A. \qquad (56a)$$

In view of (46) we have

$$A_{m+n} = -CA_m = -A_m C. \qquad (57)$$

From the commutators

$$[P_{ac}, P_{bd}] = P_{ac}P_{bd} - P_{bd}P_{ac}$$

which vanish unless exactly one congruence between the indices is satisfied, we can form $n-1$ independent linear combinations of dihedral symmetry, which may be written most compactly

$$G_m = \tfrac{1}{4} \sum_{a=1}^{n} ([P_{az}, P_{a+m,z}] - [P_{za}, P_{z,a+m}])$$
$$= \tfrac{1}{2} \sum_{a=1}^{n} (P_{az}P_{a+m,z} - P_{za}P_{z,a+m}). \qquad (58)$$

While the definition of $G_1, G_2, \cdots$ will be extended conventionally to the complete period $m = 1, 2, \cdots, 2n$, they satisfy the identities

$$G_{-m} = -G_m; \quad G_0 = G_n = 0 \qquad (58a)$$

so that there are only $n-1$ independent operators included in the set. The analog of (57)

$$G_{m+n} = -CG_m = -G_m C \qquad (59)$$

is valid. The commutators of $A_1, A_2, \cdots, A_{2n}$ with each other necessarily can be expressed in terms of the set (58). By definition

$$[A_m, A_m] = 0.$$

By (57) we have also

$$[A_m, A_{m+n}] = 0.$$

If we exclude these trivial cases, the congruences

$$a \equiv b; \quad a+k \equiv b+m$$

will not be satisfied simultaneously when we compute

$$[A_k, A_m] = \sum_{a,b=1}^{n} [P_{a,a+k}, P_{b,b+m}]$$
$$= \sum_{c=1}^{n} [P_{zc}, P_{z,c+m-k}] + \sum_{a=1}^{n} [P_{az}, P_{a-m+k,z}]$$

whence

$$[A_k, A_m] = 4G_{k-m}, \qquad (60)$$

With the convention (58a) this is also valid in the trivial case $k \equiv m$. Conversely, the commutators of $A_1, A_2, \cdots, A_{2n}$ with their own commutators $G_1, G_2, \cdots, G_{n-1}$ can be expressed in terms of the former as follows:

$$2[G_m, A_k] = \sum_{a,b=1}^{n} ([P_{az}P_{a+m,z}, P_{b,b+k}]$$
$$-[P_{za}P_{z,a+m}, P_{b-k,b}]).$$

Again if we exclude the trivial case $m \equiv 0$, the congruences $b \equiv a$, $b \equiv a+m$ will not be satisfied simultaneously. Hence,

$$2[G_m, A_k] = \sum_{a=1}^{n} ([P_{az}P_{a+m,z}, P_{a,a+k}]$$
$$+[P_{az}P_{a+m,z}, P_{a+m,a+k+m}]$$
$$-[P_{za}P_{z,a+m}, P_{a-k,a}]$$
$$-[P_{za}^n P_{z,a+m}, P_{a-k+m,a+m}]).$$

Here

$$[P_{az}P_{a+m,z}, P_{a,a+k}] = -2P_{a,a+k}P_{az}P_{a+m,z}$$
$$= -2P_{a,a+k}P_{a,a+k}P_{a+m,a+k} = -2P_{a+m,a+k},$$

$$[P_{az}P_{a+m,z}, P_{a+m,a+k+m}]$$
$$= 2P_{az}P_{a+m,z}P_{a+m,a+k+m} = 2P_{a,a+k+m},$$

etc., and we obtain

$$2[G_m, A_k] = \sum_{a=1}^{n} (-2P_{a+m,a+k} + 2P_{a,a+k+m}$$
$$+2P_{a-k,a+m} - 2P_{a-k+m,a})$$
$$= \sum_{a=1}^{n} (4P_{a,a+k+m} - 4P_{a,a+k-m}),$$

or, simply

$$[G_m, A_k] = 2A_{k+m} - 2A_{k-m} \qquad (61)$$

which is also valid in the trivial case $m \equiv 0$. It is easily shown from (60) and (61) that the commutators $G_1, \cdots, G_{n-1}$ commute with each other. We have generally

$$[G_m, [A_k, A_0]] = [[G_m, A_k], A_0] + [A_k, [G_m, A_0]].$$

Using this together with (60) and (61) we find

$$[G_m, G_k] = [(2A_{k+m} - 2A_{k-m}), A_0]$$
$$+[A_k, (2A_m - 2A_{-m})]$$
$$= 8G_{k+m} - 8G_{k-m}$$
$$+8G_{k-m} - 8G_{k+m} = 0. \qquad (61a)$$

A Set of Generating Basis Elements

The commutation rules (60) and (61) suggest the introduction of the operators X_0, X_1, $\cdots$, X_n, Y_1, $\cdots$, Y_{n-1}, Z_1, $\cdots$, Z_{n-1} defined as follows:

$$X_r = (4n)^{-1} \sum_{a,b=1}^{2n} \cos((a-b)r\pi/n)P_{ab},$$

$$Y_r = (4n)^{-1} \sum_{a,b=1}^{2n} \sin((a-b)r\pi/n)P_{ab}, \qquad (62)$$

$$Z_r = (-i/8n) \sum_{a,b=1}^{2n} \sin((a-b)r\pi/n) \times (P_{az}P_{bz} - P_{za}P_{zb}).$$

Again these definitions will apply conventionally for the entire period $r = 1, 2, \cdots, 2n$, whereby

$$X_{-r} = X_r,$$
$$Y_{-r} = -Y_r; \quad Y_0 = Y_n = 0, \qquad (62a)$$
$$Z_{-r} = -Z_r; \quad Z_0 = Z_n = 0.$$

Comparison of (62) with (56) and (58) yields directly

$$X_r = (2n)^{-1} \sum_{m=1}^{2n} \cos(mr\pi/n)A_m,$$

$$Y_r = -(2n)^{-1} \sum_{m=1}^{2n} \sin(mr\pi/n)A_m, \qquad (63a)$$

$$Z_r = (i/2n) \sum_{m=1}^{2n} \sin(mr\pi/n)G_m.$$

Conversely, from the orthogonal properties of the coefficients

$$A_m = \sum_{r=1}^{2n} (X_r \cos(mr\pi/n) - Y_r \sin(mr\pi/n)),$$
$$\qquad (63b)$$
$$G_m = -i \sum_{r=1}^{2n} Z_r \sin(mr\pi/n).$$

In particular, A and B admit the expansions

$$B = -A_0 = -\sum_1^{2n} X_r = -X_0 - 2X_1$$
$$\qquad\qquad -\cdots - 2X_{n-1} - X_n,$$

$$A = A_1 = \sum_1^{2n} (X_r \cos(r\pi/n) - Y_r \sin(r\pi/n))$$
$$= X_0 + 2(X_1 \cos(\pi/n) \qquad (63c)$$
$$- Y_1 \sin(\pi/n)) + \cdots$$
$$+ 2(X_{n-1} \cos((n-1)\pi/n)$$
$$- Y_{n-1} \sin((n-1)\pi/n)) - X_n.$$

Conversely, the operators (62) can be constructed from A and B by means of (60), (61), and (63a).

The relations (57) and (59) now become

$$(1 + (-)^r C)X_r = (1 + (-)^r C)Y_r$$
$$= (1 + (-)^r C)Z_r = 0 \quad (64)$$

so that the operators (62) belong to the algebra of even functions when r is odd, and vice versa; they annihilate even functions for even r and odd functions for odd r. We could deal with each set independently, but the two have many properties in common, and it is often convenient to keep them together.

The commutation rules for the operators (62) are easily obtained from (60) and (61). Thanks to the orthogonal properties of the trigonometric coefficients, we find

$$[X_r, X_s] = [Y_r, Y_s] = [Z_r, Z_s] = 0,$$
$$[X_r, Y_s] = [Y_r, Z_s] = [Z_r, X_s] = 0; \quad (65a)$$
$$\cos(r\pi/n) \neq \cos(s\pi/n).$$

The only commutators which do not vanish are

$$[X_r, Y_r] = -2iZ_r,$$
$$[Y_r, Z_r] = -2iX_r, \quad (r = 1, 2, \cdots, n-1) \quad (65b)$$
$$[Z_r, X_r] = -2iY_r.$$

We shall show next that the operators (X_r, Y_r, Z_r), which commute with the members of other sets (X_s, Y_s, Z_s), anticommute among themselves. For this purpose we compute

$$(X_r + iY_r)^2 = (4n)^{-2} \sum_{a,b=1}^{2n} e^{(a+b-c-d)r\pi i/n} P_{ac}P_{bd}$$

by the definition (62). In the given expansion we may interchange (a, b) and (c, d) independently without disturbing the coefficient. Taking the mean of the four alternatives we obtain

$$(8n)^{-2} \sum_{a,b=1}^{2n} e^{(a+b-c-d)r\pi i/n}$$
$$\times (P_{ac}P_{bd} + P_{ad}P_{bc} + P_{bc}P_{ad} + P_{bd}P_{ac})$$

and by the exchange rule (52), this equals

$$(8n)^{-2} \sum_{a,b=1}^{2n} e^{(a+b-c-d)r\pi i/n}$$
$$\times (1 - (-)^{D(a-b)})(1 - (-)^{D(c-d)}) P_{ac}P_{bd}.$$

The only products which contribute to this sum

are those for which $a \equiv b$ as well as $c \equiv d$, whereby according to (46) and (47)

$$P_{ac}^2 = P_{ac}P_{a+n, c+n} = 1; \quad P_{ac}P_{a, c+n} = P_{ac}P_{a+n, c} = -C$$

and we obtain

$$(X_r + iY_r)^2 = (8n)^{-2} \sum_{a, c=1}^{2n} 8e^{2(a-c)r\pi i/n}(1 - (-)^r C),$$

whence, observing (62a)

$$X_0^2 = \tfrac{1}{2}(1 - C) = R_0,$$

$$X_n^2 = \tfrac{1}{2}(1 - (-)^n C) = R_n = \begin{cases} R_0; & (n \text{ even}) \\ 1 - R_0; & (n \text{ odd}) \end{cases} \tag{66}$$

$$(X_r \pm iY_r)^2 = 0; \quad (r = 1, 2, \cdots, n-1). \tag{67}$$

The latter result implies that X_r, Y_r anticommute

$$X_r Y_r = -Y_r X_r \tag{67a}$$

and that their squares are identical

$$X_r^2 = Y_r^2 = R_r. \tag{67b}$$

Now (65b) implies

$$X_r Y_r = -Y_r X_r = -iZ_r$$

and since the product in turn must anticommute with either factor, e.g.,

$$X(YX) = (XY)X = -(YX)X,$$

we have similarly

$$Y_r Z_r = -Z_r Y_r = -iX_r; \quad Z_r X_r = -X_r Z_r = -iY_r.$$

These results together imply (67b) as well as its extension

$$Z_r^2 = iX_r Y_r Z_r = X_r^2 = Y_r^2 = R_r.$$

The operator R_r thus defined satisfies the relations appropriate to the unit of a quaternion algebra with the hermitian basis (R_r, X_r, Y_r, Z_r), viz., the real basis (R_r, X_r, Y_r, iZ_r) because

$$R_r X_r = Y_r Y_r X_r = i Y_r Z_r = X_r,$$

etc., and

$$R_r^2 = R_r X_r X_r = X_r^2 = R_r.$$

The multiplication table is accordingly

$$\begin{aligned}
R_r &= R_r^2 = X_r^2 = Y_r^2 = Z_r^2, \\
X_r &= R_r X_r = X_r R_r = i Y_r Z_r = -i Z_r Y_r, \\
Y_r &= R_r Y_r = Y_r R_r = i Z_r X_r = -i X_r Z_r, \\
Z_r &= R_r Z_r = Z_r R_r = i X_r Y_r = -i Y_r X_r.
\end{aligned} \tag{68}$$

For $r = 0$ and $r = n$ the operators Y_r and Z_r vanish; in either case we obtain a commutative algebra with a degenerate basis (R_0, X_0) or (R_n, X_n); cf. (66).

In the original quaternion algebra described by (31) the operators $R_0, R_1, \cdots, R_{n-1}, R_n$ are projections. It remains to enumerate the dimensions of these projections [in the irreducible representation (30)]. By (66), R_0 and R_n are 2^{n-1} dimensional. By (64), the others have at most that many dimensions, because R_r is contained in R_0 or in $(1 - R_0)$ as r is even or odd. Further information can be obtained from suitable explicit expansions of R_r in terms of the basis elements P_{ab}. A very compact expression is obtained by evaluating

$$R_r = \tfrac{1}{2}(X_r^2 + Y_r^2)$$

directly from the expansions (62), which yields

$$R_r = 2^{-5}n^{-2} \sum_{a, b, c, d=1}^{2n} \cos[(a-b-c+d)r\pi/n] P_{ac} P_{bd}. \tag{69}$$

To obtain an interesting modification of this formula we interchange the pairs of indices (a, b) and (c, d) independently. By (52) the mean of the four alternatives equals

$$R_r - 2^{-7}n^{-2} \sum_{a, b, c, d} (\cos((a-b-c+d)r\pi/n)$$
$$- (-)^{D(a-b)} \cos((a-b+c-d)r\pi/n))$$
$$\times (1 + (-)^{D(a-b)+D(c-d)}) P_{ac} P_{bd}.$$

The anticommuting product terms in which either $a \equiv b$ but not $c \equiv d$, or vice versa, cancel out. The terms for which neither congruence is satisfied yield

$$2^{-5}n^{-2} \sum_{a, b, c, d} \sin((a-b)r\pi/n) \sin((c-d)r\pi/n) P_{ac} P_{bd}.$$

The terms for which both congruences are satisfied yield either the identity or $-(-)^r C$ and add up to $\tfrac{1}{4}(1 - (-)^r C)$. Accordingly,

$$R_r = (1 - (-)^r C)$$
$$\times (\tfrac{1}{4} + (8n)^{-2} \sum_{a, b, c, d=1}^{2n} \sin((a-b)r\pi/n)$$
$$\times \sin((c-d)r\pi/n) P_{ac} P_{bd}). \tag{70}$$

The dimensions (character) of the projection R_r are simply enumerated from (70). When $n > 2$, we have $-1 \neq P_{ac}P_{bd} \neq 1$ unless $a \equiv b$; $c \equiv d$, so that the dimension of R_r is one fourth of the total 2^n.

$$\text{Dim}(R_r) = 2^{n-2}; \quad (r = 1, 2, \cdots, n-1);$$
$$(n > 2). \quad (71)$$

For the case $n = 2$; $r = 1$ we have from (63c) and (64)

$$B(1+C) = -4X_1; \quad (n=2).$$

By (43a), the even eigenvectors of B are the two possible even homogeneous functions: (1) and $(\mu_1\mu_2)$; the corresponding characteristic numbers of B are 2 and -2. Neither vanishes, so that in this case R_1 contains all of $1 - R_0$, and we have the "exceptional" result

$$R_1 = 1 - R_0; \quad \text{Dim}(R_1) = 2; \quad (n=2). \quad (71a)$$

Irreducible Representations

In the previous section we have completed the abstract theory of the algebra generated by the operators A and B defined by (27) and (23), which are involved in the eigenwert problem (29). Apart from illustrative references to the representation (30), our main results have been derived from the abstract commutation rules (31). Now the task before us is to apply these results to the solution of problems such as (29), which can be formulated explicitly in terms of A and B. For this purpose we shall again think of the abstract numbers $s_1, s_2, \cdots, s_n, C_1, C_2, \cdots, C_n$ as linear operators in a 2^n-dimensional vector space, defined by the matrix representation (30).

Essentially, the problem of reduction consists in finding the vector spaces which are invariant towards A and B, in the sense that neither A nor B operating on any member of an invariant set can yield anything but a linear combination of the vectors which belong to the same set. Since the set of operators (62) can be constructed from A and B and vice versa [cf. (63c)], it is evident that the invariant sets in question must be invariant towards all the operators (62). In view of the commutation rules (65), we may specify for added convenience that the matrices representing the operators (62) and their products shall be direct products of irreducible representations of the factor groups (R_r, X_r, Y_r, Z_r).

To answer the main questions dealt with in the present communication it will suffice to find the largest solution to the problem (29). The identification of the corresponding irreducible representation of the operators (62) is an easier task than the complete reduction, which we shall leave aside for future attention. We shall go a little further and identify separately for the vector spaces of even and odd functions the representations of maximal dimensions.

By (64) we may consider the sets of even and odd functions separately. The former are annihilated by (R_r, X_r, Y_r, Z_r) for even $r = (0, 2, 4, \cdots)$, the latter for odd $r = (1, 3, 5, \cdots)$. By (66), the space of even functions is formed by the projection $1 - R_0 = \frac{1}{2}(1+C)$, that of odd functions by $R_0 = \frac{1}{2}(1-C)$.

In the representation (30) the operator of (29) has a matrix whose elements are all positive. Accordingly, by Frobenius' theorem[11] the components of the "maximal" eigenvector must be all of one sign in this representation. It follows immediately that the vector in question cannot be odd. More precisely, it cannot be orthogonal to the even vector

$$\chi_0(\mu_1, \cdots, \mu_n) = \text{const} = 2^{-n/2}$$

(normalized), which happens to be the only function of degree zero. By (43a), this is an eigenvector of B corresponding to the simple characteristic number n

$$(B, \chi_0) = (B(1 - R_0), \chi_0) = n\chi_0. \quad (72)$$

Since by (63c) and (64)

$$-B = X_1 + X_2 + \cdots + X_{2n},$$
$$-B(1 - R_0) = X_1 + X_3 + \cdots + X_{2n-1}$$
$$= \begin{cases} 2X_1 + 2X_3 + \cdots + 2X_{2m-1}; \\ \qquad (n = 2m) \\ 2X_1 + 2X_3 + \cdots + 2X_{2m-1} \\ \qquad + X_{2m+1}; \quad (n = 2m+1) \end{cases} \quad (73)$$

and $X_0, X_1, \cdots, X_n$ commute with each other, they commute with B. Accordingly, operation with X_r on χ_0 will either annihilate it or convert it into some function of the same degree. But χ_0 is the only function of degree zero. Hence, χ_0 is a common eigenvector of $X_0, X_1, \cdots X_n$ satisfying

$$(X_r, \chi_0) = \xi_r\chi_0; \quad (r = 0, 1, \cdots, n).$$

By (68) and (66) the operators X_r all satisfy the identity

$$X_r^3 - X_r = X_r(X_r - 1)(X_r + 1) = 0, \qquad (74)$$

so that ξ_0, ξ_1, $\cdots$, ξ_n are to be selected from the numbers 0, 1, -1. Since χ_0 is an even vector we have

$$\xi_0 = \xi_2 = \xi_4 = \cdots = 0$$

and by comparison with (73)

$$\xi_1 + \xi_3 + \cdots + \xi_{2n-1} = -n.$$

There is just one way to satisfy this requirement

$$\xi_1 = \xi_3 = \cdots = \xi_{2n-1} = -1,$$
$$(X_r, \chi_0) = -\chi_0;$$
$$(r = 1, 3, \cdots, 2[(n-1)/2]+1). \qquad (75)$$

Having identified one common eigenvector χ_0 of X_1, X_3, $\cdots$, we proceed to construct others which belong to the same invariant set by operations with Y_1, Y_3, $\cdots$. In general if χ satisfies

$$(X_r, \chi) = \xi_r \chi; \quad (r = 1, 3, \cdots),$$

then by (68) and (65)

$$(X_r, (Y_r, \chi)) = (X_r Y_r, \chi) = (-Y_r X_r, \chi)$$
$$= -\xi_r(Y_r, \chi),$$
$$(X_s, (Y_r, \chi)) = (X_s Y_r, \chi) = (Y_r X_s, \chi)$$
$$= \xi_s(Y_r, \chi); \quad (s \neq r)$$

so that the operation with Y_r gives a new common eigenvector; the sign of the corresponding ξ_r is reversed and the others are unchanged. If ξ_r vanishes, Y_r will only annihilate χ because by (68)

$$(Y_r, \chi) = (Y_r X_r^2, \chi) = (Y_r, \xi_r^2 \chi).$$

For this reason it is useless to operate with Y_2, Y_4, $\cdots$ on χ_0. It is equally useless to operate more than once with any one Y_r, because by (68) Y_r is its own inverse in the subspace R_r,

$$Y_r^2 = X_r^2 = R_r,$$

so that the second operation at best undoes the effect of the first. Finally, since

$$Z_r = iX_r Y_r,$$

the operation with Z_r can only give us such eigenfunctions of X_r as we can obtain with Y_r alone. Accordingly, the vectors which we can obtain by operating with various combinations of $(Y_1, Y_3, \cdots)$ on χ_0—at most once apiece—form an invariant set. The vectors thus obtained can be distinguished from each other by their different sets of characteristic numbers; when necessary we shall designate them individually

$$\chi(\xi_1, \xi_3, \cdots, \xi_{2m-1}; (\mu)); \quad (n \geq 2m \geq n-1).$$

In the case of odd n we have $\xi_n = -1$ for every vector in the set; there is no Y_n. For the rest, every combination occurs exactly once, as is expressed by the following formulation of the relations derived above

$$(X_r, \chi(\xi_1, \xi_3, \cdots, \xi_{2m-1}; (\mu)))$$
$$= \xi_r \chi(\xi_1, \xi_3, \cdots, \xi_{2m-1}; (\mu)),$$
$$(Y_r, \chi(\xi_1, \xi_3, \cdots, \xi_{2m-1}; (\mu)))$$
$$= \chi(\xi_1, \cdots, -\xi_r, \cdots, \xi_{2m-1}; (\mu)), \qquad (76)$$
$$(X_n, \chi(\xi_1, \xi_3, \cdots, \xi_{n-2}; (\mu)))$$
$$= -\chi(\xi_1, \cdots, \xi_{n-2}; (\mu)).$$

A more explicit description of the functions χ is not necessary. The relations (76) in conjunction with the verified construction of the one vector

$$\chi(-1, -1, \cdots, 1; (\mu)) - 2^{-n/2} \qquad (76a)$$

suffice for definition. From the one given vector, the construction is in effect completed by means of the formal relations (68).

The representation (76) is analogous to (30); but it has only $m = [n/2]$ factors and the number of dimensions is $2^{[n/2]}$, if we denote by $[x]$ the greatest integer which does not exceed x. Moreover, denoting by Q_r a general function of the factor basis (R_r, X_r, Y_r, Z_r), the representation (76) is a direct product of the two-dimensional representations

$$\mathbf{D}_{max} = \mathbf{D}_2(Q_1) \times \mathbf{D}_2(Q_3) \times \cdots \times \mathbf{D}_2(Q_{n-1});$$
$$(n = 2m)$$
$$\mathbf{D}_{max} = \mathbf{D}_2(Q_1) \times \mathbf{D}_2(Q_3) \times \cdots$$
$$\times \mathbf{D}_2(Q_{n-2}) \times \mathbf{D}_-(Q_n); \quad (n = 2m+1)$$
$$\qquad (77)$$
$$\mathbf{D}_2(X_r) = \begin{pmatrix} -1 & 0 \\ 0 & 1 \end{pmatrix}; \quad \mathbf{D}_2(Y_r) = \begin{pmatrix} 0 & 1 \\ 1 & 0 \end{pmatrix};$$
$$\mathbf{D}_+(X_r) = (1); \quad \mathbf{D}_-(X_r) = (-1)$$

supplemented by the one-dimensional representation $\mathbf{D}_-(Q_n)$ when n is odd.

An invariant vector space may be constructed by the same procedure from any common eigenvector of $(X_0, X_1, \cdots, X_n)$. The characters of the representation will be determined by the initial ξ_0 and ξ_n, which are common to all members of an invariant set, together with the values (0 or 1) of $\xi_1{}^2, \xi_2{}^2, \cdots, \xi_{n-1}^2$. For the signs of $\xi_1, \xi_2, \cdots, \xi_{n-1}$ can be changed independently by operation with $Y_1, Y_2, \cdots, Y_{n-1}$, excepting those which vanish in the first place. To put it in geometrical language, the general type of the representation is determined by an intersection of projections

$$P_0 P_1 \cdots P_n$$

where for $r = 1, 2, \cdots, n-1$, P_r is either R_r or $1 - R_r$, while R_0 is subdivided into $\frac{1}{2}(R_0 + X_0)$ and $\frac{1}{2}(R_0 - X_0)$, and R_n is similarly subdivided. The vector space constructed above from χ_0 is the only intersection common to all of the projections $R_1, R_3, \cdots, R_{[n/2]-1}$. For every such intersection must contain an even eigenvector of B corresponding to the characteristic number n when n is even or $n-1-\xi_n$ when n is odd, in either case an even function of degree less than 2, and the only function that satisfies this specification is const. $= \chi_0$.

We have already pointed out that by Frobenius' theorem, that eigenvector which corresponds to the largest solution of (29) cannot be orthogonal to χ_0. Therefore, this eigenvector belongs to the invariant set constructed above, which contains χ_0. The expectation of a *simple* largest characteristic number is in accord with our finding that the invariant vector space which corresponds to the maximal representation (77) is unique.

In connection with the considerations of Lassettre and Howe,[13] it ia a matter of interest to identify not only the largest solution $\lambda_{\max}$ of (29), but in addition the largest solution $\bar{\lambda}_{\max}$ which belongs to an *odd* eigenvector. As it happens, this solution is the second largest in magnitude. We shall prefer a slight variation of the above procedure. From one common eigen-

---

[13] E. N. Lassettre and J. P. Howe, reference 12.

vector of the operators

$$X_r{}^* = -X_r \cos(r\pi/n) + Y_r \sin(r\pi/n);$$
$$(r = 0, 1, \cdots, n-1, n) \quad (78)$$
$$X_0{}^* = -X_0; \quad X_n{}^* = X_n$$

we shall construct others by operations with, say, $Z_1, \cdots, Z_{n-1}$, which anticommute with the corresponding $X_r{}^*$. Then instead of B we consider

$$A = \sum_{j=1}^{n} s_j s_{j+1} = -\sum_{r=1}^{2n} X_r{}^* = X_0 - \sum_{r=1}^{n-1} 2X_r{}^* - X_n,$$

$$R_0 A = \frac{1}{2}(1 - C)A \quad (79)$$

$$= \begin{cases} X_0 - 2X_2{}^* - 2X_4{}^* - \cdots - 2X_{n-2}^* - X_n; (n = 2m) \\ X_0 - 2X_2{}^* - 2X_4{}^* - \cdots - 2X_{n-1}^*; \quad (n = 2m+1) \end{cases}$$

[cf. (63c)]. By (68) the operators (78) satisfy

$$X_r{}^{*2} = R_r = X_r{}^2; \quad X_r{}^{*3} - X_r{}^* = 0. \quad (80)$$

The matrix which represents the operator

$$\exp(H'A) = \prod_{j=1}^{n} (\cosh H' + s_j s_{j+1} \sinh H')$$

in terms of the vector basis (41) which renders B diagonal consists of two 2^{n-1}-dimensional matrices which transform even functions among themselves and odd functions among themselves, respectively. In either matrix all elements are positive (provided $H' > 0$), so that Frobenius' theorem applies in the vector spaces of even and odd functions separately. The modification by the factor $\exp(H^*B)$ in (29) does not affect this conclusion because this factor is represented here by a diagonal matrix whose elements are all positive.

By (44) the largest characteristic number n of A is double and belongs to the two configurations $(+, +, \cdots, +)$, $(-, -, \cdots, -)$, which are free from alternations of sign. When the corresponding even and odd eigenvectors

$$\begin{matrix} \chi_0{}^* \\ \chi_0{}^{(-)*} \end{matrix} \Bigg\} = 2^{-\frac{1}{2}} \left(2^{-n} \prod_1^n (1 + \mu_j) \right.$$
$$\left. \pm 2^{-n} \prod_1^n (1 - \mu_j) \right) \quad (81)$$

are expanded in terms of the vector basis (41), their components are all positive, so that the representations which correspond to the characteristic numbers $\lambda_{\max}$ and $\bar{\lambda}_{\max}$ of (29) must belong to unique invariant vector spaces which contain χ_0^* and $\chi_0^{(-)*}$, respectively. The former has been constructed above from χ_0. The odd vector $\chi_0^{(-)*}$ is a common eigenvector of all X_r^* and satisfies

$$n\chi_0^{(-)*} = (A, \chi_0^{(-)*})$$

$$= -\sum_{q=1}^{n} (X_{2q}^*, \chi_0^{(-)*}) = -\left(\sum_{q=1}^{n} \xi_{2q}^*\right)\chi_0^{(-)*}$$

which requires

$$\xi_0 = -\xi_0^* = 1,$$

$$\xi_{2q}^{*2} = 1; \quad (0 < q < \tfrac{1}{2}n)$$

$$\xi_n = \xi_n^* = -1; \quad (n = 2m),$$

so that the vector $\chi_0^{(-)*}$ is included in the projection

$$\tfrac{1}{2}(R_0 + X_0)R_2R_4 \cdots R_{2[n/2]}$$

and in $\tfrac{1}{2}(R_n - X_n)$ when n is even. The "maximal odd" representation is therefore

$$\mathbf{D}_{\max}^- = \mathbf{D}_+(Q_0) \times \mathbf{D}_2(Q_2) \times \cdots \times \mathbf{D}_2(Q_{n-1});$$
$$(n = 2m+1)$$

$$\mathbf{D}_{\max}^- = \mathbf{D}_+(Q_0) \times \mathbf{D}_2(Q_2) \times \cdots$$
$$\times \mathbf{D}_2(Q_{n-2}) \times \mathbf{D}_-(Q_n); \quad (n = 2m)$$
(82)

with the notation of (77). When n is odd, $\mathbf{D}_{\max}^-$ has $2^{(n-1)/2}$ dimensions, the same number as $\mathbf{D}_{\max}$, and there can be no further representation with this number of dimensions, for there is only one pair of configurations with less than two alternations of sign. For even n the representation $\mathbf{D}_{\max}^-$ has only $2^{(n-2)/2}$ dimensions; it is still unique. However, one other representation with the same number of dimensions is compatible with the known spectrum (44) of A, namely,

$$\mathbf{D}_{\mathrm{alt}} = \mathbf{D}_-(Q_0) \times \mathbf{D}_2(Q_2) \times \cdots$$
$$\times \mathbf{D}_2(Q_{n-2}) \times \mathbf{D}_+(Q_n). \quad \text{(82a)}$$

The count of configurations with n (even) alternations yields one pair: $(+ - + - \cdots + -)$ and $(- + - + \cdots - +)$; reasoning analogous to the above shows that exactly one invariant vector space belongs to (82a).

The identification of the remaining invariant vector spaces involves more elaborate calculations; the complete theory will be given in a later publication. However, in order to settle an interesting question which will arise, we shall complete the discussion of those n vector spaces which contain the n functions of degree one

$$\chi_1(2r; \mu_1, \cdots, \mu_n) = n^{-\frac{1}{2}} \sum_{k=1}^{n} \exp(2rk\pi i/n)\mu_k. \quad \text{(83a)}$$

These functions may be classified according to the representations of the dihedral symmetry group (55). To the identical representation of (55) belongs $\chi_1(0; (\mu))$; the corresponding representation of the algebra (68) is (82). For even n, $\chi_1(n; (\mu))$ belongs to another one-dimensional representation of (55) and to the representation (82a) of (68). The remaining functions (83a) belong pairwise to two-dimensional representations of (55). To show that the pair of similar representations which belong to the pair $\chi_1(\pm 2r; (\mu))$ are described by

$$\mathbf{D}_-(Q_0) \times \mathbf{D}_0(Q_{2r}) \times \prod_{s \neq r} \mathbf{D}_2(Q_{2s}); \quad (n \text{ odd})$$
(83b)

$$\mathbf{D}_-(Q_0) \times \mathbf{D}_-(Q_n) \times \mathbf{D}_0(Q_{2r}) \times \prod_{s \neq r} \mathbf{D}_2(Q_{2s});$$
$$(n \text{ even})$$

one may verify that the vectors in question are annihilated by X_{2r}

$$(X_{2r}, \chi_1(2r; (\mu))) - (X_{2r}, \chi_1(-2r; (\mu))) = 0;$$
$$(0 < 2r < n). \quad \text{(83c)}$$

The direct calculation by means of (62) and (54) is not too laborious; since X_{2r} commutes with B, only the terms of first degree have to be computed. The remaining factors in (83b) are then determined as before by the relation

$$(B, \chi_1) = (n - 2)\chi_1.$$

SOLUTION OF THE EIGENWERT PROBLEM

It is a simple matter now to compute those solutions of the eigenwert problem (29) which belong to the representations (77), (82), and (83); we only have to solve a set of quadratic equations.

As to the "physical" significance of the solutions, we recall that the characteristic number itself yields the partition function and with it the thermodynamic functions of the crystal. Moreover, as shown by Kramers and Wannier, the individual components $\psi(\mu_1, \cdots, \mu_n)$ of the eigenvector which belongs to λ_{max} describe the statistical distribution of configurations of the last tier of atoms added to the crystal.[10] On the other hand, if we know in addition the eigenvector $\psi'(\mu_1, \cdots, \mu_n)$ which belongs to λ_{max} for the adjoint operator $V' = V_1'V_2' = V_1V_2$, then the component products

$$\psi(\mu_1, \cdots, \mu_n)\psi'(\mu_1, \cdots, \mu_n)$$

describe the statistical distribution of configurations in the *interior* of the crystal.[13]

With the latter result in mind, we first transform V to self-adjoint form by operation with $V_1^{\frac{1}{2}}$; then

$$\bar{V} = V_1^{\frac{1}{2}} V V_1^{-\frac{1}{2}} = V_1^{\frac{1}{2}} V_2 V_1^{\frac{1}{2}}$$

$$= (2 \sinh 2H)^{n/2}$$

$$\times \exp(\tfrac{1}{2}H'A) \exp(H^*B)\exp(\tfrac{1}{2}H'A) \quad (84)$$

has the same characteristic numbers as V, and the squares

$$\psi^2(\mu_1, \cdots, \mu_n) = f(\mu_1, \cdots, \mu_n)$$

of the components of its principal eigenvector will describe the distribution of configurations in the interior of the crystal.

To solve the eigenwert problem we may transform $\bar{V}$ into a function of the operators $X_0, X_1, \cdots, X_n$ alone, or equally well into a function of $X_0^*, X_1^*, \cdots, X_n^*$. The latter procedure involves slightly simpler relations; either result is readily converted into the other by means of the real orthogonal operator[14]

$$\exp\left(\sum_{r=1}^{n-1} ((n-r)/2n)\pi i Z_r\right)$$

$$= \prod_{r=1}^{n-1} (1 - (1 - \sin(r\pi/2n))R_r + \cos(r\pi/2n)iZ_r).$$

For reference we observe here the somewhat more general rule

$$\exp(i\alpha Z_r)(X_r \cos \beta + Y_r \sin \beta) \exp(-i\alpha Z_r)$$

$$= X_r \cos(2\alpha+\beta) + Y_r \sin(2\alpha+\beta) \quad (85)$$

which is easily verified by means of the multiplication table (68). The latter—and (85)—are

also valid if X_r, Y_r, Z_r are replaced throughout by

$$X_r^* = -X_r \cos(r\pi/n) + Y_r \sin(r\pi/n),$$

$$Y_r^* = X_r \sin(r\pi/n) + Y_r \cos(r\pi/n), \quad (86)$$

$$Z_r^* = -Z_r$$

(dual transformation). We now recall the expansions

$$A = -X_0^* - 2X_1^* - \cdots - 2X_{n-1}^* - X_n^*$$

$$B = -X_0 - 2X_1 - \cdots - 2X_{n-1} - X_n$$

$$= X_0^* + 2 \sum_{r=1}^{n-1} (\cos(r\pi/n)X_r^*$$
$$- \sin(r\pi/n)Y_r^*) - X_n^*.$$

Substitution in (84) yields

$$\bar{V} = (2 \sinh 2H)^{n/2} \exp((H^* - H')X_0^*)$$

$$\prod_{r=1}^{n-1} (U_r) \exp(-(H^*+H')X_n^*) \quad (87)$$

where

$$U_r = \exp(-H'X_r^*) \exp(-2H^*X_r) \exp(-H'X_r^*)$$

$$= (1 - R_r) + (\cosh 2H' \cosh 2H^*$$
$$- \sinh 2H' \sinh 2H^* \cos(r\pi/n))R_r$$
$$- (\sinh 2H' \cosh 2H^*$$
$$- \cosh 2H' \sinh 2H^* \cos(r\pi/n))X_r^*$$
$$- \sinh 2H^* \sin(r\pi/n)Y_r^*. \quad (87a)$$

By inspection of the factored form we note that transformation with $(1 - R_r + iZ_r)$ changes U_r into U_r^{-1}, which differs from U_r in the signs of the coefficients of X_r^* and Y_r^*. Adding the two together we obtain

$$U_r + U_r^{-1} = 2(1 - R_r + (\cosh 2H' \cosh 2H^*$$
$$- \sinh 2H' \sinh 2H^* \cos(r\pi/n))R_r)$$
$$= 2(1 - R_r) + 2 \cosh \gamma_r R_r$$

whence the characteristic numbers of U_r belonging to the subspace R_r are e^{γ_r} and $e^{-\gamma_r}$, of equal multiplicity. (Those which belong to $(1 - R_r)$ are trivial, all equal to unity.) Accordingly we may write

$$U_r = 1 - R_r + \cosh \gamma_r R_r - \sinh \gamma_r \cos \delta_r^* X_r^*$$
$$- \sinh \gamma_r \sin \delta_r^* Y_r^*$$

$$= \exp(-\gamma_r(\cos \delta_r^* X_r^* + \sin \delta_r^* Y_r^*)) \quad (88)$$

where γ_r and δ_r^* can be computed from $2H'$,

---

[14] This is not exactly the dual transformation; it takes a further transformation with $\prod(1 - R_r + X_r)$ to obtain (86).

$2H^*$, $\omega_r = r\pi/n$ by the rules of hyperbolic trigo-nometry, as indicated by Fig. 4. We note in particular the trigonometric relations[15]

$$\cosh \gamma = \cosh 2H' \cosh 2H^*$$
$$- \sinh 2H' \sinh 2H^* \cos \omega, \quad \text{(a)}$$

$$\sinh \gamma \cos \delta^* = \sinh 2H' \cosh 2H^*$$
$$- \cosh 2H' \sinh 2H^* \cos \omega, \quad \text{(b)}$$

$$\sin \omega / \sinh \gamma = \sin \delta^* / \sinh 2H^* \quad (89)$$
$$= \sin \delta' / \sinh 2H', \quad \text{(c)}$$

$$\cot \delta^* = (\sinh 2H' \coth 2H^*$$
$$- \cosh 2H' \cos \omega) / \sin \omega. \quad \text{(d)}$$

With the aid of (85) and (86) we find immediately the transformations which reduce U_r to diagonal form with regard to the eigenvectors of either X_r^* or X_r as a vector basis,

$$U_r = \exp(-\tfrac{1}{2}\delta_r^* i Z_r) \exp(-\gamma_r X_r^*) \exp(\tfrac{1}{2}\delta_r^* i Z_r), \quad (90a)$$

$$U_r = \exp(\tfrac{1}{2}(\pi - \omega_r$$
$$- \delta_r^*) i Z_r) \exp(-\gamma_r X_r) \exp(-\tfrac{1}{2}(\pi - \omega_r - \delta_r^*) i Z_r),$$
$$(r = 1, 2, \cdots, n-1) \quad (90b)$$

and by substitution in (87)

$$\exp\left(\sum_{r=1}^{n-1} \tfrac{1}{2}\delta_r^* i Z_r\right) \bar{V} \exp\left(-\sum_{r=1}^{n-1} \tfrac{1}{2}\delta_r^* i Z_r\right)$$

$$= (2 \sinh 2H)^{n/2} \exp\left((H^* - H') X_0^*\right.$$

$$\left. - \sum_{r=1}^{n-1} \gamma_r X_r^* - (H' + H^*) X_n^*\right), \quad (91a)$$

$$\exp\left(-\sum_{r=1}^{n-1} \tfrac{1}{2}(\pi - \omega_r\right.$$
$$\left. - \delta_r^*) i Z_r\right) \bar{V} \exp\left(\sum_{r=1}^{n-1} \tfrac{1}{2}(\pi - \omega_r - \delta_r^*) i Z_r\right)$$

$$= (2 \sinh 2H)^{n/2} \exp\left((H' - H^*) X_0\right.$$

$$\left. - \sum_{r=1}^{n-1} \gamma_r X_r - (H' + H^*) X_n\right). \quad (91b)$$

<hr>

[15] The formulas of hyperbolic trigonometry are obtained from those of spherical trigonometry by the substitution of imaginary lengths for the sides. See F. Klein, *Vorlesungen über Nicht-Euklidische Geometrie* (Springer, Berlin, 1928), p. 195.

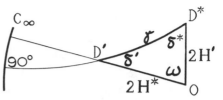

FIG. 4. Hyperbolic triangle. Stereographic projection, conformal. Circles are represented by circles, geodetics by circles invariant towards inversion in the limiting circle $C\infty$ of the projection. See F. Klein, reference 15, pp. 293–299.

The spectrum of $\bar{V}$ is now known to the extent that the common eigenvectors χ of $(X_0, X_1, \cdots, X_{n-1}, X_n)$ and the corresponding sets of characteristic numbers $(\xi_0, \xi_1, \cdots, \xi_{n-1}, \xi_n)$ are known. For every solution χ to that problem $\bar{V}$ has an eigenvector ψ with characteristic number λ given by

$$\psi = \left(\exp\left(-\sum_{r=1}^{n-1} \tfrac{1}{2}(\pi - \omega_r - \delta_r^*) i Z_r\right), \chi\right),$$
$$(\bar{V}, \psi) = \lambda\psi, \quad (92)$$
$$\log \lambda = \tfrac{1}{2} n \log(2 \sinh 2H) + (H' - H^*) \xi_0$$
$$- \sum_{r=1}^{n-1} \gamma_r \xi_r - (H' + H^*) \xi_n.$$

In the previous section we were able to show that the largest characteristic number of $\bar{V}$ belongs to the representation (77). In view of (76) and (76a) we must put $\xi_0 = \xi_2 = \xi_4 = \cdots = 0$. All the others may be taken negative, which calls for $\chi = \chi_0$, and obviously this is the best choice. Accordingly,

$$\log \lambda_{\max} - \tfrac{1}{2} n \log(2 \sinh 2H)$$
$$= \begin{cases} \gamma_1 + \gamma_3 + \cdots + \gamma_{2m-1}; & (n = 2m) \\ \gamma_1 + \gamma_3 + \cdots + \gamma_{2m-1} + (H' + H^*); & (n = 2m+1). \end{cases} \quad (93)$$

These two results can be combined if we adopt the natural conventions

$$\gamma_{-r} = \gamma_r; \quad \gamma_n = 2H' + 2H^*. \quad (94)$$

Then (93) takes the compact form

$$\log \lambda_{\max} - \tfrac{1}{2} n \log(2 \sinh 2H) = \tfrac{1}{2} \sum_{r=1}^{n} \gamma_{2r-1}$$

$$= \tfrac{1}{2} \sum_{r=1}^{n} \cosh^{-1}(\cosh 2H' \cosh 2H^*$$

$$- \sinh 2H' \sinh 2H^* \cos((2r-1)\pi/2n)). \quad (95)$$

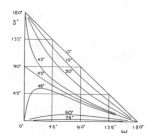

FIG. 5. Variation of δ^* with ω for the quadratic crystal, at various reduced temperatures $(k/J)T = 1/H$.

gd $2H$	$1/H$
$0°$	∞
$15°$	7.551
$30°$	3.641
$42°$	2.472
$45°$	2.269
$48°$	2.089
$66°$†	1.29
$77°$‡	0.92

† By mistake marked 60° on figure.
‡ By mistake marked 75° on figure.

The corresponding eigenvector can be described by either of the two formulas

$$\psi_{max} = \left(\exp\left(-\sum_{0<2r<n} \tfrac{1}{2}(\pi - \omega_{2r-1}\right.\right.$$
$$\left.\left. -\delta^*_{2r-1})iZ_{2r-1}\right), \chi_0\right), \quad (96a)$$

$$\psi_{max} = \left(\exp\left(\sum_{0<2r<n} \tfrac{1}{2}\delta^*_{2r-1}iZ_{2r-1}\right), \chi_0^*\right). \quad (96b)$$

Here we have omitted operations with $\exp((\text{const})Z_{2r})$; $(r=1, 2, \cdots)$, because these have no effect on χ_0 or any other even vector. The somewhat simpler form (96b) involves construction from the even δ function χ_0^* given by (81); one easily verifies that $\xi_1^*, \xi_3^*, \cdots$ are all negative for this vector.

The Principal Eigenvector

The result (96) describes explicitly the distribution of configurations (ψ^2) of one row of atoms in a crystal. The terms of the description are quite unfamiliar. We should not regret this; for among the main objects of the development of this theory is the invention of more suitable methods for the description of such distributions. Even so, we must try to establish the connection with more familiar types of description in terms

of more tangible quantities. Moreover, the result (96) fails to describe the statistical correlation of configurations in different rows of atoms. Methods of dealing with these two tasks have been perfected to a considerable degree; but the algebraic apparatus involved is elaborate enough to make a separate publication advisable.

One quite tangible result is readily obtained from (96b): The probability of a configuration free from alternations is

$$2\psi^2(++\cdots+) = 2\psi^2(--\cdots-)$$
$$= \prod(\cos^2(\tfrac{1}{2}\delta^*_{2r-1})).$$

Similarly, when n is even so that a configuration free from persistencies of sign is possible, the probability of such a configuration is

$$2\psi^2(+-+-\cdots+-) = 2\psi^2(-+-+\cdots-+)$$
$$= \prod(\sin^2(\tfrac{1}{2}\delta^*_{2r-1})).$$

Presumably, the probabilities of the 2^n-4 remaining configurations are intermediate between the given limits. One can show from the commutation rules that the probabilities of configurations with 2, 4, 6, $\cdots$ alternations of sign can be computed successively from those terms in the expansion of (96b) which contain, respectively, 1, 2, 3, $\cdots$ factors of the type $(Z_r \sin \tfrac{1}{2}\delta_r^*)$. However, the labor involved increases rapidly with the number of steps.

Of greater interest is the fact that the description (96) brings out a striking qualitative difference between those distributions which occur for $n(H'-H^*)\gg 1$ and those which occur for $n(H^*-H')\gg 1$.

The variation of δ^* with $\omega = r\pi/n$ is given analytically by (89d); but a qualitative inspection of Fig. 4 is even more convenient. Let us keep the side $OD^* = 2H'$ of the triangle OD^*D' fixed; then the positions of the vertex D' for $\omega = \pi/n$, $\omega = 3\pi/n$, $\cdots$ are equidistant points on a circle of radius $OD' = 2H^*$. Now if $2H^* > 2H'$, then $\delta^* = \pi$ for $\omega = 0$, and δ^* decreases from π to 0 as ω increases from 0 to π. On the other hand, if $2H^* < 2H'$, then $\delta^* = 0$ for $\omega = 0$. As ω increases, δ^* increases to a maximum value

$$\sin^{-1}(\sinh 2H^*/\sinh 2H') = \delta^*_{max} < \tfrac{1}{2}\pi$$

given by (89c) for $\delta' = \tfrac{1}{2}\pi$; with further increase of ω, δ^* decreases again until $\delta^* = 0$ for $\omega = \pi$.

As regards the angle $\frac{1}{2}(\pi - \delta^* - \omega)$ which occurs in (96a), the axiom

$$\delta' + \delta^* + \omega < \pi$$

of hyperbolic geometry implies

$$\pi - \omega - \delta^* > 0.$$

A graphical representation of δ^* as a function of ω at various temperatures is given in Fig. 5 for a quadratic crystal $(H' = H)$. As regards the variation of δ^* with the temperature for a fixed ω, we note that as the former increases from 0 to ∞, δ^* increases from 0 to $\pi - \omega$. *For small values of ω, nearly all of this increase takes place over a small interval of temperatures which includes the critical point.* A detailed inspection of the operators Z_r can be made to show that those for which r is small are the most effective in introducing pairs of alternations far apart, which is just what it takes to convert order into disorder.

Concerning the extreme cases of low and high temperatures, one sees easily that in the former case δ^* and in the latter case $\pi - \omega - \delta^*$ will remain small for all ω. This is neither new nor surprising; it means that the low temperature distribution consists mostly of configurations which resemble the perfectly ordered arrangement represented by χ_0^*, while the high temperature distribution resembles the completely random distribution described by χ_0.

Propagation of Order

The principal characteristic number (95) of the eigenwert problem (29) yields the partition function, and the principal eigenvector determines the distribution of configurations. Nevertheless, the remaining $2^n - 1$ characteristic numbers are of some interest. In general, these describe the propagation of order.[16] In particular, if long-range order is present, at least one of the subsidiary characteristic numbers should be very nearly equal to the principal.[17] Moreover, at the transition point where long-range order appears, one may expect a phenomenon analogous to the branching of multiple-valued analytic functions such as $z^{1/n}$, whereby the order of the branch-

---

[16] F. Zernike, Physica **7**, 565 (1938).
[17] See reference 13. See also E. Montroll, J. Chem. Phys. **10**, 61 (1942).

point bears a relation to the type of the singularity. A survey of the solutions which belong to the representations (77), (82), and (83) will answer the most important questions.

To obtain all the solutions which belong to (77b), the angles $\frac{1}{2}\delta_1^*$, $\frac{1}{2}\delta_3^*$, $\cdots$ in (96) may be replaced independently by $\frac{1}{2}\delta_1^* + \frac{1}{2}\pi$, $\frac{1}{2}\delta_3^* + \frac{1}{2}\pi$, $\cdots$; the corresponding characteristic numbers are given by the formula

$$\log \lambda - \tfrac{1}{2}n \log(2 \sinh 2H)$$
$$= \begin{cases} \pm\gamma_1 \pm \gamma_3 \pm \cdots \pm \gamma_{2m-1}; & (n = 2m) \\ \pm\gamma_1 \pm \gamma_3 \pm \cdots \pm \gamma_{2m-1} + \frac{1}{2}\gamma_{2m+1}; & (n = 2m+1) \end{cases} \quad (97)$$

where the optional signs are independent and each combination occurs once.

That part of the spectrum of V which belongs to the representation (82) is obtained similarly. The result is

$$\log \lambda - \tfrac{1}{2}n \log(2 \sinh 2H)$$
$$= \begin{cases} (H' - H^*) \pm \gamma_2 \pm \gamma_4 \pm \cdots \pm \gamma_{2m-2} + \frac{1}{2}\gamma_{2m}; \\ \qquad\qquad\qquad\qquad (n = 2m) \\ (H' - H^*) \pm \gamma_2 \pm \gamma_4 \pm \cdots \pm \gamma_{2m}; \\ \qquad\qquad\qquad\qquad (n = 2m+1). \end{cases} \quad (98)$$

For the largest of these we find the counterpart of (95)

$$\log \lambda_{max}^{(-)} - \tfrac{1}{2}n \log(2 \sinh 2H)$$
$$= (H' - H^*) + \tfrac{1}{2}\sum_{r=1}^{n-1}\gamma_{2r} =$$
$$= \tfrac{1}{2}\left(\gamma_0 \operatorname{sgn}(H' - H^*) + \sum_{r=1}^{n-1}\gamma_{2r}\right) \quad (99)$$

with the natural notation

$$\gamma_0 = \cosh^{-1}(\cosh 2H' \cosh 2H^*$$
$$- \sinh 2H' \sinh 2H^*) = 2|H' - H^*|. \quad (100)$$

First of all, this differs from (95) in the substitution of γ_0, γ_2, $\cdots$ for γ_1, γ_3, $\cdots$. Moreover, the sign of $\frac{1}{2}\gamma_0$ is mandatory and changes at the critical point. This leads to a most remarkable limiting result for large values of n. The two sums

$$\gamma_1 + \gamma_3 + \cdots + \gamma_{2n-1} = \gamma(\pi/n) + \gamma(3\pi/n) + \cdots,$$
$$\gamma_0 + \gamma_2 + \cdots + \gamma_{2n-2} = \gamma(0) + \gamma(2\pi/n) + \cdots$$

may be considered as different numerical quad-ratures of

$$(n/2\pi)\int_0^{2\pi}\gamma(\omega)d\omega$$

by the trapeze rule. For periodic, analytic functions the approximation to the integral improves very rapidly (exponentially) with increasing number of intervals. Hence, with rapid convergence

$$\lim_{n=\infty}(\lambda_{max}^{(-)}/\lambda_{max})=\begin{cases}1;&(H'>H^*)\\\exp(2H^*-2H');&(H'<H^*).\end{cases}\quad(101)$$

The distribution of the logarithms of the remaining characteristic numbers which belong to (77) and (82)

$$\log\lambda_{max}-2\gamma_1,\ \log\lambda_{max}$$
$$-2\gamma_3,\ \cdots,\ \log\lambda_{max}-2\gamma_1-2\gamma_3,\ \cdots,$$
$$\log\lambda_{max}^{(-)}-2\gamma_2,\ \log\lambda_{max}^{(-)}$$
$$-2\gamma_4,\ \cdots,\ \log\lambda_{max}^{(-)}-2\gamma_2-2\gamma_4,\ \cdots$$

becomes *dense* as $n\to\infty$. These "continua" remain distinct from the two largest characteristic numbers as long as $H'\neq H^*$. However, for $H'<H^*$ the representations (83b) give rise to a continuum

$$\log\lambda=\log\lambda_{max}^{(-)}+\gamma_0-\gamma_{2r};\quad(H'<H^*)\quad(102)$$

which contains $\lambda_{max}^{(-)}$ as a superior limit. Further complications due to representations which we have not investigated here are excluded by (43) and (63c).

At the critical point $H'=H^*$, we have $\gamma_0=0$; $\gamma_r=O(r/n)$. In this exceptional case both $\lambda_{max}^{(-)}$ and λ_{max} itself are limits of "continua." The following scheme summarizes the results

$$\lambda_{max}=\lambda_{max}^{(-)}>\text{(continuum)};\quad(H'>H^*)$$
$$\lambda_{max}=\lambda_{max}^{(-)}=\lim\text{(continuum)};\quad(H'=H^*)\quad(103)$$
$$\lambda_{max}>\lambda_{max}^{(-)}=\lim\text{(continuum)};\quad(H'<H^*).$$

At temperatures below the critical point ($H'>H^*$), we have "asymptotic degeneracy" of order 2, as a symptom of long-range order. Above the critical point, there is no degeneracy of the principal characteristic number. At the critical point we observe branching associated with an "asymptotic degeneracy" of *infinite order*.

In regard to the "propagation of order," the mean distance to which a local disturbance in the crystal is propagated is inversely proportional to $\log(\lambda_{max}/\lambda)$. We note that this "range" of the "short-range order" becomes infinite at the critical point.

Since the functions (83a) belong to different representations of the dihedral group (55), —and the same is necessarily true for their respective invariant vector spaces—, the result (102) ought to contain information about the anisotropy of the propagation of order. A tentative computation leads to the implicit formula

$$\cosh 2H\cosh 2H'-\sinh 2H\cosh(\beta\sin\varphi)$$
$$-\sinh 2H'\cosh(\beta\cos\varphi)=0\quad(104)$$

for the mean range $(1/\beta)$ of the short-range correlation in the direction φ, when $H'<H^*$.

THERMODYNAMIC PROPERTIES OF A LARGE CRYSTAL

To compute the partition function per atom

$$\lambda=\lambda_\infty=\lim_{n=\infty}(\lambda_{max})^{1/n}\quad(105)$$

for an infinite crystal we replace the sum (95) by the integral

$$\log\lambda_\infty=\tfrac{1}{2}\log(2\sinh 2H)+\frac{1}{2\pi}\int_0^\pi\gamma(\omega)d\omega\quad(106)$$

where

$$\cosh\gamma(\omega)=\cosh 2H'\cosh 2H^*$$
$$-\sinh 2H'\sinh 2H^*\cos\omega.$$

There are several ways to show that (106) actually describes a symmetrical function of H and H'. For example, with the aid of the useful identity

$$\int_0^{2\pi}\log(2\cosh x-2\cos\omega)d\omega=2\pi x\quad(107)$$

we can convert (106) into the double integral

$$\log(\lambda/2)=\tfrac{1}{2}\pi^{-2}\int_0^\pi\int_0^\pi\log(\cosh 2H\cosh 2H'$$
$$-\sinh 2H\cos\omega-\sinh 2H'\cos\omega')d\omega d\omega'.\quad(108)$$

It seems rather likely that this result could be derived from direct algebraic and topological considerations without recourse to the operator method used in the present work. Such a development might well amount to a great improvement of the theory.

A generalization of the expansion whose initial terms were given by Kramers and Wannier is easily obtained from (108). With the notation

$$2\kappa = \tanh 2H/\cosh 2H' = \sin g \cos g',$$
$$2\kappa' = \tanh 2H'/\cosh 2H = \cos g \sin g' \quad (109a)$$

we expand the logarithm in powers of κ and κ' and integrate term by term to obtain

$$\log \lambda - \tfrac{1}{2} \log(4 \cosh 2H \cosh 2H')$$

$$= \tfrac{1}{2}\pi^{-2} \int_0^\pi\!\!\int \log(1 - 2\kappa \cos \omega - 2\kappa' \cos \omega')d\omega d\omega'$$

$$= -\tfrac{1}{2} \sum_{r+s>0} (2r+2s-1)!(r!)^{-2}(s!)^{-2}\kappa^{2r}\kappa'^{2s}. \quad (109b)$$

Specialization to the case $H=H'$; $\kappa=\kappa'$ of quadratic symmetry yields

$$\log \lambda - \log(2 \cosh 2H)$$

$$= \tfrac{1}{2}\pi^{-2} \int_0^\pi\!\!\int \log(1 - 4\kappa \cos \omega_1 \cos \omega_2)d\omega_1 d\omega_2$$

$$- -\sum_{n=1}^\infty \left(\binom{2n}{n}^2 \Big/ 4n \right)\kappa^{2n} \quad (109c)$$

$$= \log(1 - \kappa^2 - 4\kappa^4 - 29\kappa^6 - 265\kappa^8 - 2745\kappa^{10}$$
$$- 30773\kappa^{12} - 364315\kappa^{14} - \cdots)$$

which confirms the result given previously to the order κ^{10} by Kramers and Wannier.[10]

These expansions converge for all values of H and H' because

$$|2\kappa \cos \omega + 2\kappa' \cos \omega'| \leqq 2\kappa + 2\kappa' = \sin(g+g') \leqq 1.$$

The limit of convergence of the series is given by the transition point,[18] and it converges even at the limit. Of course it is not very suitable for computation in the critical region; much better formulas for this purpose will be obtained.

The partition function (106) yields directly the free energy F of the crystal; the energy U and the specific heat C are obtained by differentiation with regard to the temperature T.

[18] Kramers and Wannier might well have inferred this much from the uniform sign of the terms, which locates the nearest singularity on the real axis.

For a crystal of N atoms

$$F = U - TS = -NkT \log \lambda,$$
$$U = F - T(dF/dT) = NkT^2 d(\log \lambda)/dT, \quad (110)$$
$$C = dU/dT.$$

For the purpose of differentiation it will be convenient to consider λ, given by (106), as a function of two independent variables $H=J/kT$ and $H'=J'/kT$. In this notation we have

$$U = -NJ(\partial \log \lambda/\partial H) - NJ'(\partial \log \lambda/\partial H')$$
$$= -NkT(H(\partial \log \lambda/\partial H)$$
$$\qquad\qquad + H'(\partial \log \lambda/\partial H')) \quad (111a)$$

$$C = Nk(H^2(\partial^2 \log \lambda/\partial H^2)$$
$$\qquad + 2HH'(\partial^2 \log \lambda/\partial H\partial H')$$
$$\qquad + H'^2(\partial^2 \log \lambda/\partial H'^2)). \quad (111b)$$

The two terms of (111a) are separately the mean energies of interaction in the two perpendicular directions in the crystal (remember the interpretation of temperature as a statistical parameter).

In differentiating the integral of (106) we consider H^* as a function of H which satisfies [cf. (17)]

$$dH^*/dH = -\sinh 2H^* = -1/\sinh 2H.$$

The following formulas, wherein γ, δ', and δ^* are considered as functions of H', H^*, and ω, are easily obtained by differentiation from the formulas (89); some of them are obvious by inspection of Fig. 4.

$$\partial\gamma/\partial H' = 2 \cos \delta^*,$$
$$\partial\gamma/\partial H^* = 2 \cos \delta',$$
$$\partial^2\gamma/\partial H'^2 = 4 \sin^2 \delta^* \coth \gamma, \quad (112)$$
$$\partial^2\gamma/\partial H^{*2} = 4 \sin^2 \delta' \coth \gamma,$$
$$\partial^2\gamma/\partial H'\partial H^* = -4 \sin \delta^* \sin \delta'/\sinh \gamma.$$

Comparison with (106) yields for substitution in (111)

$$\partial \log \lambda/\partial H' = \int_0^\pi \cos \delta^* \, d\omega/\pi,$$

$$\partial \log \lambda/\partial H = \cosh 2H^* - \sinh 2H^* \int_0^\pi \cos \delta' \, d\omega/\pi \quad (113a)$$

and

$$\partial^2 \log \lambda/\partial H'^2 = 2 \int_0^\pi \sin^2 \delta^* \coth \gamma \, d\omega/\pi,$$

$\partial^2 \log \lambda / \partial H \partial H' = 2 \sinh 2H^*$

$$\times \int_0^\pi (\sin \delta^* \sin \delta' / \sinh \gamma) d\omega / \pi,$$

$\partial^2 \log \lambda / \partial H^2 = 2 \sinh^2 2H^*$ $\qquad (113b)$

$$\times \left(-1 + \coth 2H^* \int_0^\pi \cos \delta' d\omega / \pi \right.$$

$$\left. + \int_0^\pi \sin^2 \delta' \coth \gamma \, d\omega / \pi \right).$$

The reduction of these integrals by means of an uniformizing elliptic substitution for the hyperbolic triangle of Fig. 4 is dealt with in the appendix. The qualitative behavior of the integrals is easily seen from Fig. 4: The integrals (113a) are continuous functions of H' and H (or H^*) for all values of these parameters, even for $H' = H^*$ (critical point). The three integrals (113b) are infinite at the critical point, otherwise finite. The singularity results from a conspiracy: In the case $H' = H^*$ we have $\gamma(0) = 0$ and at the same time $\delta'(0) = \delta^*(0) = \frac{1}{2}\pi$, although at all other temperatures

$$\gamma(\omega) \geqq \gamma(0) = 2|H' - H^*| > 0,$$

$$\sin \delta'(0) = \sin \delta^*(0) = \sin 0 = \sin \pi = 0.$$

For the special case of quadratic symmetry $H' = H$ the computation of the thermodynamic functions can be simplified considerably. The most convenient starting point is the double integral of (109c), which is converted into a single integral with the aid of (107). Using the notation

$$k_1 = 4\kappa = 2 \sinh 2H / \cosh^2 2H,$$
$$k_1'' = \pm(1 - k_1^2)^{\frac{1}{2}} = 2 \tanh^2 2H - 1; \quad |k_1''| = k_1' \qquad (114)$$

we obtain

$$\log(\lambda / 2 \cosh 2H)$$

$$= \frac{1}{2\pi} \int_0^\pi \log(\tfrac{1}{2}(1 + (1 - k_1^2 \sin^2 \varphi)^{\frac{1}{2}})) d\varphi. \quad (115)$$

Differentiation under the integral sign yields for the energy U,

$$U = -NJ \frac{d \log \lambda}{dH}$$

$$= -NJ \coth 2H \left(1 + \frac{2}{\pi} k_1'' K_1 \right) \qquad (116)$$

and for the specific heat C,

$$C = Nk \frac{d^2 \log \lambda}{dH^2} = Nk(H \coth 2H)^2 (2/\pi)$$

$$\times (2K_1 - 2E_1 - (1 - k_1'')(\tfrac{1}{2}\pi + k_1'' K_1)). \quad (117)$$

In these formulas K_1 and E_1 denote the complete elliptic integrals

$$K_1 = K(k_1) = \int_0^{\pi/2} (1 - k_1^2 \sin^2 \varphi)^{-\frac{1}{2}} d\varphi,$$

$$E_1 = E(k_1) = \int_0^{\pi/2} (1 - k_1^2 \sin^2 \varphi)^{\frac{1}{2}} d\varphi. \qquad (118a)$$

The integral (115) cannot be expressed in closed form; but rapidly convergent series can be given. With the notation

$$K_1' = K(k_1'),$$

$$\log q_1 = \pi \tau_1 i = -\pi K_1' / K_1, \qquad (118b)$$

$$G = 1^{-2} - 3^{-2} + 5^{-2} - 7^{-2} + \cdots$$

$$= 0.915\ 965\ 594 \text{ (Catalan's constant),}$$

one or the other of the following expansions (derived in the appendix) will be found suitable for computation:

$$\log \lambda = \tfrac{1}{2} \log(2 \sinh 2H) - \tfrac{1}{4} \log q_1$$

$$+ \sum_{n=1}^\infty (-)^n (2n-1) \log(1 - q_1^{2n-1}), \quad (119a)$$

$$\log \lambda = \tfrac{1}{2} \log(2 \sinh 2H) + (2/\pi)G + \frac{1}{\pi} \sum_{n=0}^\infty (-)^n$$

$$\times \frac{(1 + (2n+1)(\pi i / \tau_1) - \exp[-(2n+1)\pi i / \tau_1])}{(2n+1)^2 \sinh^2((n+\tfrac{1}{2})\pi i / \tau_1)}. \qquad (119b)$$

A singularity occurs for $H = H^* = \frac{1}{2} \log \cot \pi/8$, in which case $k_1 = 1$; $K_1 = \infty$; $K_1' = \frac{1}{2}\pi$; $E_1 = 1$. The specific heat becomes infinite at this critical point; the energy is continuous because $k_1'' = 0$. The analytic nature of the singularity is evident from the approximate formulas

$$K_1 \sim \log(4/k_1') \sim \log(2^{\frac{1}{2}}/|H - H_c|),$$

$$C/Nk \sim (2/\pi)(\log \cot \pi/8)^2 (K_1 - 1 - \tfrac{1}{4}\pi). \qquad (120)$$

The critical data are

$$H_t = J/kT_c = \tfrac{1}{2} \log \cot \pi/8 = 0.440\ 686\ 8,$$

$$-F_c/NkT_c = \log \lambda_c = \tfrac{1}{2} \log 2 + (2/\pi)G$$

$$= 0.929\ 695\ 40,\quad (121)$$

$$-U_c/NJ = 2^{\frac{1}{2}} = 1.414\ 213\ 56,$$

$$S_c/Nk = \log \lambda_c - 2^{\frac{1}{2}} H_c = 0.306\ 470$$

$$= \log 1.358\ 621.$$

The critical temperature and energy were given already by Kramers and Wannier.[10] In addition, their estimate of "about 2.5335" for λ_c was very close to the exact value

$$\lambda_c = 2^{\frac{1}{2}} e^{2G/\pi} = 2.533\ 737\ 28.$$

The specific heat of a quadratic crystal is represented in Fig. 6 as a function of $1/H = (k/J)T$; the result of the best approximate computation in the literature[10] is indicated for comparison.

Figure 7 represents the specific heat curve of a highly anisotropic crystal $(J'/J = 1/100)$, computed from Eq. (6.2) in the Appendix. The corresponding curves for the case of quadratic symmetry $(J'/J = 1)$ and for the linear chain $(J'/J = 0)$ are also shown. The abscissa is $2/(H+H') = 2kT/(J+J')$, so that the three curves have the same area.

SOME PROPERTIES OF FINITE CRYSTALS

The partition function of any *finite* crystal is an analytic function of the temperature. Perfect order and a sharp transition point occur only when the extent of the crystal is infinite in two directions. It is a matter of interest to study the approach to complete order and the sharpening of the transition point with increasing size (width) of the crystal.

Boundary Tension

First, let us compute the "boundary tension" between two regions of opposite order below the transition point. This we can do by a slight variation of the model. We simply chose a *negative* interaction energy between adjacent atoms in the same tier:

$$u = +J' \sum_1^n \mu_j \mu_{j+1}.$$

Now if n is *even*, the reversal of J' has the same trivial effect as a redefinition of the sign of every other μ_j. The interpretation of the result is a little different in that the type of order which occurs below the transition point is now a superlattice with the ideal structure indicated by Fig. 8. However, let us choose an *odd* value of n. Now a perfect alternation of signs around the polygon of n atoms is impossible; in one place at least the given superlattice must be adjacent to one of opposite order. As we build a long crystal on the odd polygonal base, we build it with one misfit "seam." The free energy difference due to this seam will equal that of a boundary between regions of opposite order.

The formula (93) is still valid, only now

$$J'/kT = -H' > H^* > 0$$

whence

$$\log \lambda_{\max} - \tfrac{1}{2} n \log(2 \sinh 2H)$$

$$= \gamma_1 + \gamma_3 + \cdots + \gamma_{2m-1} - (|H'| - H^*)$$

$$= \gamma_1 + \gamma_3 + \cdots + \gamma_{2m-1} + \tfrac{1}{2}\gamma_{2m+1} - 2(|H'| - H^*)$$

where

$$\cosh \gamma_r = \cosh 2H' \cosh 2H^*$$

$$- \sinh(-2H') \sinh 2H^* \cos(\pi - \omega_r).$$

The replacement of ω_r by $\pi - \omega_r$ has no appreciable effect on the sum; the modification of the partition function due to the seam is therefore practically equal to

$$-2|H'| + 2H^*.$$

This represents the effect of a lengthwise boundary for our original model as well, and we obtain

$$\sigma'/kT = -\log \lambda_b' = 2(H' - H^*),$$

$$\sigma' = 2J' - kT \log \coth(J/kT) \qquad (122a)$$

for the boundary tension σ' (per atom) of a longitudinal boundary. The transverse boundary tension σ is similarly given by

$$\sigma/kT = -\log \lambda_b = 2(H - H'^*). \qquad (122b)$$

The boundary tensions both vanish at the critical point, as one should expect. The formulas

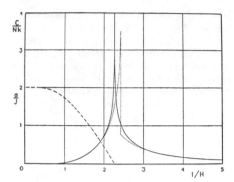

FIG. 6. Properties of quadratic crystal. — — — — — Boundary tension σ between regions of opposite order. ——————— Specific heat C. — — — — — — — — — — Approximate computation of C by Kramers and Wannier.

for the boundary entropies and energies follow

$$s_b' = -d\sigma'/dT = 2k((H/\sinh 2H)+H^*),$$
$$s_b = -d\sigma/dT = 2k((H'/\sinh 2H')+H'^*),$$
$$u_b' = \sigma'+Ts_b' = (2/\sinh 2H)J+2J', \qquad (123)$$
$$u_b = \sigma+Ts_b = 2J+(2/\sinh 2H')J'.$$

The variation of σ with the temperature for a quadratic crystal $(J=J')$ is represented graphically in Fig. 6.

Mean Ordered Length

In a crystal of moderate width n we may expect to find sections of rather well-defined order similar to that of an infinite crystal, with occasional transitions to opposite order. The result (122b) allows an estimate of the "mean ordered length" l_0 between such interruptions. By the general theory of fluctuations

$$l_0 = \lambda_b^{-n} = \exp[2n(H-H'^*)]$$
$$= (e^{2H}\tanh H')^n. \quad (124)$$

Evidently, when n is large enough we have

$$l_0 \gg n.$$

The estimate (124) is significant when this condition is fulfilled.

The Specific Heat of a Finite Crystal

Since the partition function of a finite crystal is an analytic function of the temperature, its specific heat will be finite at all temperatures. However, a maximum will occur near the

transition point of the infinite crystal; this maximum will be increased and sharpened with increasing size of the crystal.

Kramers and Wannier[10] computed the specific heats at the transition point successively for $n=1, 2, \cdots, 6$ (screw arrangement). From the results they inferred the asymptotic rule that the height of the maximum increases linearly with $\log n$. We shall verify this remarkable conjecture.

The maximum of (C/N) for a finite crystal does not occur exactly at the asymptotic critical point; but the differences in location and magnitude are only of the order $n^{-2}\log n$. This order of accuracy will suffice.

To compute the specific heat we substitute the exact partition function (95) of a finite crystal in (111b), whereby the relations (112) are useful. The specialization to the critical case

$$H'=H^*; \quad \text{gd } 2H+\text{gd } 2H'=\tfrac{1}{2}\pi;$$
$$\sinh 2H \sinh 2H'=1$$

simplifies the computation in several respects; the relations

$$\delta'=\delta^*=\delta, \quad \sinh \tfrac{1}{2}\gamma = \sinh 2H'|\sin \tfrac{1}{2}\omega|$$

are valid in this case. We write the result in the form

$$C/Nk = -2H^2\sinh^2 2H'+(H'+H\sinh 2H')^2$$
$$\times \sinh 2H \times \frac{1}{n}\sum_{r=1}^{n}\text{cosec}[(r-\tfrac{1}{2})\pi/n)]$$
$$+\frac{1}{n}\sum_{r=1}^{n}[(H'+H\sinh 2H')^2$$
$$\times (2\sin^2 \delta_{2r-1}\coth \gamma_{2r-1}-\text{cosech }\tfrac{1}{2}\gamma_{2r-1})$$
$$-4HH'\sinh 2H'\sin^2 \delta_{2r-1}\tanh \tfrac{1}{2}\gamma_{2r-1}$$
$$+H^2\sinh 4H'\cos \delta_{2r-1}]. \quad (125)$$

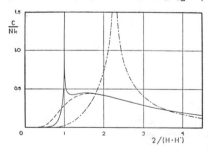

FIG. 7. Specific heats for varying degrees of anisotropy. ——————— $J'/J=1/100$; — — — — — — — — $J'/J=1$ (quadratic crystal); — — — — — $J'=0$ (linear chain).

The second summand, when considered as a periodic function of the variable $\omega_{2r-1} = (2r-1)\pi/2n$, has a differential quotient of bounded variation. Its sum may be replaced accordingly by the corresponding integral, with an error of the order n^{-2}.

We have purposely split off the first sum

$$S_1 = \frac{1}{n} \sum_{r=1}^{n} \mathrm{cosec}((r-\tfrac{1}{2})\pi/n).$$

The corresponding integral diverges. For an asymptotic estimate of the sum we expand the cosecants in partial fractions; the result is easily arranged in the form

$$S_1 = \frac{4}{\pi}\left(\frac{1}{1} + \frac{1}{3} + \cdots + \frac{1}{2n-1} - \frac{1}{2n+1} - \frac{1}{2n+3}\right.$$
$$\left. - \cdots - \frac{1}{4n-1} + \frac{1}{4n+1} + \cdots\right)$$
$$= \frac{2}{\pi} \sum_{m=0}^{\infty} (-)^m (\psi((m+1)n+\tfrac{1}{2}) - \psi(mn+\tfrac{1}{2}))$$

whereby

$$\psi(z) \equiv \Gamma'(z)/\Gamma(z) = \log(z-\tfrac{1}{2}) + O(z^{-2}).$$

From the given asymptotic estimate of $\psi(z)$ together with

$$\psi(\tfrac{1}{2}) = -2\log 2 - C_E,$$

$$C_E = 0.577\ 215\ 665, \quad \text{(Euler's constant)}$$

$$\log\left(\frac{2}{1} \cdot \frac{2}{3} \cdot \frac{4}{3} \cdot \frac{4}{5} \cdots\right) = \log\frac{\pi}{2}, \quad \text{(Wallis' product)}$$

we obtain

$$S_1 = (2/\pi)(\log(8n/\pi) + C_E) + O(n^{-2}).$$

Substitution of the given estimates in (125) yields the asymptotic formula for the specific heat maximum

$$C/Nk \sim (2/\pi)[\sinh 2H(H' + H\sinh 2H')^2$$
$$\times (\log n + \log(8/\pi \cosh 2H') + C_E)$$
$$+ \sinh 2H(H' - H\sinh 2H')^2$$
$$- 2(H'\sinh 2H)^2 \,\mathrm{gd}\, 2H'$$
$$- 2(H\sinh 2H')^2 \,\mathrm{gd}\, 2H]. \quad (126)$$

The result is not quite symmetrical with regard

FIG. 8. Superlattice at low temperatures in crystal for which the interaction energy between neighbors of opposite spin is $J > 0$ lengthwise and $-J' < 0$ transversely.

to H and H'; neither was such symmetry to be expected. Interchange of the two leads to the replacement

$$\log(n/\cosh 2H') \to \log(n/\cosh 2H).$$

For the case of quadratic symmetry

$$H = H' = \tfrac{1}{2}\log\cot \pi/8$$

the coefficient of $\log n$ attains a maximum. We find

$$C/Nk \sim (2/\pi)(\log\cot \pi/8)^2$$
$$\times (\log n + \log(2^{5/2}/\pi) + C_E - \tfrac{1}{4}\pi)$$
$$= 0.4945\log n + 0.1879. \quad (127)$$

The estimate indicated graphically by Kramers and Wannier[10] may be described by the equation

$$C/Nk = 0.48\log n + 0.21.$$

APPENDIX

We shall deal here with the evaluation of various integrals which occur in the text. Most of these can be reduced in straightforward fashion to complete elliptic integrals; only the partition function itself is of a type one step higher than the theta functions, and involves a little analysis which is not found in textbooks.

1. General Notation

One of the main reasons for setting this material apart is that the notation must be specified because there are several systems in common use. We shall adopt the notation used by Whitaker and Watson.[19]

[19] E. T. Whitaker and G. N. Watson, *Modern Analysis* (Cambridge University Press, 1927), fourth edition, Chapters 20–22. Referred to in the following as WW.

Theta Functions

$$\vartheta_3(z\,|\,\tau) = \vartheta_3(z) = \vartheta_4(z+\tfrac{1}{2}\pi)$$

$$= \sum_{-\infty}^{\infty} \exp\,(n^2\pi i\tau)\,\cos 2nz,$$

$$\vartheta_2(z\,|\,\tau) = \vartheta_2(z) = \vartheta_1(z+\tfrac{1}{2}\pi)$$

$$= \sum_{-\infty}^{\infty} \exp[(n+\tfrac{1}{2})^2\pi i\tau]\,\cos(2n+1)z,$$

$$\tag{1.1}$$

$$\vartheta_r = \vartheta_r(0); \quad \vartheta_r'' = \vartheta_r''(0); \quad (r=2,3,4).$$

Periods, Modulus, Comodulus

$$2K = \vartheta_3^2\pi; \quad 2iK' = \vartheta_3^2\pi\tau,$$

$$k = \sin\theta = \vartheta_2^2/\vartheta_3^2,$$

$$k' = \cos\theta = \vartheta_4^2/\vartheta_3^2, \tag{1.2}$$

$$\log q = \pi i\tau = -\pi K'/K.$$

We shall only have to deal with "natural" cases in which $-i\tau$ is real (not negative); $0 < k \leqq 1$; $0 \leqq k' < 1$. The ratio τ of the periods will be specified when necessary. Tables of $\log q$ are available.[20]

Jacobian Elliptic Functions

$$\mathrm{sn}(\vartheta_3^2z) = (\vartheta_3/\vartheta_2')\vartheta_1(z)/\vartheta_4(z),$$

$$\mathrm{cn}(\vartheta_3^2z) = (\vartheta_4/\vartheta_2)\vartheta_2(z)/\vartheta_4(z),$$

$$\mathrm{dn}(\vartheta_3^2z) = (\vartheta_4/\vartheta_3)\vartheta_3(z)/\vartheta_4(z), \tag{1.3}$$

$$\mathrm{am}\,u = \sin^{-1}(\mathrm{sn}\,u) = \cos^{-1}(\mathrm{cn}\,u).$$

We shall use Glaisher's notation for the quotients of the Jacobian elliptic functions

$$\mathrm{sc}\,u = \mathrm{sn}\,u/\mathrm{cn}\,u; \quad \mathrm{nd}\,u = 1/\mathrm{dn}\,u; \quad \text{etc.} \tag{1.4}$$

The Functional Relations

$$\mathrm{cn}^2\,u + \mathrm{sn}^2\,u = \mathrm{dn}^2\,u + k^2\,\mathrm{sn}^2\,u = 1 \tag{1.5}$$

are important for the uniformization of certain algebraic functions.

Elliptic Integrals of the First Kind

$$d(\mathrm{sn}\,u)/du = \mathrm{cn}\,u\,\mathrm{dn}\,u; \quad \text{etc.}$$

$$u = F(k,\,\mathrm{am}\,u) = \int_0^{\mathrm{am}\,u} (1-k^2\sin^2\varphi)^{-\frac{1}{2}}d\varphi, \tag{1.6}$$

$$K(k) = F(k,\tfrac{1}{2}\pi); \quad K'(k) = K(k').$$

---

[20] Four place tables of $\log q$ by $5'$ intervals of the modular angle are given by E. Jahnke and F. Emde, *Tables of Functions* (B. G. Teubner, Leipzig and Berlin, 1933), second edition, p. 122.

Elliptic Integrals of the Second Kind

These involve the following definitions and relations.

$$Z(u) = \vartheta_3^{-2}\vartheta_4'(\vartheta_3^{-2}u)/\vartheta_4(\vartheta_3^{-2}u),$$

$$E = \int_0^{\pi/2} (1-k^2\sin^2\varphi)^{\frac{1}{2}}d\varphi \tag{1.7}$$

$$= \int_0^K \mathrm{dn}^2\,u\,du = (-\vartheta_2''/\vartheta_2\vartheta_3^2)\pi/2,$$

$$E(u) = Z(u) + (E/K)u = \int_0^u \mathrm{dn}^2\,v\,dv.$$

Complete Elliptic Integral of the Third Kind

$$KZ(a) = \int_0^K k^2\,\mathrm{sn}\,a\,\mathrm{cn}\,a\,\mathrm{dn}\,a\,\mathrm{sn}^2\,u$$

$$\times (1-k^2\,\mathrm{sn}^2\,a\,\mathrm{sn}^2\,u)^{-1}\,du. \tag{1.8}$$

2. Uniformizing Substitution

The trigonometric functions of the angles and sides of a spherical triangle with two fixed parts can always be expressed in terms of single-valued functions by a suitable elliptic substitution. The same is true for hyperbolic (and plane) triangles. In any case the determination of the modulus is suggested by the sine proportion.

For the triangle of Fig. 4, with two given sides $2H'$ and $2H^*$, the simplest uniformizing substitution depends on which side is longer. For temperatures below the critical point we choose the modulus k, the imaginary parameter ia, and the new variable u as follows

$$k = k_0 = \sin\Theta_0 = \sinh 2H^*/\sinh 2H'$$

$$= 1/\sinh 2H \sinh 2H' < 1,$$

$$\mathrm{am}(ia) = \mathrm{am}(2iK_0'y,\,k_0) = 2iH', \tag{2.1a}$$

$$\mathrm{am}\,u = \delta'.$$

Then the trigonometric, *viz.*, hyperbolic functions of δ^*, δ', $2H^*$ and $2H'$ are given by the formulas

$$\sin\delta' = \mathrm{sn}\,u; \quad \sin\delta^* = k\,\mathrm{sn}\,u$$

$$\cos\delta' = \mathrm{cn}\,u; \quad \cos\delta^* = \mathrm{dn}\,u$$

$$\tan g' = \sinh 2H' = -i\,\mathrm{sn}\,ia$$

$$\cot g = \sinh 2H^* = -ik\,\mathrm{sn}\,ia \tag{2.2a}$$

$$\sec g' = \cosh 2H' = \mathrm{cn}\,ia$$

$$\mathrm{cosec}\,g = \cosh 2H^* = \mathrm{dn}\,ia.$$

We are using the abbreviation

$$g' = \text{gd } 2H'; \quad g = \text{gd } 2H = \tfrac{1}{2}\pi - \text{gd } 2H^* \quad (2.3)$$

for the gudermannian angles, which can often be employed to advantage in numerical computations.

For temperatures above the critical point the starred and the primed quantities exchange roles:

$$k = k_0 = \sin \Theta_0 = \sinh 2H'/\sinh 2H^*$$
$$= \sinh 2H' \sinh 2H = <1,$$

$$\text{am}(ia) = \text{am}(2iK_0'y, k_0) = 2iH^*, \quad (2.1b)$$

$$\text{am } u = \delta^*$$

and explicitly

$$\begin{aligned}
\sin \delta' &= k \text{ sn } u; \quad \sin \delta^* = \text{sn } u \\
\cos \delta' &= \text{dn } u; \quad \cos \delta^* = \text{cn } u \\
\tan g' &= \sinh 2H' = -ik \text{ sn } ia \\
\cot g &= \sinh 2H^* = -i \text{ sn } ia \\
\sec g' &= \cosh 2H' = \text{dn } ia \\
\text{cosec } g &= \cosh 2H^* = \text{cn } ia.
\end{aligned} \quad (2.2b)$$

The formulas for ω and γ are the same in the two cases. The abbreviation

$$M = \text{dn } ia \text{ dn } u - k \text{ cn } ia \text{ cn } u$$

will be convenient; then

$$\begin{aligned}
\cosh \gamma &= (\text{cn } ia \text{ dn } u - k \text{ dn } ia \text{ cn } u)/M \\
\sinh \gamma &= -ik'^2 \text{ sn } ia/M \\
-\cos \omega &= (\text{dn } ia \text{ cn } u - k \text{ cn } ia \text{ dn } u)/M \\
\sin \omega &= k'^2 \text{ sn } u/M.
\end{aligned} \quad (2.4)$$

Moreover, when ω and γ are considered as functions of u and a

$$\begin{aligned}
-\partial\omega/\partial u &= \partial\gamma/\partial a = k'^2/M, \\
\partial\gamma/\partial u &= \partial\omega/\partial a = ik'^2 k \text{ sn } ia \text{ sn } u/M.
\end{aligned} \quad (2.5)$$

The conformal mapping indicated by (2.5) may be described by the functional relation

$$\cot \tfrac{1}{2}(\omega - i\gamma)$$
$$= (1+k) \text{ sc } \tfrac{1}{2}(u+ia) \text{ nd } \tfrac{1}{2}(u+ia). \quad (2.6)$$

3. Alternative Substitution

While the substitution given above is preferred for most purposes, there are certain advantages to be gained by a Landen transformation to

modulus and periods given by

$$\begin{aligned}
k_1' &= \cos \Theta_1 = \coth(H' + H^*) \tanh|H' - H^*| \\
&= |\cos(g + g')|/\cos(g - g') = (1 - k_0)/(1 + k_0),
\end{aligned}$$

$$\begin{aligned}
k_1 &= \sin \Theta_1 = 2(\sinh 2H \sinh 2H')^{\frac{1}{2}}/ \\
&\qquad (1 + \sinh 2H \sinh 2H'),
\end{aligned} \quad (3.1)$$

$$K_1 = (1 + k_0)K_0,$$
$$K_1' = \tfrac{1}{2}(1 + k_0)K_0' = K_0'/(1 + k_1'),$$
$$\tau_1 = iK_1'/K_1 = \tfrac{1}{2}\tau_0 = \tfrac{1}{2}\tau.$$

The parameter

$$ia_1 = 2yiK_1' = \tfrac{1}{2}(1 + k_0)ia_0 \quad (3.2)$$

and the variable u_1 are given by

$$\begin{aligned}
\text{am}(ia_1, k_1) &= i(H' + H^*), \\
\text{am}(u_1, k_1) &= \tfrac{1}{2}(\delta' + \delta^*)
\end{aligned} \quad (3.3)$$

and the following relations hold

$$\begin{aligned}
\text{am}(K_1 - ia_1) &= \tfrac{1}{2}\pi - i|II' - II^*|, \\
\text{am}(K_1 - u_1) &= \tfrac{1}{2}\pi - \tfrac{1}{2}|\delta' - \delta^*|, \\
\text{am}(u_1 \pm ia_1) &= \tfrac{1}{2}(\pi - \omega \pm i\gamma).
\end{aligned} \quad (3.4)$$

Here the functions of the double parameter are of some interest

$$\begin{aligned}
\text{sc}(2ia_1) &= \text{sc}(4iyK_1') = i \cos(g - g'), \\
\text{nc}(2ia_1) &= \text{nc}(4iyK_1') = \sin(g - g'), \\
\text{dc}(2ia_1) &= \text{dc}(4iyK_1') = \sin(g + g').
\end{aligned} \quad (3.5)$$

This substitution shows up the analytic nature of the parameter through the critical point, and (2.6) is replaced by a simpler equivalent in (3.4). Moreover, formulas valid for all temperatures can be constructed by furnishing a sign for the comodulus

$$k_1' = |k_1''|;$$
$$k_1'' = \coth(H' + H^*) \tanh(H' - H^*). \quad (3.6)$$

That advantage together with the previous usage of Kramers and Wannier[10] decided the choice of elliptic substitution in the text.

4. Discussion of the Parameter

The imaginary parameter ia given by (2.1) can be evaluated as a real elliptic integral of the first kind. For that purpose we use Jacobi's imaginary transformation, described by the relations

$$i \, cs(iu, k) \, sn(u, k') = cn(iu, k) \, cn(u, k')$$

$$= dn(iu, k) \, cd(u, k') = 1 \quad (4.1)$$

and the parameter can be determined by the formula

$$am(a_0, k_0') = am(2yK_0', k_0') = gd \, 2H' = g',$$
$$a_0 = 2yK_0' = F(\cos \Theta_0, g'). \quad (4.2)$$

In tables this elliptic integral is often denoted by $F(90° - \Theta_0, g')$. Alternative formulas can be obtained by Landon transformations, whereby the "elliptic angle" πy is invariant. From (3.4) and (3.5) we obtain, respectively,

$$am(a_1, \cos \Theta_1) = am(2yK_1', k_1')$$
$$= gd(H' + H^*), \quad (4.3)$$
$$am(2a_1, \cos \Theta_1) = am(4yK_1', k_1')$$
$$= \tfrac{1}{2}\pi - g + g'. \quad (4.4)$$

Further Landen transformations are best described by the quotients of theta functions. The following are suitable for temperatures not too far removed from the critical point

$$\tan^2 \tfrac{1}{2}\Theta_2 = \sin \Theta_1$$
$$= (\tan 2g \tan 2g')^{\frac{1}{2}}/\cos(g - g'), \quad (4.5a)$$

$$(\cos \Theta_2)^{\frac{1}{2}} sn(a_2, \cos \Theta_2)$$
$$= \frac{\vartheta_1(2\pi y \mid -4/\tau)}{\vartheta_4(2\pi y \mid -4/\tau)} = \frac{\sinh |H' - H^*|}{\cosh(H' + H^*)}$$
$$= |\tan \tfrac{1}{2}(g + g' - \tfrac{1}{2}\pi)|, \quad (4.5b)$$

$$(\cos \Theta_2)^{\frac{1}{2}} sn(2a_2, \cos \Theta_2) = \frac{\vartheta_1(4\pi y \mid -4/\tau)}{\vartheta_4(4\pi y \mid -4/\tau)}$$
$$= \sin(g - g') |\cot(g + g')|. \quad (4.5c)$$

For temperatures not too close to the critical point we may take

$$k_{-1} = \sin \Theta_{-1} = \tan^2 \tfrac{1}{2}\theta_0, \quad (4.6a)$$

and the theta series involved in the following formula (imaginary argument) will converge rapidly:

$$-ik_{-1}^{\frac{1}{2}} sn(iyK_{-1}', k_{-1}) = \frac{\vartheta_1(\pi\tau y \mid 2\tau)}{i\vartheta_4(\pi\tau y \mid 2\tau)}$$
$$= \begin{cases} e^{-2H}; & (H' > H^*) \\ \tanh H'; & (H' < H^*). \end{cases} \quad (4.6b)$$

One or the other of the two formulas (4.5) and (4.6) will yield at least 3 significant figures per term of the theta series. When a table of $\log q$ is available (a table of complete elliptic integrals will do), it is not necessary to evaluate θ_2 or Θ_{-1} explicitly.

As regards the variation of the parameter a with the temperature, we note first of all that

$$y = a_0/2K_0' = a_1/2K_1'$$

is analytic at the critical point. This is evident from (3.1) and (3.3); the elliptic integrals which determine a_1 and K_1' are analytic functions of $\cos^2 \Theta_1$, which has no singularity there. In this sense the singularities of the various functions at the critical point are completely described by the degeneration of the period parallelogram

$$i/\tau = K/K' = \infty; \quad (H' = H^*).$$

For extreme temperatures, with

$$H = J/kT; \quad H' = J'/kT$$

the asymptotic estimates

$$\lim_{T=0} (2y) = H'/(H + H') = J'/(J + J'),$$
$$\lim_{T=\infty} (2y) = \tfrac{1}{2} \quad (4.7)$$

can be derived from (2.1).

The parameter is an unsymmetrical function of H and H'; the nature of the asymmetry is evident from either of the formulas (4.4) and (4.5c), which involve the double parameter: Interchanging the roles of H and H' replaces y by $\tfrac{1}{2} - y$, and the parameter ia_0 is replaced by its elliptic complement $i(K_0' - a_0)$. Thus (2.1a) and (2.1b) have the counterparts

$$am(iK' - ia, k_0) = am(iK_0'(1 - 2y), k_0)$$
$$= \begin{cases} 2iH; & (H' > H^*) \\ 2iH'^*; & (H' < H^*). \end{cases} \quad (4.8)$$

It is evident that if $J' < J$, we shall have

$$y < \tfrac{1}{4}; \quad a < \tfrac{1}{2}K'; \quad (J' < J) \quad (4.9)$$

for all temperatures, and in the special case of quadratic symmetry

$$\nu = \tfrac{1}{4}; \quad a = \tfrac{1}{2}K'; \quad (J = J'). \quad (4.10)$$

This is the reason why the results for that case

can be expressed by the complete integrals $K(k)$ and $E(k)$ alone, as given in the text.

5. Reduction of Integrals

By the substitution (2.2), the integrals of (113) in the text are readily evaluated as complete elliptic integrals. For example,

$$\int_0^\pi \cos \delta' d\omega = \int_0^{2K} k'^2 \operatorname{cn} u (\operatorname{dn} ia \operatorname{dn} u$$
$$- k \operatorname{cn} ia \operatorname{cn} u)^{-1} du; \quad (H' > H^*).$$

If we replace u by $2K - u$, the sign of $\operatorname{cn} u$ is changed; that of $\operatorname{dn} u$ is preserved. Accordingly, that part of the integrand which is odd with regard to $\operatorname{cn} u$ may be omitted, which yields

$$\frac{k \operatorname{cn} ia \operatorname{cn}^2 u}{1 - k^2 \operatorname{sn}^2 ia \operatorname{sn}^2 u} = k \operatorname{cn} ia - \frac{k \operatorname{cn} ia \operatorname{dn}^2 ia \operatorname{sn}^2 u}{1 - k^2 \operatorname{sn}^2 ia \operatorname{sn}^2 u}$$

whence

$$\int_0^\pi \cos \delta' d\omega = 2K(k \operatorname{cn} ia - k^{-1} \operatorname{ds} ia Z(ia));$$
$$(H' > H^*).$$

The other integrals (113) are treated similarly. The results are

$$\int_0^\pi \cos \delta^* d\omega = -2iK(k \operatorname{sn} ia \coth 2H^*$$
$$- Z(ia) \coth 2H'),$$
$$\int_0^\pi \cos \delta' d\omega = -2iK(k \operatorname{sn} ia \coth 2H'$$
$$- Z(ia) \coth 2H^*),$$
$$(5.1)$$
$$\int_0^\pi (\sin \delta^* \sin \delta'/\sinh \gamma) d\omega$$
$$= 2(K - E)(i/k \operatorname{sn} ia),$$
$$\sinh^2 2H' \int_0^\pi \sin^2 \delta^* \coth \gamma \, d\omega$$
$$= \sinh^2 2H^* \int_0^\pi \sin^2 \delta' \coth \gamma \, d\omega = -2iKZ(ia).$$

The mixed notation is designed to throw the distinction between the cases (2.2a) and (2.2b) entirely into the choice of modulus and parameter; so that the formulas (5.1) are equally valid for $H' < H^*$ and for $H' > H^*$.

The Z function of imaginary argument which enters into (5.1) can be expressed in terms of the tabulated functions of real argument $Z(a, k')$ as

follows:

$$Z(iu, k) = \operatorname{dn}(iu, k) \operatorname{sc}(iu, k)$$
$$- iZ(u, k') - i(\pi u/2KK'). \quad (5.2)$$

For substitution in (113) it is convenient to make use of the function

$$Z(iK' - ia) = -\operatorname{dn} ia \operatorname{cs} ia - i(\pi/2K) - Z(ia) \quad (5.3)$$

together with $Z(ia)$; this leads to more symmetrical formulas. The connection between the corresponding functions of real argument is

$$Z(a, k') + Z(K' - a, k')$$
$$= k'^2(-i) \operatorname{sc} ia \operatorname{nd} ia. \quad (5.4)$$

6. Derived Thermodynamic Functions

In assembling the terms of the formula (113a) for the energy it is best to deal with the cases $H' > H^*$ and $H' < H^*$ separately. The results are

$$U = -NJ'(\partial \log \lambda/\partial H')$$
$$- NJ(\partial \log \lambda/\partial H), \quad (6.1a)$$
$$\partial \log \lambda/\partial H' = \coth 2H'(2y + (2K/\pi)Z(2yK', k')),$$
$$(H' > H^*) \quad (6.1b)$$
$$\partial \log \lambda/\partial H = \coth 2H$$
$$\times (1 - 2y + (2K/\pi)Z((1 - 2y)K', k'))$$

valid for temperatures below the critical point; above the critical point we have instead

$$\partial \log \lambda/\partial H' = \coth 2H'$$
$$\times (2y - (2K/\pi)Z((1 - 2y)K', k'))$$
$$(H' < H^*). \quad (6.1c)$$
$$\partial \log \lambda/\partial H = \coth 2H$$
$$\times (1 - 2y - (2K/\pi)Z(2yK', k'))$$

The energy is continuous at the critical point because, when u is real [cf. (5.4)]

$$0 \leq Z(u, k') \leq 1 - k.$$

For the specific heat we obtain from (113b) and (5.1), in mixed notation valid for all temperatures

$$\frac{C}{Nk} = \frac{4}{\pi} \left(-iKZ(ia) \left(\frac{H'}{\sinh 2H'} \right)^2 \right.$$
$$- iKZ(iK' - ia) \left(\frac{H}{\sinh 2H} \right)^2$$
$$\left. + 2(K - E) \left(\frac{\operatorname{sn} ia}{i \sinh 2H'} \right) HH' \right). \quad (6.2)$$

Every term is positive. The specific heat is infinite at the critical point, where $k=1$. This is caused by the factor K; the general case is substantially like the special case of quadratic symmetry treated in the text.

7. The Partition Function

To compute the partition function (106) in terms of τ and y we first differentiate under the integral sign

$$\frac{\partial}{\partial a}\int_0^\pi \gamma\, d\omega = \int_0^{2K}\left(-\frac{\partial\gamma}{\partial a}\frac{\partial\omega}{\partial u} - \gamma\frac{\partial^2\omega}{\partial a\partial u}\right)du.$$

Integration by parts yields with the aid of (2.5)

$$\int_0^{2K}\left(-\frac{\partial\gamma}{\partial a}\frac{\partial\omega}{\partial u} + \frac{\partial\gamma}{\partial u}\frac{\partial\omega}{\partial a}\right)du$$

$$= \int_0^{2K}\left(\left(\frac{\partial\gamma}{\partial a}\right)^2 + \left(\frac{\partial\gamma}{\partial u}\right)^2\right)du$$

$$= \int_0^{2K}\left(\frac{\mathrm{dn}^2\, ia\, \mathrm{dn}^2\, u + k^2\, \mathrm{cn}^2\, ia\, \mathrm{cn}^2\, u}{1 - k^2\, \mathrm{sn}^2\, ia\, \mathrm{sn}^2\, u}\right)du.$$

The integral is easily evaluated; then considering that

$$\gamma(0, u) = \cosh^{-1}(1) = 0$$

for $a=0$, we have

$$\int_0^\pi \gamma\, d\omega = 2K\int_0^a \Phi(iu)du, \qquad (7.1)$$

$$\Phi(u) = 2\,\mathrm{dn}^2\, u - k'^2 - 2\,\mathrm{dn}\, u\, \mathrm{cs}\, u\, Z(u).$$

The function $\Phi(u)$ is even and periodic with the period $2K$. Its singularities are poles at the points

$$u = 2mK + niK'; \quad (m, n \text{ integers}; n \neq 0).$$

Jacobi's imaginary transformation (4.1), (5.2) yields

$$\Phi(iu) = -k'^2 + \mathrm{dc}(u, k')\, \mathrm{ns}(u, k')$$
$$\times (2Z(u, k') + (\pi u/KK')). \qquad (7.2)$$

Accordingly, the poles of $\Phi(u)$ are simple; the residues are

$$R(2mK + niK') = R(niK') = (-)^n n\pi i/K. \qquad (7.3)$$

It is easily seen that $\Phi(u)$ satisfies the standard

conditions for development in partial fractions *qua* function of $\cos(\pi u/K)$. The residues (7.3) together with the condition

$$\Phi(K) + \Phi(K+iK') = k'^2 - k'^2 = 0$$

suffice to determine the expansion

$$\Phi(u) = \left(\frac{\pi}{2K}\right)^2\left(1 - 8\sum_{n=1}^\infty (-)^n n\right.$$

$$\left.\times\frac{(q^n \cos(\pi u/K) - q^{2n})}{(1 - 2q^n \cos(\pi u/K) + q^{2n})}\right). \qquad (7.4)$$

The integration required by (7.1) is readily performed, and the formula (106) of the text becomes

$$-\frac{F}{NkT} = \log\lambda = \tfrac{1}{2}\log(2\sinh 2H) + \frac{1}{2\pi}\int_0^\pi \gamma\, d\omega$$

$$= \tfrac{1}{2}\log(2\sinh 2H) - \tfrac{1}{2}y\log q$$

$$- \sum_{n=1}^\infty (-)^n n\log\left(\frac{1 - q^{n+2y}}{1 - q^{n-2y}}\right). \qquad (7.5)$$

Specialization to the case $y = \tfrac{1}{4}$ yields the formula (119a) of the text, where $q_1^2 = q$.

The series of (7.5) converges rapidly except in the immediate neighborhood of the critical point. To obtain an expansion suitable for computation in that region we note from (7.2) that

$$\Phi(iu) - (\pi u/2KK')i\, \mathrm{cs}\, iu\, \mathrm{dn}\, iu$$

$$= 2\mathrm{dc}(u, k')\, \mathrm{ns}(u, k')Z(u, k') - k'^2$$

is a periodic function of u with the real period $2K'$. Its Fourier series is easily derived with the aid of the identity

$$K(2\,\mathrm{dc}\, a\, \mathrm{ns}\, a\, Z(a) - k^2)$$

$$= \int_0^K k^2(\mathrm{sn}(u-a)\,\mathrm{sn}(u+a) - \mathrm{cn}(u-a)\,\mathrm{cn}(u+a))du$$

from those of the Jacobian elliptic functions.[21] The result is

$$\Phi(iu) - \frac{\pi u}{2KK'}i\, \mathrm{cs}\, iu\, \mathrm{dn}\, iu$$

$$= 2\left(\frac{\pi}{K'}\right)^2\sum_{n=0}^\infty \frac{\cos((2n+1)\pi u/K')}{\sinh^2((2n+1)\pi K/K')}.$$

---

[21] WW, reference 19, 22.6.

Moreover,[22]

$$iu \text{ cs } iu \text{ dn } iu - \frac{d}{du}(u \log(-ik^{\frac{1}{2}} \text{ sn } iu))$$

$$= -\log(k^{\frac{1}{2}} \text{ sc}(u, k')) = \log \cot(\pi u/2K')$$

$$+ \sum_{n=0}^{\infty} \frac{2e^{-(2n+1)\pi K/K'} \cos((2n+1)\pi u/K')}{(2n+1) \sinh((2n+1)\pi K/K')}.$$

Termwise integration of the Fourier series yields the alternative expansion for the partition function

$$- F/NkT = \log \lambda = \tfrac{1}{2} \log(2 \sinh 2H)$$

$$+ \frac{K}{\pi} \int_0^a \Phi(iu)du = y \log(2 \sinh 2H')$$

$$+ (\tfrac{1}{2} - y) \log(2 \sinh 2H) + 2 \int_0^y \log \cot (\pi z)dz$$

$$+ \frac{1}{\pi} \sum_{n=0}^{\infty} \frac{(1 + (4n+2)(\pi i/\tau) - \exp[-(4n+2)\pi i/\tau])}{(2n+1)^2 \sinh^2((2n+1)\pi i/\tau)}$$

$$\times \sin((4n+2)\pi y). \quad (7.6)$$

The formula (119b) of the text is obtained by specialization to the case $H' = H$; $y = \tfrac{1}{4}$, with $\tau = 2\tau_1$, whereby

$$\int_0^{\pi/4} \log \cot x \, dx = 1^{-2} - 3^{-2} + 5^{-2} - 7^{-2} + \cdots = G.$$

[22] WW, reference 19, 22.5, example 3.

8. Critical Data

The transition temperature is given by the condition

$$\sinh 2H \sinh 2H'$$

$$= \sinh(2J/kT) \sinh(2J'/kT) = 1; \quad (T = T_c). \quad (8.1)$$

Then

$$\tau = 0; \quad K = \infty; \quad K' = \tfrac{1}{2}\pi$$
$$\pi y = a = \text{gd } 2H' = \tfrac{1}{2}\pi - \text{gd } 2H. \quad (8.2)$$

The formula (6.1) for the energy becomes

$$U_c = -(2/\pi)(NJ' \text{ gd } 2H' \coth 2H'$$
$$+ NJ \text{ gd } 2H \coth 2H) \quad (8.3)$$

and the formula (7.6) for the partition function simplifies to

$$- F_c/NkT = \log \lambda_c = \tfrac{1}{2} \log 2 + (\tfrac{1}{2} - 2y) \log \cot \pi y$$
$$+ 2 \int_0^y \log \cot \pi z \, dz. \quad (8.4)$$

The specific heat becomes infinite; the approximate formula

$$C/Nk \sim (4/\pi)(K(H + H' \sinh 2H)^2 \sinh 2H'$$
$$- H^2 \sinh^2 2H' \text{ gd } 2H - 2HH'$$
$$- H'^2 \sinh^2 2H \text{ gd } 2H') \quad (8.5)$$

where

$$K \sim \log(4/k') \sim \tfrac{1}{2}[\log(4T/|T - T_c|)$$
$$- \log(H \coth 2H + H' \coth 2H')]$$

is valid for temperatures near the critical point.

NOTES

PREFACE

1. Reed and Simon, *Methods of Modern Mathematical Physics*, p. vii.

2. So, for example, "Teller's proof [of the no-binding theorem for Thomas-Fermi atoms] is a very clever one, employing a fair amount of chitchat but almost no mathematics, but the proof is not without weaknesses" (Spruch, "Pedagogic Notes on Thomas-Fermi Theory," p. 191).

3. Simon, *Statistical Mechanics of Lattice Gases*, p. 17.

4. *Zustandsumme* is, it seems, Planck's term, in papers of 1921 and 1924. But Gibbs developed the concept somewhat earlier (1902). See Brush, *Statistical Physics*, pp. 77–78, 152–153.

5. Krieger, *Marginalism and Discontinuity*, chap. 1. See note 3, p. 312.

6. Wightman (introduction to Israel, *Convexity in the Theory of Lattice Gases*, p. lii) points out, following Gibbs and Uhlenbeck, that a finite-sized object in contact with a heat bath is our model system. The problem then lies in what is exchanged with the bath. Hence, rather than an infinite N, what we need is the notion of bulk matter.

7. See, for example, Simon, *Statistical Mechanics of Lattice Gases*, chaps. 3 and 4. See also Georgii, *Gibbs Measures and Phase Transitions*, pp. 1–7. The crucial figures here are R. L. Dobrushin, O. E. Lanford, and D. Ruelle (DLR).

8. I engage in this presumption throughout the book. For a lovely statement of the comparativist's position, see Zeitlin, introduction to Vernant, *Mortals and Immortals*, p. 22.

9. See Baxter, "Exactly Solved Models in Statistical Mechanics" (1985), for example, and note how differently this reads than Baxter's 1982 book with the same title.

10. Baxter, "Solvable Eight-Vertex Model."

11. Maillard, "Introduction," p. v.

12. Stephen Brush and a volume edited by Lillian Hoddeson provide the initial forays for a history of the Ising model; Walter Thirring provides a brief survey for the stability of matter (Brush, "Lenz-Ising Model"; Hoddeson et al., *Out of the Crystal Maze*, pp. 518–533; Thirring, introduction to Lieb, *Stability of Matter*). For a lovely bibliographical essay, especially about more recent work, see Georgii, *Gibbs Measures and Phase Transitions*, pp. 450–454.

13. Here I am thinking in particular of Michael Fisher, who has made major contributions to both problems that concern me in this book. See his "Free Energy of a Macroscopic System," for example, as well as his many papers on phase transitions.

14. It is perhaps a coincidence that antisymmetry is needed to account for the stability of matter and for the existence of phase transitions (in two

dimensions, at least). And it is perhaps another such coincidence that the fundamental work on both of these problems was done first by Lars Onsager, in 1939 and 1944, respectively. It is just these coincidences that provided an extra pleasure in working out this book.

15. Newell and Montroll, "Theory of the Ising Model," p. 362.

16. Here I am thinking of Buchwald's *From Maxwell to Microphysics*, and Kuhn's *Black-Body Theory and the Quantum Discontinuity*. Of course, we see such care in analyses of Copernicus or Ptolemy or the Babylonians, in the history of astronomy.

CHAPTER ONE

1. I believe novelists do much the same as do other modelers; they find themselves constrained by the facts of characters and plots.

2. Krieger, *Marginalism and Discontinuity*.

3. It should be noted that the bargains may be cooperative and collusive, that competitive equilibrium implies that the individuals do not collude, and that, in such a competitive market, prices are not negotiated but are taken as given, individuals deciding what to buy and sell based on alternatives available to them. For a brief survey, see Hildenbrand, "Cores"; Roberts, "Large Economies."

In general, the convergence we shall be concerned with is an asymptotic convergence, a convergence to a certain form or shape. The central limit theorem speaks of convergence to a Gaussian; here the asymptotic convergence, at a rate $1/N$, is to a competitive equilibrium. One should not confuse the norming constants $(1/\sqrt{N})$ of the central limit theorem with measures of rates of convergence (such as the Berry-Esseen theorem). When talking about the infinite volume limits in the constitution of matter, the relevant rate of convergence would perhaps be a measure of the surface area enclosed divided by the volume ($\approx V^{1/3} \approx 1/N$, for a fixed density). See note 11, where agents are points on a real line and the $1/N$ convergence measure is put aside for the advantage of more precisely modeling the negligible influence each agent has on prices.

4. Here I am thinking of the work of Jed Buchwald on Maxwellian electrodynamics, James Cushing on S-matrix theory, Thomas Kuhn on black-body radiation, and S. S. Schweber on quantum field theory. Cushing, in *Theory Construction*, is quite explicit about this: "My claim is that the origin of methodologically interesting ideas and questions, at least in modern physics, lies in the (highly) *technical* details of practice" (p. xv).

5. Of course, mathematics' modes of observation are somewhat different (Tragesser, *Husserl and Realism in Logic and Mathematics*).

6. McCoy and Wu, *Two-Dimensional Ising Model*, pp. 43, 248, 402.

7. Simon, *Statistical Mechanics of Lattice Gases*, p. 153, notes that only for the Ising model do we rigorously know that there is just one critical temperature: for correlations, for spontaneous magnetization, and for the specific heat. See Lebowitz and Martin-Löf, "Uniqueness of the Equilibrium State"; Abra-

ham and Martin-Löf, "Transfer Matrix." Below the critical point there is no unique state, since there are many directions of alignment. And the decay of correlations depends on the analyticity of the free energy (the "pressure," in their terms), which of course does not hold at phase transition.

8. Here I am thinking of Samuel's use of Grassmann integration ("Use of Anticommuting Variable Integrals"), Baxter's various methods, and the renormalization-group formulations by a number of authors.

9. Of course, theoretical physicists may use formalism as such to push through their work. But such formalism almost always turns out to be meaningful physically. In a defense of the freedom of the theoretical physicist, Steven Weinberg in *Dreams of a Final Theory* attacks the positivists and Bridgman in particular for their operationalism. But that we do not require that formalism be about the physics does not then prevent its being about the physics on its own.

10. These limits are asymptotic limits, limits that reflect the functional shape, the emergent form, of the sum.

11. See note 3, especially Roberts, "Large Economies." The analogy to the problem of infinite volume limit is quite precise, although, for the physicists, statistical averaging does the work of going to negligibly small influence, the role performed by going from a countable infinity to the real line in Aumann's model of markets.

12. Baxter, *Exactly Solved Models*. Baxter distinguishes exactly solved models from rigorously solved ones, for one might well guess the answer to a model without having a rigorous proof of it. There are some wonderful examples of this, for which see Baxter, "Exactly Solved Models," in which a well-guessed solution by Zamolodchikov is eventually shown by Baxter to be provably correct.

13. Yet, in some important work, such as Yang's derivation of the spontaneous magnetization, we are still left feeling that formalism and mathematics rather than physics seems to have motivated the flow, at least after the initial insight.

14. For example, Baxter's eight-vertex model incorporates and illuminates the Ising model.

15. It makes perhaps for inadequate philosophy and bad history, but my interest is rather more in meaning than in analysis or in story.

16. Here I an thinking of the work of Nancy Cartwright (*How the Laws of Physics Lie*) and her considerations of Pierre Duhem and the meaning of mathematical symbolism in scientific work, and of Hartry Field in *Science without Numbers*, which is surely not science without models, as well as work on mathematical fictionalism discussed by Liston ("Taking Mathematical Fictions Seriously"). Livingston's description of the "Lebenswelt pair" (in *Ethnomethodological Foundations of Mathematics*) strikes me as close to actual practice, in which a model allows us to see the world anew, the model being a guide to the world, an indexical device.

I remain agnostic about mathematical realism as interpreted by philosophers. Mathematicians in their work are naive realists, and believe mathematics is true and mathematical objects are real. (See Hersh, "Some Proposals.") This is not probative for the philosophers. I actually have some sympathy for Hartry Field's notion that for physics mathematics needs to be good (and not true) in the sense that it is a "conservative" instrument or auxiliary theory needed to do one's work, preserving the truth of physical theories so that any physical argument with mathematics can be replaced by one without: "underlying every good extrinsic explanation [physics with mathematics] there is an intrinsic explanation [physics alone]" (Field, *Realism, Mathematics, and Modality*, p. 44; see also Field, *Science without Numbers*; Maddy, *Realism in Mathematics*, pp. 159–163). I believe in the physics. Field would also allow for mathematics as a discovery device, my main concern.

I think the curious artifact that causes more trouble than good is the distinct separation by many philosophers of mathematics and physics.

17. Schultz, Mattis, and Lieb, "Two-Dimensional Ising Model," p. 867.

18. Schrödinger, *Statistical Thermodynamics*, pp. 36–37.

19. Onsager, "Ising Model in Two Dimensions," p. 9. See also Baxter, *Exactly Solved Models*, p. 103, who refers to the elliptic functions as "delightfully easy" to use.

20. Dyson, preface to Lieb, *Stability of Matter*, p. v.

21. Wilson, "Renormalization Group," p. 805.

22. I should note that I am not considering how philosophical or ontological presuppositions are taken by mathematicians as formative of their work, for examples of which see Hankins, *William Rowan Hamilton*, on the Kantian-Coleridgian influence on Hamilton; Crowe, *History of Vector Analysis*, on both Hamilton and Grassmann. From Gibbs's testimony as well, it is clear that Grassmann deserves to be a better-sung hero than he is.

23. Dyson, "Missed Opportunities," p. 646.

24. James Cushing emphasizes the importance of what he calls technical advances to the progress of theoretical physics, in *Theory Construction*.

25. See, for example, Dyson, *Origins of Life*.

26. Whether it be about abundances of elements or the cosmological constant or Ω in the big bang or the saltiness of the oceans, these issues are addressed by the anthropic principle, for which see Barrow and Tipler, *Anthropic Cosmological Principle*.

27. Dyson, "Missed Opportunities"; Peierls, *More Surprises in Theoretical Physics*.

28. See Krieger, *Marginalism and Discontinuity*; Krieger, *Doing Physics*.

29. Yang, *Selected Papers*, p. 12.

30. Newell and Montroll, "Theory of the Ising Model," p. 377.

31. Onsager, "Crystal Statistics." An Ising model assigns to each vertex of a grid a "spin" that takes two values, and a spin–adjacent-spin interaction, $-J\sigma_1\sigma_2$—which is taken to model a ferromagnet or an alloy mixture.

32. Newell and Montroll, "Theory of the Ising Model," sec. 4.2 in particular. See also Onsager, "Ising Model in Two Dimensions." See chapter 4 for my own attempt.

33. Here I am thinking of Samuel's use of Grassmann integration ("Use of Anticommuting Variable Integrals") and Zamolodchikov's techniques, which Baxter describes in "Exactly Solved Models."

34. Pickering, *Constructing Quarks*; Pickering, *Mangle of Practice*. Another variety of example is provided by Willis Lamb's attempt to account for what we now call the "Lamb dip," as recounted by Cartwright in *Nature's Capacities and Their Measurement*, sec. 2.2, "Causes in Mathematical Physics." At first, Lamb's derivation did not accord with experiment. Carefully going through the mathematical derivation, he tries to find the presumed technical error, an error that might lead him to understand the physics. Eventually, Lamb discovers that he had done an integral too soon, corresponding to actual physics that would preclude such a dip.

35. See Onsager, "Ising Model in Two Dimensions."

36. See ibid., p. 8; Yang, *Selected Papers*, p. 13.

37. Yang, *Selected Papers*, p. 15.

38. Kramers and Wannier, "Statistics of the Two-Dimensional Ferromagnet"; Montroll, "Statistical Mechanics of Nearest Neighbor Systems."

39. Baxter, *Exactly Solved Models*, pp. 404–409, quoted from p. 409.

40. Andrews, *q-Series*.

41. Buchwald, *From Maxwell to Microphysics*; Siegel, *Innovation in Maxwell's Electromagnetic Theory*.

42. Wheeler, *Journey into Gravity and Spacetime*.

CHAPTER TWO

1. For a nice statement concerning multiple rederivations, in the context of mathematics, see Rota, "Pernicious Influence."

2. Lieb, "Stability of Matter," p. 554. See note 2, p. 311.

3. Dyson, preface to Lieb, *Stability of Matter*. I should note that Fisher and Ruelle ("Stability of Many-Particle Systems") are credited with setting up this research program for proving the stability of matter. See also Fisher, "Free Energy of a Macroscopic System." See Thirring's introduction to Lieb, *Stability of Matter*, for a history. The more perspicuous solution is due to Lieb and Thirring, "Bound for Kinetic Energy of Fermions," and Lebowitz and Lieb, "Existence of Thermodynamics."

4. Lieb and Lebowitz, "Constitution of Matter," pp. 364–377, summarized in Lieb, "Stability of Matter," pp. 567–568. As for the electric charge, either the surface charge in the limit is negligible, or it dominates and there is no thermodynamic limit, or there is just a fixed surface charge whose contribution to the free energy is separable from the bulk. As for bosons, see note 15 below —the basic point is that bosons can get close enough to each other to lead to instability, while fermions cannot.

5. Lieb, "Stability of Matter," p. 563.

6. Ibid., p. 553.

7. Ibid., p. 555. Note that the Sobolev inequality is dimension dependent.

8. Lieb develops a technically weaker but sufficient result, using the Holder inequality, in which there is no external power on the integral, that is, the kinetic energy is bounded by an integral of the density to the $5/3$ power (rather than the cube root of the integral of the density to the third power, as in Sobolev), which as he points out makes for a nice extensive additivity of the effect of density peaks: $\int \rho^{5/3}$ versus $[\int \rho^3]^{1/3}$.

9. In the context of the "relativistic" Schrödinger equation, $-Z\alpha < 2/\pi$ for stability, where $\alpha \approx 1/137$ (Lieb, "Stability of Matter: From Atoms to Stars," p. 35).

10. Ibid., p. 13. I have replaced some symbols with words.

11. The Thomas-Fermi approximation, suited to atoms with large numbers of electrons (high Z), is a $Z = \infty$ model (just as the hydrogen atom is a $Z = 1$ model) and is the source of this limit. The Thomas-Fermi atom provides a lower limit on the energy, E: $E_{matter} > E_{Thomas-Fermi} > \Sigma^N E_{i,Thomas-Fermi}$ for N atoms. See Spruch, "Pedagogic Notes on Thomas-Fermi Theory." It should be noted that the Thomas-Fermi model is said to be statistical in the sense that it works with average properties, namely, the average density, rather than because it works within the realm of quantum statistical mechanics.

12. Lieb, "Stability of Matter," pp. 556–557.

13. Ibid., p. 557. Note that only the electrons need be fermions; the nuclei could presumably be bosons. See note 15.

14. Lieb takes this quote from Dyson, "Ground-State Energy," who gets it from Ehrenfest, Collected Scientific Papers, p. 617.

15. See Dyson, "Ground-State Energy," pp. 1538–1541. In fact, if we have a nonrelativistic Hamiltonian, with no exclusion principle, then

> The binding energy of a macroscopic number of charges. . . would be at least of the order of. . . 1 megaton. . . . In particular, any material containing hydrogen would be a nuclear explosive. . . . This argument shows that the decisive factor in making matter without exclusion principle unstable is the cooperative effect of many particles in screening each other. The charge cloud around each particle is composed not of one or two nearest neighbors, but of a large number, $N^{2/5}$, of slightly distorted wavefunctions. This enables the charge clouds to be produced with a very small expenditure of kinetic energy. The exclusion principle makes matter stable by forbidding such a cooperation of many particles with small momentum.

So Dyson also accounts for the $-N^{7/5}$ in the ground-state energy of bosons.

16. Lieb, "Stability of Matter: From Atoms to Stars," pp. 22–23.

17. Ibid., p. 21.

18. Lieb, "Stability of Matter," p. 563.

19. "Unpleasant" is Lieb's term.

20. Lieb, "Stability of Matter," p. 563.

21. Lieb, "Stability of Matter: From Atoms to Stars," p. 23.

22. Ibid., p. 27.

23. Ibid., pp. 22–23.

24. See Thirring, introduction to Lieb, *Stability of Matter*; Lebowitz and Lieb, "Existence of Thermodynamics."

25. Schweber has emphasized how pervasive this hierarchization has become, effective Lagrangians justified by renormalization-group arguments, separating out the scales of physical phenomena in a very deep way, the infinite volume limit at each scale leading to condensed matter having its own sort of properties at the next scale ("Physics, Community, and the Crisis in Physical Theory").

26. Lieb, "Stability of Matter: From Atoms to Stars," p. 15.

27. Lieb points out that this is a small-r effect still, and depends on a "no-binding" (among atoms) theorem in the Thomas-Fermi theory of large atoms. Lieb and Thirring, in their proof, much improved over Dyson and Lenard's proof of the stability of matter, uses this no-binding to provide space for each atom. Dyson suggests in his "Ground-State Energy" paper of 1967 that the Thomas-Fermi model would be effective in getting a much tighter limit (at least for bosons) than he and Lenard achieved, but "[w]e have no idea at present how the conjecture might be proved" (p. 1545).

For a nice survey, see Spruch, "Pedagogic Notes on Thomas-Fermi Theory," pp. 190–194.

28. Here I am thinking of both the Darwin-Fowler derivation of the canonical ensemble (see chapter 3), where the steepest descent sum occurs at the well-defined temperature, and the Stirling's approximation derivation, where technical features such as the Lagrange multipliers pick out the temperature variable. Note that N/V's being constant appears in the Darwin-Fowler derivation by the fact that the energies of the states remain unchanged as N goes to infinity, there being no interaction due to "greater density" (greater number of systems).

See Parisi, *Statistical Field Theory*, pp. 1–2, 8, on averaging, and not much fluctuation, and thermal diffusion evening things out.

29. The canonical and grand canonical ensembles are equivalent here.

CHAPTER THREE

1. Spruch, "Pedagogic Notes on Thomas-Fermi Theory," p. 191.

2. Yang and Lee, "Statistical Theory of Equations of State," p. 405 n. 4.

3. Darwin and Fowler, "On the Partition of Energy," pp. 450–451.

4. Huang, *Statistical Mechanics*, sec. 8.1; Schrödinger, *Statistical Thermodynamics*, pp. 27–41; Feynman, *Statistical Mechanics*, pp. 1–6.

5. See McCoy and Wu, *Two-Dimensional Ising Model*, p. 13; Huang, *Statistical Mechanics*, p. 212. The quote is from Darwin and Fowler, "On the Partition of Energy," p. 450.

6. See Feynman and Hibbs, *Quantum Mechanics and Path Integrals*, p. 271.

7. Here I follow, in somewhat technical detail, Huang and Schrödinger, who are close to Fowler's presentation in his text. (The source, it would seem, is Fowler, *Statistical Mechanics* [1936], rather than Darwin and Fowler's 1922 paper, "On the Partition of Energy.") McCoy and Wu offer a mathematically different derivation, one that employs an integral representation of a delta function rather than a generating function and a Cauchy integral, and that is in the complex-β plane. They fulfill a promise in Darwin and Fowler's original paper, "On the Partition of Energy," p. 465. For our purposes, McCoy and Wu's derivation exhibits many technical phenomena similar to Huang's (Huang, *Statistical Mechanics*, pp. 206–213; Schrödinger, *Statistical Thermodynamics*, pp. 27–41; McCoy and Wu, *Two-Dimensional Ising Model*, pp. 9–18).

8. If $A(s) = \sum_i a_i s^i$, then $A(s)$ is a generating function for the series $\{a_i\}$. The grand partition function, *GPF*, the partition function for the grand canonical ensemble, is a generating function for the n-particle partition functions, Q_n: $GPF(z) = \sum_n z^n Q_n$. The analogy with G is manifest. And the partition function, $Q(\beta) = \int \exp - \beta E \, \rho(E) \, dE$, is a generating function for the moments, in E, of the density of states, $\rho(E)$—merely expand the exponential in a power series—as well as being its Laplace transform.

9. β real and positive is thermodynamics, β imaginary is quantum mechanics, and β real and negative applies to systems with inverted populations, as in nuclear magnetic resonance or in lasers, but these, too, are unstable.

10. See McCoy and Wu, *Two-Dimensional Ising Model*, pp. 13–14; Schrödinger, *Statistical Thermodynamics*, p. 31. Note that the E_i are independent of N_S by definition. But in the particle-as-system interpretation of the Darwin-Fowler derivation, so that N_S refers to the number of particles, this is not true in fact. See Landau and Lifshitz, *Statistical Physics*, p. 29; Feynman, *Statistical Mechanics*, chap. 1, presents a particularly nice physical account.

11. One wants g to be monotone increasing and be greater than one, and the assumption that the ground state has energy zero and that the energies are rationally incommensurable is crucial for this. See Huang, *Statistical Mechanics*, pp. 209–210; Schrödinger, *Statistical Thermodynamics*, pp. 29–30; McCoy and Wu, *Two-Dimensional Ising Model*, pp. 12–13, where it would seem that condition 3 is not directly invoked.

12. Schrödinger, *Statistical Thermodynamics*, pp. 28–29, 31–33; Darrigol, *From c-Numbers to q-Numbers*, p. 305. By the way, Dirac was Fowler's student.

13. If the contour is the circle $r \exp i\theta$, and our units are such that $r = 1$, then the contributions to the contour integral are $\sum_j \exp i\theta E_j$. The absolute value of this sum goes as $\sqrt{N}$ because of the mutual incommensurability of the energies, except on the real line ($\theta = 0$) where the sum goes as N. Recall Feynman's showing how quantum mechanics becomes classical in the short wavelength limit when interference kills the amplitude except on the classical trajectory (which is in effect a "saddle line") (Feynman, Leighton, and Sands, *Feynman Lectures on Physics*, vol. 2, chap. 19).

14. There are no terms z^{E_i}, where E_i is negative (Huang, *Statistical Mechanics*, equation 10.16).

15. McCoy and Wu, *Two-Dimensional Ising Model*, pp. 15–18, especially equation 3.37.

16. Van der Waerden, "Die lange Reichweite"; Kac and Ward, "Combinatorial Solution"; Kastelyn, "Dimer Statistics and Phase Transitions."

17. Schrödinger, *Statistical Thermodynamics*, p. 23.

18. Darwin and Fowler, "On the Partition of Energy: Part 2," pp. 840–841.

19. Fourier is presumably the source of this. See Lawden, *Elliptic Functions and Applications*, pp. 1–3. The nome q is $\exp - 4\alpha t$ (where α is the time constant and t is the time), essentially the decay factor for the heat distribution.

20. Whittaker and Watson, *Modern Analysis*, pp. 462–463. See also Griffiths, "Microcanonical Ensemble," for a more recent consideration of the inversion issue.

21. McCoy and Wu, *Two-Dimensional Ising Model*, pp. 7–8.

22. Hence, many derivations implicitly use a concept of entropy. See also Simon, *Statistical Mechanics of Lattice Gases*, on states.

23. As, for example, in Ruelle, *Statistical Mechanics*, or in Georgii, *Gibbs Measures and Phase Transitions*. More generally, what one does is, rather than concern oneself with each limit process separately, as it appears, one studies operators and variables in which the limit itself is well defined. The $\exp - \beta E$ conditional probability is assumed to start with.

24. Lebowitz, "Statistical Mechanics," p. 389.

25. Schweber, "Physics, Community, and the Crisis in Physical Theory."

26. Once $N_S \to \infty$, there is likely to be a pole (between the two saddle points) within the contour, so the derivation would not go through in its current form.

27. Huang, *Statistical Mechanics*, pp. 209–211. Curiously, were G a partition function of a system, which it is for a system of noninteracting components, the analysis of Lee and Yang, ("Statistical Theory of Equations of State") suggests that only one such temperature or value of an intensive variable is a critical temperature between two phases. The existence of two or more minima indicates there are other phases of such matter.

28. McCoy and Wu, *Two-Dimensional Ising Model*, pp. 12–13, especially equations 3.16–3.18.

29. See equation 10.25 in Huang, *Statistical Mechanics*—the "$1/\sqrt{N_S}$" corrections to the partition function are now to the free energy ($\approx - \ln Q$), and are actually $\approx (1/2N_S) \ln N_S$, where N_S is the number of systems. But N_S might be interpreted here as the number of particles, each system being a particle. All of this is reminiscent of higher-order terms in a central limit theorem.

30. See equations 10.24 and 10.25 in ibid., where N_S (M in Huang) nicely separates out—the crucial fact being that we are at a saddle point.

31. Feynman, *QED*.

32. I take this argument from Parisi, *Statistical Field Theory*, pp. 2, 8, 9. I have discussed diffusion, equilibrium, and the central limit theorem in *Doing Physics*, pp. 21, 22, and notes therein.

33. This adherence of physics to objects includes the experimental apparatus, and is true for much of scientific practice. See, for example, Frank, "Instruments, Nerve Action."

34. Kuhn, *Structure of Scientific Revolutions*, chap. 11. Of course, there are physicists, such as Fermi and Feynman, who go out of their way to hide the mathematics, including extensive calculations they themselves have performed, and in this fashion make the physics central to the discovery.

35. Thirring (introduction to Lieb, *Stability of Matter*, pp. 3–5) shows that there are other such limits, nonlinear ones, which define a meaningful infinite volume limit (for example, $N^{-7/3}E$) but do not give one stable matter or mechanical stability. See the last section of this chapter.

36. Dresden, "Kramers's Contributions to Statistical Mechanics."

37. Lieb and Lebowitz, "Constitution of Matter."

38. Ibid., p. 322.

39. Uniform convergence of continuous functions means that their limit is a continuous function. In general, we expect the infinite volume limit of the thermodynamic functions to be smooth, except perhaps at phase transitions. Uniform convergence—$|h_n(z) - h(z)| < \epsilon'$, if $n > n(\epsilon')$ for all z—is convergence over a *range* of the independent variable z, $n(\epsilon')$ being independent of z.

40. Van Hove, "Quelques propriétés générales"; Yang and Lee, "Statistical Theory of Equations of State." I should also mention Fisher, "Free Energy of a Macroscopic System" (which on p. 378 indicates flaws in van Hove's and Lee and Yang's proofs), and Fisher and Ruelle, "Stability of Many-Particle Systems."

41. So one sees references to van Hove and Fisher sequences of volumes: Lieb and Lebowitz, "Thermodynamic Limit for Coulomb Systems," p. 152; Lieb and Lebowitz, "Constitution of Matter," p. 351; Simon, *Statistical Mechanics of Lattice Gases*.

42. The sources for the following argument are Fisher and Ruelle, "Stability of Many-Particle Systems"; Lieb and Lebowitz, "Constitution of Matter."

43. As good extensive variables, we would expect the free energies of independent systems to just add; then the partition functions should multiply—although this is also clear from writing out the terms.

44. Huang, *Statistical Mechanics*, pp. 458–459. A similar argument can be made for the partition function.

45. Ibid., pp. 321–328, 458–464.

46. See ibid., equation 15.33.

47. When the canonical ensemble and partition function are considered, then uniform convergence depends on the fact that corridor depth (the range of the force) rather than the volume of the bulkette determines the convergence rate (ibid., p. 324).

48. Lieb, "Stability of Matter," pp. 563–564.

49. Ibid., p. 564.

50. As r gets larger, the density of the surface charge (where all the excess charge is) decreases.

Onsager had the original insight, that the total electrostatic field energy is positive ("Electrostatic Interaction of Molecules," pp. 189–190).

Consider all the particles of our assembly immersed one at a time in our conducting fluid. When all the particles are immersed, the energy of the whole system equals $-\Sigma u_i$ [the energies released by immersing a charged particle in the fluid], regardless of the arrangement of the particles. Now the removal of the fluid cannot release any more energy; on the contrary, some energy must in general be expended to accomplish it. Thus, however the particles may be arranged, we know a lower bound for the energy: $U = \Sigma u_{ik}$ [the Coulomb interaction energies] $\geq -\Sigma u_i$.

See the appendix for a reprint of this article.

For an interpretation, see Lieb and Lebowitz, "Thermodynamic Limit for Coulomb Systems," p. 140; Dyson and Lenard, "Stability of Matter," p. 426; Spruch, "Pedagogic Notes on Thomas-Fermi Theory," p. 191—and the interpretations no longer sound so chemical as they are particulate.

51. Lebowitz and Lieb, "Existence of Thermodynamics"; Lieb, "Stability of Matter," p. 566.

52. Lieb, "Stability of Matter," p. 566, equation 88.

53. Ibid., p. 565.

54. Lebowitz and Lieb, "Existence of Thermodynamics," p. 633; Lieb, "Stability of Matter," pp. 564–566.

55. Lieb, "Stability of Matter," p. 565, equation 69. This is not the case at phase transition.

56. Ibid., p. 565, equation 79.

57. This part of the proof invokes the Peierls-Bogoliubov inequality: if $A \to \operatorname{Tr} A$ is convex, then $\operatorname{Tr} \exp(A + B) \geq \operatorname{Tr} \exp A \exp\langle B\rangle$, where $\langle B\rangle = (\operatorname{Tr} B \exp A)/(\operatorname{Tr} \exp A)$. Only if the balls are independent in the sense that their operators are independent (that is, A is a tensor product of separate bulkette operators) is the averaging as stated. Here, $A = -\beta H_{\text{bulkettes}}$ and $B = -\beta H_{\text{interbulkette}}$, where B is a perturbation on A (Lieb, "Stability of Matter," p. 565).

58. Lebowitz and Lieb, "Existence of Thermodynamics," p. 633; Lieb and Lebowitz, "Thermodynamic Limit for Coulomb Systems," pp. 151–154. The excess charge goes to the surface of a ball or bulkette. One creates a corridor or shell that contains that excess charge, and an outer shell that contains a counter charge; one then shows that the corridor and shell terms are negligible in the infinite volume limit.

59. Put differently, the convergence rate is independent of the volume, and depends only on a/r_0, as in the last formula of this paragraph, where $c = c(a/r_0)$.

60. Parisi, *Statistical Field Theory*, p. 42.

61. Note that a nonnegative compressibility is the same as the equivalence of the canonical and grand canonical ensembles, as van Hove showed in "Quelques propriétés générales." Also, I use the notion of thermodynamic stability to refer to nonnegative compressibility and nonnegative specific heat.

62. Thirring, introduction to Lieb, *Stability of Matter*, p. 5.

63. I have taken this from the wonderful essay by Wightman, introduction to Israel, *Convexity in the Theory of Lattice Gases*, pp. xiv–xv.

64. Lieb, "Stability of Matter," p. 567.

65. Ibid.

66. Widom, "Equation of State." Conversely, by the way, it may require a great deal of experimental device and care, at least in the initial experiments, for a system to exhibit a delicate limiting property. So, for example, scaling phenomena near the critical point (that properties depend on $|T - T_c|^\alpha$) were not so easy to see initially.

67. I speak of "moves" deliberately, to avoid claiming that every arithmetic step has such a meaning, which is probably asking too much.

Taking properties, here given real numbers, to be linked by arguments, where $\rightarrow$ means "corresponds to," $\Rightarrow$ means mathematical proof or argument, and $\leftrightarrow$ means a physical argument, we might say diagrammatically (figure N.1), physical argument $AD \approx$ mathematical argument $ABCD$.

So we might say PhysicalProperty$_1$ $\rightarrow$ MathematicalProperty$_1$ $\Rightarrow$ MathematicalProperty$_2$ $\rightarrow$ PhysicalProperty$_2$. Moreover, I have been suggesting that many of the mathematical properties deduced along the way, let us call one of them

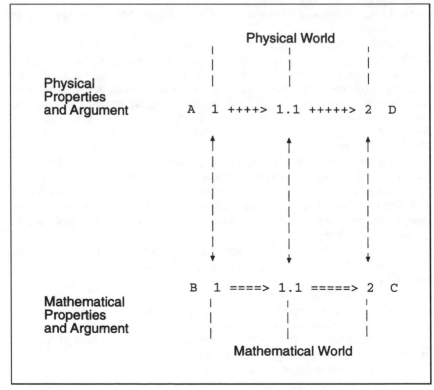

FIGURE N.1 Mathematics and physics and the world.

MathematicalProperty$_{1,x}$, correspond to a significant PhysicalProperty$_{1,x}$. And presumably there are $\leftrightarrow$ arrows connecting PhysicalProperty$_{1,x}$ to PhysicalProperty$_{1,x+e}$ and to PhysicalProperty$_{1,x-e}$ (and $\Rightarrow$ arrows connecting the corresponding mathematical properties, indicated by $\rightarrow$ arrows).

68. See Livingston, *Ethnomethodological Foundations of Mathematics*, on what he calls the "Lebenswelt pair."

69. Tragesser, *Husserl and Realism in Logic and Mathematics*, chap. 4.

70. So figure N.1 might be altered a bit, and if one believes that the world observed is the same for mathematics and physics, then one has something like a cylinder: the mathematical world and the physical world are taken as phenomenological aspects of the same world.

71. Lebowitz and Lieb, "Existence of Thermodynamics," pp. 632–633.

72. Lieb, "Stability of Matter," pp. 567–568; Lebowitz and Lieb, "Existence of Thermodynamics," pp. 632–633; Lieb and Lebowitz, "Constitution of Matter," pp. 364–377. If we turn off the charge before going to the infinite volume limit, there is no reason for the density to be uniform, and, for example, the electrons and the nuclei might well be separated from each other. If we turn off the charge afterward, then on the way to the infinite volume limit the long-range effect of the Coulomb force, namely screening and local neutrality, provides for a uniform density.

73. Thirring, introduction to Lieb, *Stability of Matter*, discusses Landsberg's "Is Equilibrium Always an Entropy Maximum?"

74. Thirring, introduction to Lieb, *Stability of Matter*. Lévy-Leblond ("Nonsaturation of Gravitational Forces") points out that, without quantum mechanics, if gravity is the only force present, it is surely the case that matter would crush itself (since there is no shielding in this case). Of course, nongravitational interactions become important at some density (hence shining stars). He also points out that in actual matter

the spatial distribution of the nuclei is much the same as that of the electrons. In particular, even when the nuclei obey Bose statistics, the exclusion principle operating on the electrons to limit their density is "transmitted" to the nuclei by the interplay of Coulomb forces [due to net electrical charge neutrality]. Due to this adjustment of the spatial distributions for both types of particles, electrons and nuclei also have the same momentum distribution. The essential contribution to the kinetic energy then is furnished by the electrons, because of their much smaller mass. On the other hand, the gravitational potential energy comes essentially from the mutual interaction between the heavier particles, the nuclei. (p. 809)

75. Lieb, "Stability of Matter," p. 564. I use "effective" since the sequence is not strictly decreasing.

76. As Spruch points out in "Pedagogic Notes on Thomas-Fermi Theory," the Thomas-Fermi theory builds in the uncertainty principle (quantum mechanics), the Pauli principle (fermions), and Coulomb forces (electrons and protons).

77. As far as I can tell, the reason for the ideal gas term is that the Thomas-Fermi bounds are computed for fixed nuclei (and hence, in the relative coordinates, the mass of the electrons is off by a factor of two), and when one adds in the motions of the nuclei, the ideal gas term does the needed work. Lieb, "Stability of Matter," p. 564, gives the proof but does not explicitly mention this fact, one that motivates splitting up the kinetic energy into two halves.

78. For example, a property of M_i, P_i, converges in the following sense: $|P_{i+1} - P_i| \le C/2^i L$, where the relevant size in M_i is $2^i L$, as in Parisi, *Statistical Field Theory*, p. 42, and Lee and Yang, "Statistical Theory of Equations of State," p. 408. Note that some models are of course not convergent approximations.

79. In effect, asymptotic approximations are uniform: what we mean by an asymptotic form is that, if our variable is large enough (here it is N), N is a function of ϵ, but not of ρ or β (Whittaker and Watson, *Modern Analysis*, pp. 44–45).

Chapter Four

1. As indicated in the preface, although I have mentioned the names of authors of papers, in what follows I have not been systematic about priority or credit, since my concerns here are conceptual. More precise historical details about priority and the like are provided by Brush, "Lenz-Ising Model," and by Heims in Hoddeson et al., *Out of the Crystal Maze*, pp. 518–533.

2. The most concrete sort of argument for geometry's role in general relativity is to be found in Wheeler's *Journey into Gravity and Spacetime*, although the relationship of mathematical devices to physical content is a major theme in philosophic work. Another very popular example is group theory with respect to crystallography or the perfect solids (tetrahedron, cube, etc.).

3. See Tragesser, *Husserl and Realism in Logic and Mathematics*, chap. 4.

4. See Maxwell, "Address to the Mathematical and Physical Sections," his reflections on Thompson on the heat equation.

5. Temperley, "Two-Dimensional Ising Models," p. 250.

6. The reduction of a previous solution to a new method is an explicit theme in Kac and Ward, "Combinatorial Solution"; Montroll, Potts, and Ward, "Correlations and Spontaneous Magnetization"; and Stephen and Mittag, "New Representation." An interesting variation on this is provided by Samuel, "Use of Anticommuting Variable Integrals," who argues that his method is different from the predecessors' methods and could not be reduced to them.

For a comparatively recent (1985) grand summary, see the text in chapter 8 at notes 15 and 16 (D'Ariano, Montorsi, and Rasetti, preface to *Integrable Systems in Statistical Mechanics*, p. 2).

7. Temperley, "Two-Dimensional Ising Models," p. 250.

8. Some of the constitutions I shall present were at least at first not exact, and were convergent approximative schemes. Other such schemes do not look like candidates for becoming exact.

9. But see Barouch, "Ising Model in Presence of Magnetic Field," for how the difficulty is moved over to another (albeit unsolved) arena, distribution of the prime numbers.

10. Onsager's original paper, "Crystal Statistics," Kaufman, "Crystal Statistics," and McCoy and Wu, *Two-Dimensional Ising Model*, give substantial attention to the finite-size case. See Ferdinand and Fisher, "Bounded and Inhomogeneous Ising Models."

11. See Schweber, "Physics, Community, and the Crisis in Physical Theory."

12. Huang (*Statistical Mechanics*, p. 324) does a rather different sort of building up out of cubes, the volume, V, being proportional to N, being divided into cubes of volume proportional to $\sqrt{N}$.

13. See Kadanoff et al., "Static Phenomena near Critical Points," pp. 403–404, fig. 2; Krieger, "Theorems as Meaningful Cultural Artifacts." At the critical point, and below, boundary conditions are crucial for determining what is going on. This is the import of the Peierls argument for the existence of a phase transition, which I shall discuss in chapter 5.

14. Lieb, "Stability of Matter," p. 564.

15. See Kadanoff and Houghton, "Numerical Evaluations of the Critical Properties," pp. 378–379, figs. 1–3, for a nice distinction between decimation and block transformation, and the combination of the two (in this latter case no block shares spins with another block).

16. Wilson, "Problems in Physics with Many Scales of Length," provides a lovely exposition of these points, including the example that follows, on scaling up by a factor of three.

17. Parisi (*Statistical Field Theory*, pp. 41–42) proves the existence of stability for the Ising lattice. See also Griffiths, "Proof That the Free Energy," for the thermodynamic limit.

18. Wilson, "Renormalization Group."

19. If the block size is 3^l, then the number of spins is 3^{2l} (again we are in two dimensions, and our cubes or blocks are actually squares), the error on that is about 3^l (the square root), and if we want to distinguish a ratio of $0.5 + \epsilon$, then we need an l such that $3^{2l}\epsilon \gg 3^l$, at least.

20. The cluster sizes are a probability distribution, and so this statement needs to be made more precise. Clustering usually falls off exponentially, but at the critical point it falls off as a power law.

As for the central limit theorem, see Iagolnitzer and Souillard, "Lee-Yang Theory and Normal Fluctuations," which connects the statistical independence of the block-spin variables to the absence of zeros of the partition function (for which see chapter 5). More generally, many of these results follow from work on correlated random fields, for which see Georgii, *Gibbs Measures and Phase Transitions*. See Krieger, "Theorems as Meaningful Cultural Artifacts," for a simple example.

In this context, the infinite volume limit is much like a "tail event," its probability or measure (that is, its macroscopic variables) determined by an infinite tail of random variables (one at each scale, say). (See Georgii, *Gibbs*

Measures and Phases Transitions, pp. 119–120.) Will there be a unique state, independent of boundary conditions, or, rather, the possibility of ordered states (very much dependent on boundary conditions), as a limit? This question is much like that answered by the Kolmogorov zero-one law. See Krieger, "Could the Probability of Doom Be Zero or One?"

21. Fisher, "Theory of Equilibrium Critical Phenomena"; Kadanoff et al., "Static Phenomena near Critical Points." Of course, early experiments never see these phenomena so clearly.

22. See Widom, "Equation of State." It would seem that Kadanoff first realized the effectiveness of block-spin techniques.

23. Baxter and Enting, "399th Solution of the Ising Model."

24. Anderson, *Basic Notions of Condensed Matter Physics*, pp. 49–51.

25. These observations depend on dimension and on whether the symmetry is continuous or not.

26. Onsager ("Electrostatic Interaction of Molecules") and van Hove ("Sur l'intégrale de configuration") assume hard cores in their bounds and thermodynamic limits, respectively, and so can use arithmetic constitutions. Dyson and Lenard, and Lenard and Dyson ("Stability of Matter") use what they call "binary" cubes filled with single particle cubicles, a geometric construction, for Coulombic matter. See Lieb and Lebowitz, "Constitution of Matter," p. 322. (Fisher and Ruelle ["Stability of Many-Particle Systems"] use a geometric construction even for their noncoulombic matter.)

27. Anderson, *Basic Notions of Condensed Matter Physics*, pp. 11–19.

28. Kramers and Wannier, "Statistics of the Two-Dimensional Ferromagnet"; Montroll, "Statistical Mechanics of Nearest Neighbor Systems"; Onsager, "Crystal Statistics," showed how this might be done for the Ising model.

29. Parisi (*Statistical Field Theory*, pp. 41–42) points out that this precludes $f \to \infty$ as $N \to \infty$.

30. See Parisi (ibid.) who employs the lower bound on the energy (when all the spins are aligned), and the vanishing smallness of surface terms, to provide a nice proof of the existence of the infinite volume limit. Peierls ("On Ising's Model of Ferromagnetism") provided the early topological proof that there ought to be a transition point; Griffiths ("Peierls Proof of Spontaneous Magnetization") corrects some of Peierls's argument. See also Weng, Griffiths, and Fisher, "Critical Temperature of Anisotropic Ising Lattices."

31. In effect, we have reduced the dimensionality of the problem. So if V adds on a row to a two-dimensional lattice, we have a one-dimensional problem when we diagonalize V.

32. I have put aside various questions concerning boundary conditions, technically quite important.

33. Huang, *Statistical Mechanics*, p. 351.

34. Smaller degeneracies, like a twofold one, have implications for correlation functions and the existence of a long-range order. The off-diagonal elements of V tell about the spontaneous magnetization, splitting otherwise degenerate states.

Again, if Λ_{max} is finite (V's dimensions are 2^N, so this is not obvious in the infinite volume limit), and this is shown by direct calculation, then there really is an infinite volume limit.

35. Next-nearest-neighbor interactions complicate the formalism substantially, and the methods do not work so readily if one wants an exact solution.

36. The one-line explanation is that these situations lead to nonplanar graphs.

37. Itzykson ("Ising Fermions," p. 450) suggests that the horizontal interaction is the potential energy, the vertical one is the kinetic energy (and the time direction). Insofar as the vertical interaction is seen as a disorder operator, as is Onsager's B, then this is quite reasonable.

38. See chapter 8, "The Mathematics and the Foundations." See also Baxter, "Solvable Models in Statistical Mechanics," p. 8.

39. Kramers and Wannier, "Statistics of the Two-Dimensional Ferromagnet," p. 264; Montroll, "Statistical Mechanics of Nearest Neighbor Systems," pp. 709–711.

40. Onsager, "Crystal Statistics," pp. 136–137, and fig. 5 of his text. Stephen and Mittag ("New Representation") offer another representation of these disorder operators (in effect expanding the exponentials of the Zs), and a plane triangle rather than the spherical triangle that Onsager employs is the diagrammatic device.

41. Baxter, "Exactly Solved Models," p. 7.

42. Montroll, "Statistical Mechanics of Nearest Neighbor Systems," is one source. But see Kramers and Wannier's work, and Lassettre and Howe's. As I shall point out in later chapters, the transfer-matrix formulation is based on the Hamiltonian, and it brings to mind quantum mechanical techniques all the time.

43. Wilson and Kogut, "Renormalization Group and the ϵ Expansion."

44. Or classical mechanics expressed in terms of action-angle variables and characteristic functions, for which see Goldstein, *Classical Mechanics*.

See chapter 8, where action-angle variables are the essence of the "inverse scattering method" for solving the Ising model. Diagonalizing the transfer matrix, and obtaining its largest eigenvalue, then allows one to have $V^N \approx \exp(N \ln \Lambda_{max})$, which is just the form we would want for the angle analogy. Presumably, some function of k would be the action. See Dolan and Grady, "Conserved Charges from Self-Duality," pp. 1590–1591.

The original connections with Euclidean field theory are credited to Schwinger ("On the Euclidean Structure of Relativistic Field Theory") and to Symanzik ("Euclidean Quantum Field Theory").

45. For further demonstrations of the formal equivalences among the various constitutions, and for more on the S-matrix analogy, see Stephen and Mittag, "New Representation." See also Hurst, "New Approach to the Ising Problem," for an early exposition.

46. In effect, they are both Green's functions.

47. Wilson and Kogut, "Renormalization Group and the ϵ Expansion," sec. 10, p. 145: "The transfer function is an hermitian operator since the statistical

mechanical hamiltonian is real and symmetric. Therefore, V^N could be chosen to be a hamiltonian of some quantum mechanical system. However, an acceptable hamiltonian must satisfy the additional requirement of locality." Note that $it/\hbar = \beta$, and $\psi(s') = \int V^N(s', s)\psi(s)\,ds$, where $\psi^2(s)$ is the probability that a particular row has spin configuration s, and $V(s', s) = \exp - \beta H_{sm}(s, s')$.

48. Parisi, *Statistical Field Theory*, p. 224.

49. See Feynman and Hibbs, *Quantum Mechanics and Path Integrals*, chap. 10. See also Wilson and Kogut, "Renormalization Group and the ϵ Expansion," p. 147.

50. Schultz, Mattis, and Lieb, "Two-Dimensional Ising Model." Their operator formulation actually starts out with the s and C operators of Onsager's 1944 paper. See also Nambu's 1950 paper, "Eigenvalue Problem in Crystal Statistics." Nambu uses a Jordan-Wigner transformation to create fermion operators. See also Srednicki ("Hidden Fermions in $Z(2)$ Theories." p. 2878) who points out the more general character of the strategy of fermionization. The Ising model is "a system of locally coupled fermions. After a canonical transformation, these fermions become noninteracting. . . . This is reminiscent of the Hilbert-space description of fermions, in which each occupation number is either "one" or "zero." The problem is turning the commuting objects of the spin description into anticommuting objects appropriate to a fermion description." He refers to "fermionic theories in disguise." It took from the mid-1960s to 1980 for the Schultz, Mattis, and Lieb work to become bread-and-butter technology (although their work employs some earlier technology, that of the Jordan-Wigner transformation [1928] for turning the commuting objects into anticommuting ones). See also Itzykson, "Ising Fermions"; Samuel, "Use of Anticommuting Variable Integrals." The tone of both Samuel's and Srednicki's papers are remarkably similar in their bravado and delight. See also Plechko, "Simple Solution of Two-Dimensional Ising Model," which also reviews the various Grassmann-integral solutions.

I should note that I will use the notion of "free" particles (here free fermions) even if the mass matrix is not diagonal—but can be diagonalized. A more restricted notion of a noninteracting or free particle is fulfilled at the critical point, when the correlation length is infinite and there is no mass gap.

51. Itzykson, "Ising Fermions."

52. As indicated in chapter 6, Onsager was explicitly aware that he was describing just such an orderly system, and that his work was in part meant to develop techniques for describing such orderliness. What I am interested in here is the particle interpretation of such a description.

53. So the pseudo-Hamiltonian is written as Column $+ k^{-1}$Row; k is small for low temperatures, large for high temperatures; and $k = 1$ at the critical point. From the point of view of perturbation theory, the critical point is obviously the least tractable. See Onsager, "Ising Model in Two Dimensions"; Stephen and Mittag, "New Representation." As we shall see, the good particles are mixtures of orderly and disorderly states.

54. Note as well that $1.76 \sin q/2$ is a good approximation to the critical spectrum (which is $\mathrm{acosh}(2 - \cos q)$).

55. Hurst and Green, "New Solution of the Ising Problem," p. 1060. See also Itzykson, "Ising Fermions."

56. Schultz, Mattis, and Lieb, "Two-Dimensional Ising Model"; Hurst and Green, "New Solution of the Ising Problem"; Samuel, "Use of Anticommuting Variable Integrals."

57. See Savit, "Duality in Field Theory and Statistical Systems"; Gruber, Hintermann, and Merlini, *Group Analysis*.

58. Technically, in one dimension, if $S(x)$ denotes the spin at each point, then a directed derivative of S, toward larger indices, ∇^+, measures differences between adjacent spins, and the energy of a configuration, E, goes as $J\int\{2(\nabla^+ S)^2 - 1\}\,dV$, in effect a kinetic energy term (like p^2). The contribution to the partition function goes as $\exp - \beta J\int\{2(\nabla^+ S)^2 - 1\}\,dV$, a nice field theoretic form. (Note that $|\nabla^+ S|$ is either zero or one; the square is as good as an absolute value.) Now, the absolute value of gradient of the dual of S is just $1 - |\nabla^+ S|$, and the energy of the dual configuration is $E + 2KN$.

59. Kramers and Wannier, "Statistics of the Two-Dimensional Ferromagnet." Dresden (*H. A. Kramers*) points out that Kramers and Wannier do not see the more general features of this transformation, as did Onsager several years later.

60. Van der Waerden, "Die lange Reichweite."

61. Onsager ("Ising Model in Two Dimensions") points out that the operator form of the dual, what he calls $L_{\mu\mu'}$, says the alternation of spins in one direction or row is associated with the average in the other direction, or column.

62. Onsager, "Crystal Statistics," pp. 123–124.

63. Baxter, *Exactly Solved Models*, chaps. 6 and 7; much of this was known to Onsager in his 1944 paper ("Crystal Statistics").

64. Baxter, *Exactly Solved Models*, pp. 118–120 on correlation lengths, p. 94 on invertibility (and equation 7.3.15 with the $i\pi/2$ added to the coupling), and p. 95 on symmetry. See also Kramers and Wannier, "Statistics of the Two-Dimensional Ferromagnet," p. 260. Given the scattering interpretation of the Ising model, these features have even more physical meaning, invertibility being unitarity, for example.

More precisely, if unitarity is denoted by $S(\theta)\cdot S^T(-\theta) = 1$, where S is the S-matrix and $\theta/2$ is the center of mass rapidity, then analogously, we have an almost-inverse of $V(k, u)$ in $V(k, -u)$. $\{K_1, K_2\}$ becomes $\{K_2 + i\pi/2, -K_1\}$. (See note 71 below for a definition of u.) Symmetry is much the same as crossing relations, where for the transfer matrix exchanging K_1 with K_2 means that u goes to $I' - u$, while θ goes to $i\pi - \theta$ for the S-matrix. See Baxter, *Exactly Solved Models*, pp. 93–95; Shankar, "Simple Derivation of the Baxter-Model Free Energy." p. 1178. See also Pokrovsky and Bashilov ("Star-Triangle Relations"), who do Baxter's solution in S-matrix terms.

65. Onsager was aware of all of this in his 1944 paper. By the way, Baxter had to do a good deal of work on his functional equation deduced from the formal symmetries, for it to yield the partition function—along the way using elliptic functions to parametrize the constraints and the solutions.

66. Baxter and Enting, "399th Solution of the Ising Model."

67. If the couplings are not equal, then their ratio is altered by the two transformations, so the high- and low-temperature hexagons are different. See Houtappel, "Order-Disorder in Hexagonal Lattices," p. 434.

68. Onsager, "Ising Model in Two Dimensions," p. 3. As Thacker points out (in "Exact Integrability," p. 284), the Jordan-Wigner transformation is used to convert interacting bosons to free fermions, the quantum inverse scattering method goes from interacting variables to decoupled action-angle variables, and duality goes from order to disorder. See note 44 above.

69. Kramers and Wannier, "Statistics of the Two-Dimensional Ferromagnet," equation 26a; Onsager, "Crystal Statistics," equation 2.1a.

70. Onsager has two definitions for k, above and below the critical point, so that k remains in the unit interval, allowing for canonical definitions of the relevant elliptic functions. Note that Onsager's ia (what he calls "the imaginary parameter") corresponds to what I call here iu, the argument of the elliptic functions. See note 71 below on u. And he defines ia differently above and below the critical point (Onsager, "Crystal Statistics," p. 146).

71. I have left out a second needed variable, what Baxter calls u (and what Onsager calls a, the "imaginary parameter" [see note 70]), when we are not at the critical point and the difference between the horizontal and vertical coupling constant actually matters. sn $iu_1 = \sin 2iK_1$ and $\Sigma u_i = I'$ (the quarter-period imaginary magnitude of the elliptic functions), where the sn function's dependence on k is suppressed in this notation. What Baxter calls symmetry is bound up with the fact that a partition sum over the neighbors of a fixed spin, which is expressed as $\cosh \sigma L$, can be converted again to an exponential bilinear in the σ.

Baxter's u equals Onsager's a, below the critical point. (Onsager's u is not directly related to Baxter's u.)

72. In the infinite volume limit, the glitches in those symmetries, surface terms, are of vanishing importance, for most purposes. See McCoy and Wu, *Two-Dimensional Ising Model*, on when and how surface terms do count. See also Ferdinand and Fisher, "Bounded and Inhomogeneous Ising Models,"

73. Here the Griffiths inequalities come to mind. "The average magnetization of any spin, $\langle \sigma_i \rangle$, considered as a function of the βJ_{kl} [k and l referring to two different sites] at fixed z [magnetic fugacity, $\exp 2\beta B$] or B, where B is the magnetic field, cannot decrease as any βJ_{kl} is made larger and $B \geq 0$" (Lebowitz, "Statistical Mechanics," p. 411).

74. Think of the Cauchy distribution, in which the variance is infinite, and so extra measurements do not improve the mean—although they do improve the median.

75. Recall that there are infinite-variance random variables whose normed sum also asymptotically converges, but here the norming constant is different than $1/\sqrt{N}$. These are the so-called Lévy stable distributions. And they might remind us of bosons whose normed sum is asymptotic, but the norming constant is $N^{-7/5}$. See note 20 above for more on the central limit theorem.

76. Krieger, "Theorems as Meaningful Cultural Artifacts."

77. See Ferdinand and Fisher, "Bounded and Inhomogeneous Ising Models," for exact results for finite-sized lattices.

78. See Schweber, "Physics, Community, and the Crisis in Physical Theory," for a discussion of the renormalization group, hierarchy, fundamentalness, and so forth, both as in physics and as in a hierarchy of sciences. The crucial figure here is P. W. Anderson, perhaps the most particle-physics-like figure in condensed matter theory, who argues for emergent principles at each level. See also Cao and Schweber, "Conceptual Foundations."

79. But see Haymet, "Freezing," for how such models might work in a density functional context.

80. In an essay, "Mathematical Beauty," G.-C. Rota points out that *beauty* is often used as a term of art in mathematics, to mean enlightening.

81. Wilson, "Renormalization Group," pp. 773–774. See Krieger, *Doing Physics*, p. 132, for the relevant quotation.

82. Technically, there is one well-known way of doing such a cumulation, an example of which is the discrete Fourier transform. A lattice of spins is a weighted sum of its spatial Fourier components, and the Hamiltonian of adjacent-spin interactions can be shown to be a weighted sum of the product of adjacent Fourier components.

83. Note that such a scheme would seem to require that we be able to exponentiate a function of a sum of spins, $\cosh \Sigma \tau L \approx \exp - H(\tau)$. This is essentially a star-triangle relation, for which see Baxter, *Exactly Solved Models*, pp. 81–82. More generally, we see similar phenomena in the definition of the star-triangle relation, in the definition of the free energy, $\exp - \beta N f = \Sigma \exp - \beta H$, and in the high-temperature expansion of the Ising partition function, where we linearize $\exp K \sigma \sigma = \cosh K (1 + w \sigma \sigma)$, where $w = \tanh K$. All of these schemes would seem to depend on the two-state nature of the spins, σ.

84. Hurst, in "Applicability of the Pfaffian Method," shows that what is crucial is that there be no crossed bonds, and then it is possible to go from bilinears to exponentials—and the criterion for doing so is expressible as a Pfaffian (or "antisymmetric") sum, which in essence says that we may reduce the problem to sums of two-body interactions. Hence, in giving structure to a vertex, so that bonds are independent of each other at that vertex (since there are now a multiplicity of terminals at each vertex), leads to factorizability into two-body interactions.

85. Wilson, "Renormalization Group," pp. 797–800.

86. Note that this model would not work if the variance per spin, or the interaction energy per spin, were proportional to N, just what might happen at the critical point. And, like all the constitutions, it in fact tells us about approaching the critical point. For the critical point is perhaps terra incognita, an infinite place in the infinite volume limit—although in the finite-N limit it is quite knowable, not unlike any other place. By the way, the holographic constitution (or any of the others) could not work for constituting gravitational matter because there is no screening and so no effective locality.

87. Hurst, "Applicability of the Pfaffian Method," discusses the requirement of bipartiteness in quite gritty detail. As we shall see, bipartiteness is also a matter of the factorizability of an associated scattering matrix into two-body interactions.

88. We end up with a quartic rather than a quadratic field Hamiltonian. Note that the projection of a three-dimensional system onto a plane requires crossed interactions.

89. Onsager, "Crystal Statistics," p. 124.

90. Baxter and Enting, "399th Solution of the Ising Model." There is an intimate connection between the renormalization group and the star-triangle relation, for which see Knops and Hilhorst, "Exact Differential Renormalization." But these are still nascent observations, for an essential account in terms of, say, field theory was only achieved much later historically. As I said earlier, the steps along the way are physically illuminating about the constitution of matter. A one-line Whiggish answer is not so illuminating.

91. The usual analogy is watching a digitized television picture, then averaging at a scale, and then upping the contrast (or lowering the contrast, as the case might be), so that the picture still looks the same.

92. Knops and Hilhorst, "Exact Differential Renormalization," and the papers they discuss work out the renormalization-group equations in this context. Padé approximants have a similar role in the high- and low-temperature expansions in the topological constitution. Presumably, they get at the same asymptotic form, and so the approximations need many fewer terms for any level of accuracy.

93. The chains presumably exert a mildly attractive force at a distance, mediated by other chains, and a reasonably repulsive force close up (since they have strict rules about how they are to share bonds). They are said to interact only when they cross each other.

See Brydges, Fröhlich, and Spencer, "Random Walk Representation of Classical Spin Systems." They refer to Symanzik as the source of the polymer model, a gas of random walks or loops (see note 94 below).

The polymer constitution also appears in Feynman's path-integral account of the quantum mechanical system of superfluid liquid helium 4. Feynman, *Statistical Mechanics*, pp. 343–350, provides an accessible version.

The quantum mechanical system can be represented by a classical system of cross-linked ring polymers (in imaginary time, it). The polymers are an expression of the delocalized character of the helium atoms at and below the phase transition, and the length of the polymer is just the permutation exchanges of these more and more (quantum mechanical) indistinguishable bosons. Imaginary time appears here, as it does in Feynman's path-integral account of statistical mechanics, because one is putting together the various configurations (in the partition function, the sum over states) in that imaginary time (classical β corresponds to $it/\hbar$ quantum mechanical). Long cycles linking N atoms are the coherent superfluid state. Note that at actual low temperatures, the polymer is most disordered (since then the boson symmetric permutation exchanges are most likely), violating our usual intuitions about what should happen at low temperatures—because in fact these are polymers in imaginary time and so their kinetic energy ($v^2 \approx 1/i^2 = -1$) has a reversed sign compared to the usual real-time value.

Ceperley, ("Path Integrals in the Theory of Condensed Helium") develops these ideas, brings the discussion up to date, and focuses on actual computation employing Monte Carlo simulations (especially pp. 279-281). He has a very fine discussion of the issues involved in pressing the analogy or model of bosonic symmetric exchange to polygons in imaginary time. By the way, Feynman asks explicitly how these polymers differ from the Ising polymers (p. 350); we might note that in computing the contribution of the Ising polymer we take into account not only its directedness but also the angles through which it turns as we complete the ring.

Ceperley points out that the polymer techniques do not apply, it would seem, to the interactions of the quasiparticle excitations of liquid helium (phonons and rotons). And it would be of some interest to connect the quasiparticle model with the polymer one.

Random walks may be conceived of as Ising models. But rather than the state of each site being known by a real number (say, $+1$ and -1) or as a vector (as in the Heisenberg ferromagnet), dimensionality, d, equals one and two, respectively—the random walk can be shown to be a $d = 0$ Ising model, in the formal sense that formulas for such models describe random walks if d is set equal to zero.

94. The original formulation in terms of loops is due to van der Waerden, "Die lange Reichweite."

95. I should note that the topology of bonds in three dimensions is such that most of the two-dimensional devices stop working at this point –the lattice is no longer planar or orientable or bipartite.

96. Kac and Ward, "Combinatorial Solution," p. 1332.

97. These correspond to Baxter's P and Q, which draw horizontal and vertical links (*Exactly Solved Models*, p. 84), and to Kaufman's P_r and P_{r+1}, spinor rotations ("Crystal Statistics").

98. The exact solution taken as the generating function, for finite N, is used to guess the form of the elements of the matrix whose determinant will be the partition function or the generating function. Since the determinant is the product of the eigenvalues of the matrix, and the finite-N generating function for the polygons (effectively, the partition function) is also expressed in the Kaufman-Onsager form as a product of factors, Kac and Ward might begin to guess the matrix elements (p. 1336). We have here a problem in combinatorial topology and intuition.

99. Ferdinand and Fisher, "Bounded and Inhomogeneous Ising Models," p. 837; Baxter, *Exactly Solved Models*, p. 107.

100. It would seem that the original work of Kac and Ward on the Ising model did not have the Feynman sum-of-histories notion in mind, and was a tour de force of algorithmic combinatorics. Yet it is notable that Kac did work

very similar to Feynman's on quantum mechanics, but in the realm of random walk and probability theory—the Feynman-Kac sum-of-histories formula being the epitome of this coincidence. The determinantal form of Kac and Ward explicitly enumerates and formally sums a sequence of more and more complex polygons (indexed by their perimeter length, l). That sequence is meant to be a series approximation to the partition function (the high-temperature expansion) or, by analogy, to the Feynman-Kac formula. I mention all of this without at all claiming that such a coincidence (of Kac and Ward, Feynman, and Kac) is more than just that. But it is interesting that the determinantal form and the Feynman-Kac formula, suitably applied, are each ways of enumerating random walks on the lattice. (See note 93 above.) Moreover, Feynman (*Statistical Mechanics*, pp. 137–148) eventually develops another solution to the Ising model that works out that enumeration, a working out of his and J. A. Wheeler's sum-of-histories idea in this context.

More generally, the partition function and the sum-of-histories formulas are generating functions, here for random walks, elsewhere for partitions of an interval.

More explicitly, the polymer model of the constitution of matter, insofar as we see it as focusing on the number of bonds involved (the length l), is perhaps not so alien to another way physicists have constituted matter in general. Each bond represents an interaction between atoms, multiple links represent multiple interactions, and in this sense matter is built up out of a perturbation series, each higher order of bond multiplicity, as l increases, being hopefully of lesser significance—but since there are perhaps more such different configurations of polygons as l increases, we are not so sure. Such series may well be divergent, but some of the time, by device, they may actually be summed—here, to bulk matter.

In effect, we have as our "input" a very large number of atoms, and as our "output" the sum of all their various interactions with each other (formally arranged in a hierarchy): such is a scattering theory, recalling our earlier observations, the in-states being the bare atoms, the out-state being bulk matter composed of the interaction of those constituents.

Within the tradition of statistical mechanics, the cluster expansions of Mayer and Montroll are reminiscent of what I am doing here.

101. Feynman's solution to the Ising model is even lovelier in pressing this analogy (*Statistical Mechanics*, pp. 136–150). See also Feynman and Hibbs, *Quantum Mechanics and Path Integrals*, chap. 10.

102. Kastelyn, "Dimer Statistics and Phase Transitions"; Montroll, "Statistical Mechanics of Nearest Neighbor Systems"; Caianiello, *Combinatorics and Renormalization*.

103. Caianiello, *Combinatorics and Renormalization*.

104. It would be interesting to see if the three other constitutions would lead to illuminating proofs of the thermodynamic limit and the stability of matter.

105. See Stephen and Mittag, "New Representation."

106. The renormalization group and its ideas link the geometric and the holographic constitutions. Transfer functions and their expansion in terms of a perturbation series link the arithmetic and topological constitutions. The geometric and the holographic constitutions are linked to the arithmetic and topological constitutions by the variable k (duality and the star-triangle relation), which appears most clearly in the Baxter and Enting ("399th Solution of the Ising Model") derivation using correlation functions and in the Baxter (and originally Onsager) derivation employing commuting transfer matrices (*Exactly Solved Models*): k is the right variable to use to scale up and down, it is the conserved quantity that allows transfer matrices to commute, and it defines duality.

For other versions of this story, see Hurst and Green, "New Solution of the Ising Problem"; Stephen and Mittag, "New Representation"; Temperley, "Two-Dimensional Ising Models."

107. Tragesser, *Husserl and Realism in Logic and Mathematics*.

108. See Samuel, "Use of Anticommuting Variable Integrals," on using anticommuting variables for three-dimensional approximations, and Baxter, "Exactly Solved Models," on Zamolodchikov's work.

CHAPTER FIVE

1. Dresden, *H. A. Kramers*, pp. 321–324; Dresden, "Kramers's Contribution to Statistical Mechanics."

2. Onsager, "Crystal Statistics," p. 142. Ferdinand and Fisher ("Bounded and Inhomogeneous Ising Models," pp. 834–835) suggest modifications of this.

3. However, there is a tradition of numerical approximation surrounding the topological formulation, namely, the high- and low-temperature expansions. As indicated in chapters 4, 6, and 7, there are quite exact formal studies of the renormalization-group equations in the holographic constitution (Hilhorst, Schick, and van Leeuwen, "Exact Renormalization Group Equations"; Knops and Hilhorst, "Exact Differential Renormalization").

4. Uhlenbeck and Ford, *Lectures in Statistical Mechanics*, pp. 67–68. The "more rigorous" approach (see preface, note 7) deals with these questions even more directly, because it builds in the infinite volume limit. Phase transitions are here marked by, for example, the emergence of a nonunique equilibrium state (as in freezing, where the direction of crystallization is arbitrary). See note 15 below.

5. Cartwright, *How the Laws of Physics Lie*; Franklin, *Neglect of Experiment*; Galison, *How Experiments End*. However, it is perhaps not so simple as Cartwright would seem to suggest. Fisher ("Nature of Critical Points," p. 16) points out that the experimental results, at that time (1965), seemed to accord better with some approximate solutions rather than with the exact one.

6. "Spontaneous" is here used to mean without a direct and deliberate influence, on its own, arising without a necessarily given direction. A permanent magnet exhibits a spontaneous magnetization because there is no external

magnetic field present to maintain the alignment of the atomic spins, although some external field may have been present when that magnetization was set up. More generally, spontaneous symmetry breaking occurs when an otherwise arbitrary orderliness arises in a system, without its being maintained by a direct external influence.

7. In the following discussion, I have mostly considered second-order phase transitions, where there is no heat of transition, as in a bar of iron becoming permanently magnetized below about 1043 K or losing its magnetization above that temperature. First-order transitions (as in freezing, when there is such a heat of transition) become second-order at the critical point. (It should also be noted that the second-order Ising transition can be made isomorphic to a first-order lattice gas transition—the partition function of the first being formally the same as the grand partition function of the second. Hence, for our purposes, the distinction is not so crucial.)

8. At room temperature, in time scales of the order of 10^{65} years, ordinary matter will melt and flow—due to quantum mechanical tunneling rather than statistical ergodic effects (Dyson, "Time without End," pp. 451–452). The statistical mechanical effects are $\exp - \Delta F/k_B T$ probabilities.

9. In general, the times of observation have to be short compared to the decay times, and long compared to the energy resolutions, $\Delta E \approx h/\Delta t$, that we need in order to ascribe particular masses to objects. See Anderson, *Basic Notions of Condensed Matter Physics*, p. 72, on Fermi liquids. See also Feynman, *Statistical Mechanics*, p. 1; Feynman, *Lectures on Gravitation*, p. 21—both quoted in my *Doing Physics*, pp. 14–15.

10. Parisi, *Statistical Field Theory*, p. 16.

11. Ibid.

12. Feynman, *Statistical Mechanics*.

13. Arthur, "Competing Technologies."

14. To anticipate later discussion, the $\mathbb{N}$ operator is $z\partial/\partial z$, while the $\mathbb{I}$ operator is $\lim_{N \to \infty} 1/V$, where z is the activity (or fugacity), a measure of a reaction rate or evanescence. If ρ is the density, equal to $\partial P/\partial \mu)_T$, where μ is defined by $z = \exp \beta\mu$, and *GPF* is the grand partition function, then $\rho_{\mathrm{IN}} = \mathbb{IN} \ln GPF$ and $\rho_{\mathrm{NI}} = \mathbb{NI} \ln GPF$, where $\beta P = \mathbb{I} \ln GPF$. We can show that $\mathbb{I} \ln GPF$ always exists, but only under uniform convergence to the infinite volume limit will the infinite volume limit be guaranteed to be analytic (smooth) in z (as *GPF* already is for finite V)—and so if there is no uniform convergence, then the $\mathbb{N}$ operator will not be well defined here.

15. See, for example, the chapters on states in Simon, *Statistical Mechanics of Lattice Gases*. For example, the "DLR equations" presume the canonical ensemble and then seek states that are in equilibrium under it.

16. I should note that, in this formulation, in the infinite volume limit the infinity of the degree of the polynomial that is the grand partition function ($\Sigma_{i=0 \text{ to } \infty} z^i Q_i$) is a matter of arithmetically adding on one more particle, ad infinitum, and as one does so, the zeros of the partition function could get

closer to the real axis, hitting it "when" $N = \infty$ and so allowing for phase transitions.

17. I take this example from Parisi, *Statistical Field Theory*, p. 15.

18. If Q_n were constant, then summing the geometric series (assuming $|z| \leq 1$), $GPF \approx (1 - z^{N+1})/(1 - z)$. The solutions for z are $N + 1$st roots of unity. If $Q_n \approx \exp n$, then $GPF \approx (1 - (ez)^{N+1})/(1 - ez)$.)

19. Fisher, ("Nature of Critical Points," pp. 58–59) nicely shows how these appear in the complex tanh K plane.

20. Yang and Lee, "Statistical Theory of Equations of State"; Lee and Yang, "Statistical Theory of Equations of State." See chapter 4, note 20, for the connection between Lee-Yang notions and the central limit theorem.

21. There should be no big changes for small changes in properties. But this is much too rough. The proper definition of a phase is no mean achievement. See Wightman, introduction to Israel, *Convexity in the Theory of Lattice Gases*; Simon, *Statistical Mechanics of Lattice Gases*.

22. Huang, *Statistical Mechanics*.

23. See ibid., pp. 317–318 and figures 15.3 and 15.5, for the relationship of the P-V curve to the P-z and V-z curves.

24. Ibid., p. 316.

25. If there are no real zeros, then there is uniform convergence, and then continuity and smoothness and commutativity of the operators. If there are zeros, then there may not be uniform convergence, smoothness, and commutativity. And if there is not commutativity or smoothness, then there is not uniformity and then there are real zeros (the contrapositive of the first sentence). In terms of Fourier series, one is concerned with points or their neighborhoods that are infinitely slow to converge. Modern descriptive set theory was invented to describe these problematic point sets.

26. $GPF_N(z)$ formally converges to $GPF_\infty(z)$, but only does so usefully if there are no real zeros of the grand partition function in the region.

27. Ferdinand and Fisher, "Bounded and Inhomogeneous Ising Models."

28. Huang, *Statistical Mechanics*, pp. 315–316.

29. Parisi, *Statistical Field Theory*, pp. 15–16.

30. This is reminiscent of Nancy Cartwright's arguments in *How the Laws of Physics Lie*.

31. At the critical point, not only are there drops of water and vapor, but those drops are of all sizes. See note 21 above.

32. Lieb, "Stability of Matter," p. 568. Note that here I am talking about uniform convergence in the context of the thermodynamic limit for the free energy, and so multiple phases might make sense. The previous paragraph talks about uniform convergence in the context of second-order transitions, in which only one phase at one time is possible (except, ambiguously, at the critical point, where the order parameter is about to become nonzero; in the case of the magnetization, the order parameter arises with an infinite derivative at that point).

33. Lee and Yang, "Statistical Theory of Equations of State," pp. 415–416. One can also show that the free energy is effectively the potential for point charges located at the z_r in a complex plane, $\sum \ln(1 - z_r/z)$.

34. If we are in the temperature, or the magnetic field, complex plane, the zeros might not be located on such a circle, and in fact exhibit quite curious behavior, occupying areas in the plane. See Fisher, "Nature of Critical Points"; Stephenson and Couzens, "Partition Function Zeros"; Saarloos and Kurtze, "Location of Zeros in the Complex Temperature Plane."

In general, one writes the partition function as $Q_N = C^N(K)\mathscr{P}_N$, where $\mathscr{P}_N$ is a polynomial in the variable of interest, and its constant term is one. The "normalized" partition function per spin, $Q_N^{1/N}/C = \mathscr{P}_N^{1/N}$ need be only just less than one at the critical point for the polynomial to have a zero there. Say we use the infinite-N partition function per spin computed by Onsager (which requires an asymptotic approximation of a sum by an integral, and this is what shall be crucial). $\mathscr{P}_N = (1 - z/z_N)\prod^{N-1}(1 - z/z_i)$, where the z_i change as N changes, and the Nth root of the polynomial is just the geometric mean of the factors times $(1 - z/z_N)^{1/N}$. Let z_N be the root closest to the real line. For finite N and $z = z_c$, the critical value of z, the last factor is very close to one, since there are no real zeros for finite N and the exponent does the required work. An asymptotic approximation, as used by Onsager, preserves this fact and so the normalized value is not just less than one, but rather that geometric mean ($= 1/2 \exp 2G/\pi$), where G is Catalan's constant. In approximating the finite-N polynomial by the infinite volume limit, we lose some information.

Baxter (*Exactly Solved Models*, pp. 100–101, 106–108) searches for zeros of the partition function in the arguments, u_j, of his elliptic functions, which, schematically, are $\text{sn}(i(u - u_j))$, Baxter's equation 7.8.32. The elliptic functions are multiplied to get the partition function. At the critical point ($k = 1$), $\text{sn}(iu)$ can be replaced by $\sin u$ (this is not an identity—the correct identity is that at $k = 1$, $\text{sn}(iu) = i \tan u$—but a convenient fact in this context), so $Q \approx \prod \sin(u - u_j)$. The zeros, u_j, equal $\mp \pi/4 - i\phi_j$, at $k = 1$.

The Lee-Yang account would be in terms of polynomial (Baxter's equation 7.7.12), where the zeros of the polynomial can be shown (for $k = 1$) to be equal to $\exp 2iu_j = \exp 2i(\mp \pi/4 - i\phi_j) = \mp i \exp 2\phi_j = \mp i \exp(\ln \tan \theta_j/2)$, where θ_j linearly indexes the zeros and is real. Hence, one has a real zero only if $\theta/2 = 0$. And this is possible only in the infinite volume limit. The minimum value of $\theta \approx 1/\sqrt{N}$; and hence there is a real zero, at least as an accumulation or limit point of zeros, associated with the minimum value of θ in the infinite volume limit. Note that the ϕ_j (or the u_j) are effectively phase shifts.

In relativistic kinematics, the rapidity of a particle is defined as $\text{arctanh}\, p_z/E$, the ratio of its forward momentum to its energy. The pseudorapidity is defined as $-\ln \tan \theta/2$, where θ is here the scattering angle, and is a good approximation to the rapidity for high-energy particles. When we give a particle a velocity boost, v, its rapidity increases by $\text{arctanh}\, v/c$, a constant (where c is the speed of light).

By analogy, the us are nowadays sometimes called rapidities, suggested here by their similar definition to the pseudorapidity. This is only the beginning of a sequence of analogies between statistical mechanics and particle scattering, the partition function and the scattering matrix—seen in Baxter's derivation of the partition function. See Pokrovsky and Bashilov, "Star-Triangle Relations." Also, recall from chapter 4 that the integral of the ϵ_q, at $k = 1$, is proportional to Catalan's constant, G. G is defined by an integral of $\ln \cot \theta$, and so the integral of the ϵ_q is in effect equal to an integral of the rapidity (or u).

35. Note that the only real zero is when $B = 0$. If $B > 0$, there is no spontaneous magnetization.

36. See Uhlenbeck and Ford, *Lectures in Statistical Mechanics*, pp. 66–67, for further elaboration on this.

37. In one expansion, $z = \tanh K$, so a complex z would represent a complex coupling constant.

38. More generally, complex coupling constants play curious roles. For the diagonal-to-diagonal transfer matrix, $K_1 = -K_2 = \pm i\pi/4$ is the dual transformation (the other root of $1 = \sinh^2 2K$) (Stephen and Mittag, "New Representation"). For a pure imaginary magnetic field ($= i\pi/2$, $z = \exp - 2B$, in $k_B T$ units, $= -1$), one can exactly solve for the free energy of the two-dimensional Ising model (Lee and Yang, "Statistical Theory of Equations of State," p. 417, equation 48).

39. Technically, the grand partition function for a lattice gas can be shown to correspond to the partition function for an Ising model (when the latter is written in terms of the number of paired or unpaired spins, and so the complex value of the interaction energy is just a lifetime for such a pair), and then the Gibbs potential, μ, equals $\alpha J - 2B$, where B is the magnetic field and α is the coordination number of the lattice (Huang, *Statistical Mechanics*, pp. 332–334; Lee and Yang, "Statistical Theory of Equations of State," p. 412).

40. Mattis (*Theory of Magnetism*, part 2, pp. 144–145) points out that the coefficients become more and more different from binomial coefficients as N become larger.

41. Although I shall not do it here, it would be useful to examine how the Lee-Yang theorems are proved, to further bring out some of these points.

42. Lieb, "Stability of Matter," p. 565. I have taken this quote out of its context, in the middle of another proof and have made slight changes in nomenclature.

43. Anderson, *Basic Notions of Condensed Matter Physics*, p. 14.

44. "Nonrigorously" by his own account (ibid., pp. 15, 29). But we might hope from our discussion of thermodynamic stability that we might be able to make it rigorous.

45. Clearly, if the unit is not substantially neutral electrically, then surface charges may have important effects, negating the $n^{2/3}$ effect, as we saw in chapter 3.

46. Lenard and Dyson, "Stability of Matter," p. 711.

47. Landau and Lifshitz, *Statistical Physics*, pp. 446–447.

48. Haymet, "Freezing."

49. Montroll ("Statistical Mechanics of Nearest Neighbor Systems") makes this point beautifully. The largest eigenvalue below the critical point, and the largest above, are equal to each other at that point.

50. Anderson, *Basic Notions of Condensed Matter Physics*, p. 29.

51. Phase space is no longer connected for this system; the system is said to be no longer ergodic. The modern literature on states is an attempt to make this insight precise. See, for example, Simon, *Statistical Mechanics of Lattice Gases*, chaps. 3 and 4.

52. Peierls, "On Ising's Model of Ferromagnetism"; corrected by Griffiths, "Peierls Proof of Spontaneous Magnetization"; and Weng, Griffiths, and Fisher, "Critical Temperature of Anisotropic Ising Lattices." For a modern presentation, see also Simon, *Statistical Mechanics of Lattice Gases*, pp. 470–486. Gruber, Hintermann, and Merlini (*Group Analysis*, p. 60) say, "[W]hat appears in Peierls' argument is not a discontinuity of the thermodynamic functions, but rather the fact that the properties at a given point inside the system can be strongly influenced by what happens at the boundary even if this boundary is infinitely remote." Boundary conditions are the whole story, for large β or small temperatures.

Note that the Peierls argument might be considered as marking a phase transition by a change in symmetry.

53. In Onsager, the observation was implicit in the star-triangle relation, which eventually is a foundation for the block-spin transformation and decimation. In Kadanoff, Fisher, and Wilson, the observation was explicit in the fixed-point transformation.

54. Simon, *Statistical Mechanics of Lattice Gases*, p. 279, has a lovely discussion of all the things the entropy is.

55. Ibid., p. 153, points out that the Ising model is the only one in which we can rigorously show that the critical points are the same for each of these processes. That they are necessarily equivalent is not proven, as far as I know.

56. Griffiths, "Microcanonical Ensemble," p. 1448. If the small system is in thermal contact with a temperature bath and in equilibrium, then the smallness issue is finessed for an issue about what it would mean to be "in contact."

57. Fisher points out somewhere that, if we are measuring $\ln N$, then Avogadro's number is perhaps not so large.

58. All of what follows is taken from Schultz, Mattis, and Lieb, "Two-Dimensional Ising Model," and from Griffiths, "Spontaneous Magnetization in Idealized Ferromagnets." Modern developments, especially from field theory, try to make canonical the need for external fields to split the degeneracy.

When one reads the early paper by Ashkin and Lamb, "Propagation of Order in Crystal Lattices" (1943), these issues are in effect beside the point. The problem becomes one of figuring out something justifiable that might work and be calculable.

59. See Newell and Montroll, "Theory of the Ising Model," pp. 375–376, for

an especially nice discussion. See also Schultz, Mattis, and Lieb, "Two-Dimensional Ising Model," p. 868; Baxter, *Exactly Solved Models*, pp. 21–24.

60. Here I am thinking of Montroll, Potts, and Ward ("Correlation and Spontaneous Magnetization"), who suggest they provide a more perspicuous method, and of Schultz, Mattis, and Lieb ("Two-Dimensional Ising Model") and Griffiths ("Spontaneous Magnetization in Idealized Ferromagnets"). See also McCoy and Wu, *Two-Dimensional Ising Model*, p. 25; Baxter, *Exactly Solved Models*, p. 119; Samuel, "Use of Anticommuting Variable Integrals," p. 2816.

In McCoy and Wu, *Two-Dimensional Ising Model*, pp. 177–178, 284–285, 372–376, these problems are dealt with rather more straightforwardly, the modes of calculation having by then justified the definition, careful consideration of what is open being an affordable luxury. Simon (*Statistical Mechanics of Lattice Gases*, p. 153) puts the current status nicely: "We note that neither Yang [who computes a free energy] nor Montroll, Potts, and Ward [a correlation function] directly computed the limit as the magnetic field goes to zero of the difference between the free energies with and without a magnetic field divided by the field, a restatement of the definition; for many years there was a gap in the rigorous proof in that these authors computed things that should equal M but which weren't known to be. . . . Thus, with recent progress, it is known that Yang's formula really gives M." (This is not quite a quote, but a paraphrase in part.) The "recent progress" is to be found in G. Benettin et al., "On the Onsager-Yang Value of the Spontaneous Magnetization."

61. Schultz, Mattis, and Lieb, "Two-Dimensional Ising Model," p. 867.

62. Ibid. I have made some slight changes in notation.

63. McCoy and Wu, *Two-Dimensional Ising Model*, p. 25.

64. That is why, in calculations, one may want to compute not the magnetization, but rather a projection of it in one direction or the other.

65. Yang, "Spontaneous Magnetization," especially p. 809 on the long-range order; Montroll, Potts, and Ward ("Correlations and Spontaneous Magnetization") use such a definition. That the long-range order is related to the spontaneous magnetization is shown in Newell and Montroll, "Theory of the Ising Model," p. 379.

66. McCoy and Wu, *Two-Dimensional Ising Model*, pp. 177–178, where they are concerned with $M^2 = \lim_{N \to \infty} \langle \sigma_{0,0} \sigma_{0,N} \rangle$, namely, the spontaneous magnetization is the long-range order at zero field. According to Griffiths ("Spontaneous Magnetization in Idealized Ferromagnets"), there is no proof that the limits needed for m_c exist, or that m_c and m_2 are in general the same. But see Simon, *Statistical Mechanics of Lattice Gases*, p. 153, for a more recent assessment.

67. Schultz, Mattis, and Lieb, "Two-Dimensional Ising Model," p. 867 (I have made minor changes in notation). For an argument, see Newell and Montroll, "Theory of the Ising Model," p. 376.

68. Newell and Montroll, "Theory of the Ising Model," pp. 375–376, equation 6.3. Depending on whether the spin flips are connected or not to the number of up spins, say, the partition function is either an $\exp|B|$ or a $\cosh B$

(one gets the exponential if one can drop one of the exponentials in the hyperbolic cosine), and in the latter case the lowest term in B is quadratic, in the former it is linear.

69. Schultz, Mattis, and Lieb, "Two-Dimensional Ising Model," p. 867, notation slightly altered.

CHAPTER SIX

1. Tragesser, *Phenomenology and Logic*; Tragesser, *Husserl and Realism in Logic and Mathematics*; Rota, "Pernicious Influence."

2. Cushing, *Theory Construction*.

3. For theorems, see Ruelle, introduction to *Thermodynamic Formalism*.

4. Rota, "Pernicious Influence."

5. Newell and Montroll, "Theory of the Ising Model," p. 361.

6. Onsager, "Ising Model in Two Dimensions."

7. Unfortunately, there is no archival evidence, as far as I can ascertain.

8. Krieger, "The Physicist's Toolkit," in *Marginalism and Discontinuity*; Krieger, *Doing Physics*, especially chap. 5; Pickering, *Constructing Quarks*.

9. It is said that Onsager treated Whittaker and Watson's *Modern Analysis* as a bible. His doctoral dissertation was on the Mathieu functions. I am not sure where he picked up his quanternion algebra (or Lie algebra). See Longuet-Higgins and Fisher, "Lars Onsager."

10. Hoddeson et al., *Out of the Crystal Maze*, p. 573. The quaternion algebra is that of the infinitesimal generators of these spatial rotations.

11. Steiner, "Applicability of Mathematics to Natural Science"; Rotman, *Ad Infinitum*.

12. See Krieger, *Doing Physics*, chap. 1, drawing from the first few chapters of Adam Smith's *Wealth of Nations* (1776).

13. Here I am thinking of Hessen's arguments for the practical foundations of Newton's work ("Social and Economic Roots").

14. Here I am thinking of Julian Schwinger's World War II work on waveguides, which led to his S-matrix formulation of quantum electrodynamics (Schwinger and Saxon, *Discontinuities in Waveguides*).

15. Hacking, *Representing and Intervening*; Krieger, *Doing Physics*.

16. See Itzykson and Drouffe, *Statistical Field Theory*, part 1, pp. 349–352, for a textbook presentation of duality in terms of complexes and homology.

Technically, recall that the energy of a system is the sum of the bonds' energies, and these bonds are just the spins or particles on the dual lattice—so duality "linearizes" the energy: $H_{\text{dual}} \approx \Sigma - J\sigma_{\text{dual}}$, versus $H \approx \Sigma - J\sigma_i\sigma_{i+1}$. One has to decorate the lattice to get away with this transformation. Otherwise the dual spins are not independent.

17. Date et al., "Solvable Lattice Models," p. 304–305.

18. The energy is not invariant if $q = 0$.

19. For the geometry and elliptic functions, see Lawden, *Elliptic Functions and Applications*, p. 104.

20. Hurst, "Applicability of the Pfaffian Method," points out that crossing of bonds is equivalent to the nonfactorizability of an analogous S-matrix into two-body interactions (one has quartic rather than quadratic terms).

21. Baxter, *Exactly Solved Models*, pp. 118–120.

22. However, the imputed free energies for the infinite volume limit will be different for the different transfer matrices—so we shall have a very strange mixed system.

23. Stephen and Mittag, "New Representation," p. 1951.

24. Krieger, *Doing Physics*, chap. 1; Krieger, *Marginalism and Discontinuity*, chaps. 1 and 4.

25. Lieb, "Stability of Matter," p. 564, theorem 12.

26. In Lebowitz and Lieb's proof of the stability of matter (ibid., p. 565–566) the Peierls-Bogoliubov inequality along the way shows how one might conceive of these bulkettes as "molecules" in an ideal gas.

27. Here I am thinking of the economic planner's task of "getting the prices right" for an efficient market.

28. See the nice brief survey by Lieb and Mattis, in *Mathematical Physics in One Dimension*, pp. 339–341.

29. Weinberg, "Why the Renormalization Group Is a Good Thing," p. 16.

30. Here I follow Schultz, Mattis, and Lieb, "Two-Dimensional Ising Model," and Wannier's gloss provided in his *Statistical Physics*, pp. 356–357. Note that this strategy reflects earlier work by this threesome on related problems, for which see Lieb and Mattis, *Mathematical Physics in One Dimension*. See Lieb, Schultz, and Mattis, "Two Solvable Models of an Antiferromagnetic Chain," pp. 409–416, 452–454. See also Nambu, "Eigenvalue Problem in Crystal Statistics."

Some further details: Starting out with the original spins, the strategy converts them into Pauli-spin operators, as did Onsager. Then it makes these operators into fermions, essentially by counting the number of spin flips and putting a sign on the operator to reflect even or odd numbers of flips (Jordan-Wigner transformation). Then it transforms these fermions into running waves, $\exp iqx$, still fermions, apparently reflecting the lattice's translation symmetry. And, finally, it decouples the q and $-q$ particles into standing waves by means of a rotation or Bogoliubov-Valatin transformation (from the theory of superconductivity) into $\{q, -q\}$ pairs. See chapter 8 at note 42. Those particles, as standing waves, are mixtures of order and disorder (as we might expect from having to simultaneously diagonalize Onsager's A and B operators). The mixing angle depends on the temperature and the wave number. See figure 7.12, where the disorder amplitude for K^* behaves as does the mixing angle for K. See "Phase Transitions in Ising Matter," below.

31. Onsager, albeit in 1971, says: "The eigenwert problem of the former [his operator A_0] defines a simple system of scalar spin waves (Fermi statistics, Bose symmetry) [see Schultz, Mattis, and Lieb, "Two Dimensional Ising Model," p. 861, equation 3.1]. . . . Other operators of the algebra generated by A and B can add or remove "Cooper pairs," whichever is possible" (Onsager, "Ising Model in Two Dimensions," p. 9).

Note that here pairs are created, at low temperature, by *adding* energy, the reverse of the superconductivity case.

32. The decoration of the lattice that Hurst and Green employ (in "New Solution of the Ising Problem") is just what is needed for there to be only pairs of adjacent spins, those pairings allowing for just one possible link between each pair. Kastelyn, "Dimer Statistics and Phase Transitions," shows how this is a dimer problem, with again the same decoration of the lattice.

33. Hurst and Green, "New Solution of the Ising Problem," p. 1060.

34. Samuel, "Use of Anticommuting Variable Integrals," p. 2817.

35. Onsager, "Crystal Statistics," p. 121; Kaufman, "Crystal Statistics," p. 1234. See also Schultz, Mattis, and Lieb, "Two-Dimensional Ising Model as a Soluble Problem of Many Fermions," whose title gives away the story. I imagine it is possible to avoid anticommuting objects completely, by device, but their recurrent appearance is impressive.

36. Hacking, *Representing and Intervening*.

37. Our real concern should be the Fermi surface, not simply the Boltzmann formula. For the *XY* model, there is a particle-hole picture for the excitations, the ground state has no elementary fermions, and the 'Fermi surface' consists of the points $q = \pm \pi/2$ (Lieb, Schultz, and Mattis, "Two Soluble Models," pp. 414–415).

38. Mattis, *Theory of Magnetism*, part 2, p. 128.

39. This is just what we might expect of quasiparticles in a Fermi liquid. The particles' original energies and charges are altered, but their other symmetries are preserved.

40. What seems to be crucial is that the diagonal-to-diagonal transfer matrix commutes with a pseudo-Hamiltonian, $B + k^{-1}A$, Onsager's observation. See Stephen and Mittag, "New Representation." So there is a Hamiltonian (not the original Ising Hamiltonian) whose eigenvectors or eigenstates are the eigenstates of the transfer matrix. The Hamiltonian is one-dimensional, the other dimension is time, and the energies of the particles are constants of the motion.

We might say, if a gas of fermions consists of many noninteracting fermions, we have a Fermi liquid consisting of many (weakly) interacting fermions whose energies now affect each other. Their "weak" interaction, at low temperatures, is just the columnar interactions within the thermal field, for the pseudo-Hamiltonian is in effect Column $+ k^{-1}$Row. Each fermion is at the same temperature. What Landau appreciated was that one could describe the liquid in terms of altered particles ("quasiparticles"), in their symmetry properties very much like the noninteracting fermions or particles, with perhaps a changed mass (or energy) and charge (but with no crossing of the energies or masses of states of similar symmetry). Here, again, the quasiparticles are the various configurations or patterns of rows of bonds, but their energies are no longer a sum of the spin-spin interaction energies, no longer simply a multiple of $2J$ units above the ground state. As we might expect, the interaction energy should increase with q, since then there are more spin flips per lattice separation.

Note what we have just done. To repeat, we have a two-dimensional thermodynamic system described by a partition function, $\Sigma \exp - \beta E$. We then discover one-dimensional quasiparticles—namely, the particles we discover when we try to diagonalize the transfer matrix (not the original Hamiltonian, although we are diagonalizing the pseudo-Hamiltonian), their energies or masses being dependent on the temperature. At each temperature, we now have a Fermi gas, a system of independent particles (row configurations), that system characterized directly not by a temperature but rather by the energies of its particles (which in fact are temperature dependent; namely, they are expressed in $k_B T$ units, and the coupling constants K_1 and K_2 are presumably $1/k_B T$ dependent). We have gone from a thermodynamic system characterized by a partition function to a seemingly nonstatistical one-dimensional system characterized by a Hamiltonian (namely, no explicit temperature, no partition function in view). In a moment, I shall again treat this system as a thermodynamic system, as a classical gas of independent fermions, the relevant energies being the row-particle energies.

More generally, the transfer matrix for two-dimensional lattices are sometimes simply related to a one-dimensional Hamiltonian: one takes a derivative with respect to u of the logarithm of the transfer matrix. (The two-dimensional eight-vertex model becomes a one-dimensional XYZ Hamiltonian, the two-dimensional Ising model becomes a one-dimensional Ising Hamiltonian in a transverse magnetic field.) If we think of the transfer matrix, V, as being $\exp - \Sigma N_q E_q$, and we know that the N_q are a function of k alone since the eigenvectors are only functions of k, the partial derivative operates only on the E_q, and we have an apparent one-dimensional Hamiltonian. Note that the eight-vertex model is sometimes equivalent to an even-site and an odd-site two-dimensional Ising lattice; namely, the transfer function for the eight-vertex model equals the product of the transfer functions of the two Ising models. So we find that the derived Hamiltonian from the former is the sum of the derived Hamiltonians of the latter. See Grady, "Infinite Set of Conserved Charges in the Ising Model"; Dolan and Grady, "Conserved Charges from Self-Duality." It seems that the two-dimension to one-dimension story depends on the self-duality of the two-dimensional Ising Hamiltonian: the dual just interchanges the vertical and the horizontal interactions. And, as I indicate elsewhere in the text, the symmetry that Newell and Montroll, and Wannier refer to to explain the Fourier transform's applicability is based on that duality (rather than on translational invariance).

As for the Fermi gas, this is a gas of scalar spin waves (Bose symmetry, Fermi statistics), where the Fermi surface is at $q = \pm \pi/2$. See Lieb, Schultz, and Mattis, "Two Soluble Models of an Antiferromagnetic Chain," p. 414.

41. I have made minor adjustments of horizontal and vertical coupling constants.

42. Because of the different interpretations in each solution, $\gamma(0)$ and ϵ_0 differ in sign. This has no effect on the maximal eigenvalue or the ground-state

energy, since the relevant formulas in each case compensate for the sign difference. Note that these are the first excited states, the odd combination of all up and all down: $|\uparrow\rangle - |\downarrow\rangle$. The ground state is the even combination—both being parity eigenvectors.

43. Technically, here the transfer matrix becomes $V \approx \exp - \Sigma \epsilon_q (N_q - 1/2)$. V, of course, is not the Ising Hamiltonian, and the Hamiltonian defined by $\ln V$ is also not the Ising Hamiltonian (Wilson and Kogut, "Renormalization Group and the ϵ Expansion," p. 145).

44. Schultz, Mattis, and Lieb, "Two-Dimensional Ising Model," p. 863, equation 3.29, where the first term is more than two but not by much for temperatures near the critical point, and the second term is just $\cos q$ for a square lattice.

45. Onsager, "Crystal Statistics," p. 137.

46. Griffiths, "Microcanonical Ensemble," p. 1447.

47. It is curious that almost none of the accounts of the Ising model that I have seen make use of this spectrum in great detail. Onsager provides the most hints connecting the physics to the quasiparticles' energies, although he does not think in terms of particles but, rather, in terms of disorder operators. All the other solutions are for the most part focused on the technical formalism. One exception I found after having written this section is Mattis (*Theory of Magnetism*, part 2, pp. 115–116), who notes that the occupation of the zero-energy states have no effect on the partition function. See Pfeuty ("Ising Chain in a Transverse Field") who employs plots similar to the ones I provide here and is the source of Mattis's discussion. For purposes of illustration, a square lattice will do as well as a rectangular lattice in showing the effects of the energy spectrum. The spectrum should also account for $c_v \approx \ln N$.

48. Onsager, "Crystal Statistics," p. 140.

49. The correlation lengths do differ by a factor of two (it is twice as large above T_c than below), reflecting the degeneracy below the critical point. Also, near the critical point, $\ln k \approx (K^* - K) \approx$ the mass gap.

50. Onsager, "Crystal Statistics," pp. 138–139; Kaufman, "Crystal Statistics," p. 1235; Wannier, *Statistical Physics*, p. 364; Baxter, *Exactly Solved Models*, p. 110.

51. Put differently, the partition function is the generating function for the density of states, $\rho(E)$: $Q(\beta) = \int \exp - \beta E \, \rho(E)$. It is a Laplace transform, and Tauberian theorems relating the high-energy behavior of the density to the low β (high-temperature) behavior of the partition function, and vice versa, might be thought to be relevant here. The Lee and Yang work on the unit-circle theorem (the density of zeros on the circle), Griffiths's work on the density of states and its relationship to the possibility of phase transitions, and early work by Poincaré and subsequent work by Fowler on seeing if the partition function sufficiently determines the density of states (and so the Planck deduction of the blackbody spectrum would be probative for the quantum hypothesis) are all investigations in this Laplacian light. Recall, as well, that the mathematical partition function of Euler becomes in the hands of Jacobi a theta function (or elliptic modular function). See Feller, *Introduction to Probability Theory*, vol. 2,

chap. 13, on Laplace transforms, Tauberian theorems, and generating functions; Lee and Yang, "Statistical Theory of Equations of State"; Griffiths, "Microcanonical Ensemble"; Fowler, "Simple Extension of Fourier's Integral Theorem"; Darwin and Fowler, "On the Partition of Energy," part 2; Ehrenpreis and Gunning, *Theta Functions*, p. ix.

52. Kac and Ward, "Combinatorial Solution." The relevant analogies are with random walks and with sums over quantum histories, the Feynman-Kac formula. See also Brydges, Fröhlich, and Spencer, "Random Walk Representation of Classical Spin Systems." Cellular automata, as one-dimensional objects, can be used to generate a two-dimensional Ising lattice, much as I discuss here. See Wolfram, *Theory and Application of Cellular Automata*.

53. Another way of putting this is to say that a sum of Gaussians is a Gaussian, or that a Gaussian is "infinitely divisible" into component Gaussians.

54. Here I follow Kastelyn, "Dimer Statistics and Phase Transitions." It turns out that there are some solvable models of larger genus, a more recent discovery.

55. Ibid., p. 287.

56. For ordinary random walks, the problem of genus is not the issue. It is the peculiar character of the random walk associated with the Ising model (its self-avoiding character, for example) that would seem to create difficulties. See also Hurst, "Applicability of the Pfaffian Method."

57. That Kac and Ward's account required further work by topologists and physicists (Sherman, Burgoyne) to clean up their solution, showing that it actually took into account the various cases as it should, is no indictment. Rather, this is as it should be, given the various competences of the members of the community of scientists. An equally illuminating example is provided by Baxter in his discussion of the motivations for his solution of the hard-hexagon model, in his distinction of exact from rigorous solutions, and in his willingness to say: "It ought to be shown, for instance, that the [infinite] matrix products are convergent. I have not done this, but believe that they are (at least in a sense that enables the calculation to proceed), and that as a result (13.7.21) is exactly correct" (Baxter, *Exactly Solved Models*, pp. vi). Andrews provided Baxter with assistance with the Rogers-Ramanujan identities involved here. See Andrews, *q-Series*; Andrews, Baxter, and Forrester, "Eight-Vertex SOS Model."

58. Feynman, *Statistical Mechanics*, pp. 136–150; Landau and Lifshitz, *Statistical Physics*, pp. 498–506. Here I have followed Landau and Lifshitz's account of Vdovichenko's work.

59. When the initiators Hurst and Green ("New Solution of the Ising Problem") invoke Pfaffians, they refer to Caianiello's work on field theory and combinatorics, and not only to books on matrix algebra. Essentially, the Wick theorem is the secret.

60. The fact that no site is to be used twice on this decorated lattice, or that no bond is to be used twice on the original lattice, might be said to be an $x^2 = 0$ fact, a statement about an exterior algebra. This leads naturally to Pfaffians, in expansions of exponentials of such anticommuting variables (Hurst, "Applicability of the Pfaffian Method").

61. It is a curious fact that Kac is responsible, with Feynman, for the Feynman-Kac formula concerning such random processes—although this does not figure in Kac's solution. It is central to the philosophy of Feynman's solution.

For the practicing physicist, a connection with Feynman diagrams is nowadays taken as saying what is going on.

As for the Pfaffian forms, Stephen and Mittag ("New Representation") show how they appear in a natural expansion. And Samuel ("Use the Anticommuting Variable Integrals") shows how "everything you ever wanted to know about the Ising model is expressible as a Pfaffian" (p. 2816). What makes Samuel's method of interest is its automaticity. McCoy and Wu, *Two-Dimensional Ising Model*, is a tour de force of Pfaffian methods.

62. Schultz, Mattis, and Lieb, "Two-Dimensional Ising Model," p. 861; Kastelyn, "Dimer Statistics and Phase Transitions," pp. 287–288; Samuel, "Use of Anticommuting Variable Integrals."

63. Here I am thinking of quantum field theory insights, modular functions, and the work of Zamolodchikov. For the latter, see Baxter, *Exactly Solved Models*, pp. 452–454; Baxter, "Exactly Solved Models."

64. For approximations, see Samuel, "Use of Anticommuting Variable Integrals," part 3. It would seem that the elliptic functions that appear ubiquitously in these solutions reflect the demands of the star-triangle relation, rather than a matter of two dimensionality of the lattice and the two period of those functions. They will be important for three-dimensional problems, too.

65. Baxter, *Exactly Solved Models*, pp. v–vi.

CHAPTER SEVEN

1. Gruber, Hintermann, and Merlini (*Group Analysis*, pp. xii–xiii) put it nicely: "[The mathematical structure] consists essentially of two abelian groups associated respectively with the configuration space and the interactions. . . these two groups play a 'dual' role. . . . Physically it appears that this framework reflects certain symmetries of the systems—either euclidean or internal."

The dual might be thought of as a transformation of horizontal to vertical bonds (in effect drawing normals to each original bond) or as a transformation of faces into points or vertices. Perhaps the best-known dualities to physicists are in electromagnetism, between electric fields and potentials (for example, conformal transformations in two dimensions) or between electric and magnetic fields (Srednicki, "Hidden Fermions in $Z(2)$ Theories," pp. 2883–2884). The star-triangle relation transforms vertices and the bonds attached to them (now thought of as $+s$) into diamonds, Ys into Δs.

2. There is another, similar symmetry, whose origin has not been clarified in what I have seen: above and beyond the periodicity of the lattice, there is a cyclic symmetry of Onsager's A_k matrices. See Newell and Montroll, "Theory of the Ising Model," pp. 368–369; Wannier, *Statistical Physics*, p. 358, equation 16.30. Following some hints in McKean, "Kramers-Wannier Duality," and

Gruber, Hintermann, and Merlini, *Group Analysis*, I suspect that the symmetry is due to duality.

3. The so-called odd and even spaces reflect that for circular boundary conditions an alternation of signs may or may not be consistent, depending on whether we have an even or odd number of atoms in the circle.

4. I put in "local interactions," because with less-local interactions there can be a finite-temperature phase transition. See Dyson, "Existence of a Phase Transition."

5. Kramers and Wannier, "Statistics of the Two-Dimensional Ferromagnet"; Onsager, "Crystal Statistics," pp. 123–124.

6. The dual can be seen to be a rotation of the lattice by π/m into its dual lattice, where m is two for a rectangle, three for the triangle or hexagon, and four for the square. For a more formal account, see Gruber, Hintermann, and Merlini, *Group Analysis*; Merlini and Gruber, "Spin-1/2 Lattice System."

7. Onsager, "Crystal Statistics," pp. 120–121, concerning his s_i and C_i.

8. Ibid., pp. 123–124.

9. Onsager, "Ising Model in Two Dimensions," p. 3.

10. Ibid., pp. 3–4; Stephen and Mittag, "New Representation," p. 1945.

11. It may be useful to rephrase the duality property in rather more everyday terms of mechanics. Recall chapter 4, note 58, where the energy is a kinetic energy term and duality operates on the gradient of S, taking $|\nabla^+ S|$ and converts it to $1 - |\nabla^+ S|$. Namely, duality operates in a curious tangent space (or bundle) of this field, order-to-disorder being merely complementation there.

12. I have been speaking here rather roughly, since it is not yet clear to me how to show that the transformations involved operate geometrically on k. See Wilson, "Renormalization Group," pp. 797–800, for a specific toy example. In this case, it would seem that $k_i = k_{i-1}^2$ (toy model on p. 799). For actual work in which the renormalization-group transformation is related to the star-triangle relation, see Hilhorst, Schick, and van Leeuwen, "Exact Renormalization Group Equations."

13. See Feller, *Introduction to Probability Theory*, 2:173–176.

14. Knops and Hilhorst, "Exact Differential Renormalization," diagram on p. 3690, which takes off from Hilhorst, Schick, and van Leeuwen, "Exact Renormalization Group Equations" and "Differential Form of Real-Space Renormalization."

15. The star-triangle relation is differently employed by Baxter and Enting ("399th Solution of the Ising Model") to do another such topological lattice transformation. See figure 7.8.

16. Onsager, "Crystal Statistics," p. 118.

17. It is clear from Onsager's later remarks that, at this time, he appreciated some of the implications of the star-triangle relation, in terms of the commutativity of the transfer matrix. But none of this is conveyed in the paper, except for the hint that they were a "great help to the discovery of the present results" (ibid.).

18. Baxter, "Solvable Eight-Vertex Model"; Baxter, *Exactly Solved Models*; Baxter, "Solving Models in Statistical Mechanics"; Baxter, "Solvable Models in Statistical Mechanics."

19. Note that, if the couplings differ, then their ratios will differ for the star from those of the triangle, and so the dual lattice (namely, the high-temperature triangle for the low-temperature triangle) will not have couplings in the same ratios as in the original lattice.

The method applies as well to rectangles and other lattices, as long as we might make each into a bipartite lattice, although we gain nothing new over the duality transformation unless we allow all four couplings—up, down, right, left —to be different.

20. Wannier, "Antiferromagnetism," p. 358. As indicated in note 19 above, if the three coupling constants differ, then after the duality and star-triangle transformations, the "dual" hexagon, say, has different ratios of its coupling constants than the original one.

21. If we have a square lattice with equal interaction constants, then $\sinh^2 2K = 1/k$, and the relationship to a temperature is perhaps most apparent.

22. Onsager, "Crystal Statistics," p. 142. Here $J \neq 0$.

23. Mattis, *Theory of Magnetism*, part 2, p. 107.

24. Montroll, Potts, and Ward, "Correlations and Spontaneous Magnetization." Essentially, what happens is that, rather than w, equal to $\tanh K$, appearing in the high-temperature expansion of the partition function, $1/w$ appears. Since $w = \exp - 2K^*$, we have $w \to 1/w$ corresponding to $K^* \to -K^*$, essentially introducing frustration on the dual lattice, corresponding to correlation on the original lattice. See Itzykson, "Ising Fermions," p. 452.

Note that the transformation of K, $\to K \pm i\pi/2$, preserves k and corresponds to $u \to -u$, for

$$\sin(2i(K \pm i\pi/2)) = \sin(2iK \mp \pi) = -\sin 2iK = \text{sn}(-iu).$$

25. Kramers and Wannier, "Statistics of the Two-Dimensional Ferromagnet," p. 261. κ is an elliptic modular transformation of k, the Gauss transformation. See figure 8.3.

26. Onsager, "Ising Model in Two Dimensions," p. 7.

27. Wannier, "Antiferromagnetism," p. 359.

28. See Baxter, *Exactly Solved Models*, p. 118; Onsager, "Crystal Statistics," p. 142. Baxter notes (without pointing out the fluctuation-dissipation theorem) the relationship after calculating both, while Onsager explicitly uses the fluctuation-dissipation relationship to find the correlation length, given the interfacial tension. Baxter (p. 242) also points out that the relationship seems to be derived from scaling considerations.

29. As in note 1 above, one might view duality as a process that converts the interactions around the face of a lattice into a vertex (and then factorizability is an obvious question).

30. Cushing, *Theory Construction*, pp. 189–193. k might be seen to be such an energy, while the variable u, defined by a measure of the asymmetry $(K_1 \neq K_2)$ of lattice interaction, might be in effect a momentum variable. Actually u is a rapidity. See chapter 5, note 34.

31. Chandler, *Introduction to Modern Statistical Mechanics*, chap. 8, provides a particularly lucid exposition; Landau and Lifshitz, *Statistical Physics*, sec. 123, provides greater detail, connecting fluctuation-dissipation relations to Kramers-Kronig relations as well.

32. This is the *Ansatz* that informs the Bethe-Peierls approximate solution to the Ising model, for which see Huang, *Statistical Mechanics*, p. 342.

33. The strongest interaction together with one of the weaker ones will build up a long-range order in competition with the long-range order it would build with the other of the weaker ones. See Houtappel, "Order-Disorder in Hexagonal Lattices," pp. 450–451, 454.

34. Baxter and Enting, "399th Solution of the Ising Model." See also Pais, "On a Plane Ising Lattice," which anticipates this theme of connecting the nearest and next-nearest neighbors. In effect, Pais does a decimation, what is called now a "star-square" transformation. See Baxter, *Exactly Solved Models*, p. 257, equation 10.13.7.

35. Baxter and Enting, "399th Solution of the Ising Model," p. 2465.

36. Hilhorst, Schick, and van Leeuwen, "Exact Renormalization Group Equations," p. 2760.

37. See, for example, McGuire, "Study of Exactly Soluble One-Dimensional N-Body Problems," p. 631, fig. 7.

38. Baxter, *Exactly Solved Models*, p. 93. See, for example, Savit on duality ("Duality in Field Theory and Statistical Physics") to see which issues were in the air at this time.

39. Cushing, *Theory Construction*, pp. 226–230, discusses how causality comes to have this particular technical definition.

40. Feynman and Hibbs's exposition of statistical mechanics (*Quantum Mechanics and Path Integrals*, chap. 10) makes this analogy quite precise; $i\hbar\beta$ is formally like a time variable.

41. I do not know how and when to connect equilibrium to analyticity. But I do know that the analyticity assumptions, if incorrect, will lead to incorrect results, as Baxter has demonstrated. My argument can be nicely summarized by figure N.2.

42. One also needs to specify the initial or final conditions, that for these processes time is unidirectional, that heat dissipates or flows to infinity. See Wheeler and Feynman, "Interaction with the Absorber."

43. See Hurst, "Applicability of the Pfaffian Method," for a suggestion that the crucial insight is a matter of the expansion of an exponential of anticommuting operators into multilinear operators, a problem in exterior algebra. (Appropriately, Gibbs was an admirer of Grassmann.) The bipartite character of the lattice allows for a two-body factorization, and that factorizability may be the crucial physics here. See Pokrovsky and Bashilov, "Star-Triangle Relations," p.

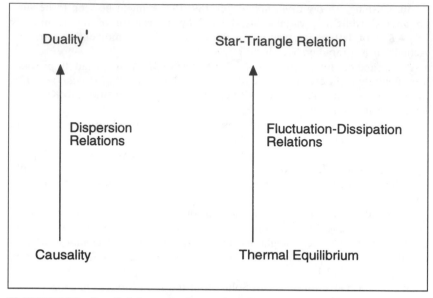

FIGURE N.2   Parallels between fluctuation-dissipation relations and dispersion relations.

107. In the case of the block-spin transformation, it would appear that greater than two spin interactions start becoming insignificant.

44. Mattis (*Theory of Magnetism*, part 2, p. 129) makes this point by showing how repeated application of the transfer matrix forces any nonoptimal eigenvector to become optimal (in the sense of being the vector with the largest eigenvalue), simply by definition of the eigenvalues and eigenvectors. V^N is a projection operator that pulls out from any vector the part parallel to the eigenvector associated with the largest eigenvalue (as long as the original vector is not orthogonal to the maximum eigenvector, and that will not be the case if all its elements are positive, which is the case for probabilities, the source of the transfer matrix).

45. The ratio of the largest to the next-largest eigenvalue is a measure of the rate of relaxation.

46. Onsager, "Crystal Statistics," p. 139. In this sense, Onsager is aware that he is dealing with the degrees of freedom in a system when he is integrating γs.

47. In some methods of solution, the star-triangle relation is not manifestly invoked or needed. I am not sure it is absent. See Baxter, "Solving Models in Statistical Mechanics," pp. 101–103.

48. Wannier, "Antiferromagnetism," and Houtappel, "Order-Disorder in Hexagonal Lattices," begin to note some of the curious formulas that arise from the star-triangle relation.

49. As Baxter shows explicitly, if one gives each of the four sides arbitrary interaction constants, the star-triangle relation constrains them so there is a temperature that is consistent with those values—they cannot be so arbitrary.

50. Baxter, "Exactly Solved Models," shows some other devices, namely, the corner-transfer matrices, as well as new, more direct modes of using commuting transfer matrices and the "matrix inversion trick," as he calls it.

51.

$$\partial/\partial K \ln \sum_{\{\sigma_i\}} \exp K \sum \sigma_i \sigma_{i+1}$$

$$= \sum_{\{\sigma_i\}} \sum \sigma_i \sigma_{i+1} \exp K \sum \sigma_i \sigma_{i+1} \bigg/ \sum_{\{\sigma_i\}} \exp K \sum \sigma_i \sigma_{i+1},$$

which is within a factor $(1/N\,2^N)$ just $\langle \sigma_i \sigma_{i+1} \rangle$, the correlation of nearest-neighbor spins to each other.

52. It would be interesting to study the various sorts of transfer matrices and how each builds in the physics of Ising matter—including the corner-transfer matrices of Baxter. The corner transfer matrix is in effect a Lorentz boost (H. B. Thacker, "Corner Transfer Matrices," *Physica* 18D (1986): 348–359).

53. Baxter, *Exactly Solved Models*, pp. 88–89. For a more recent account, see Baxter, "Solvable Models in Statistical Mechanics"; Baxter, "Exactly Solved Models."

54. See chapter 4 for more on Baxter's method. See also Baxter, "Exactly Solved Models," pp. 28, 41. For the S-matrix equivalents of these properties, see Deguchi, Wadati, and Akutsu, "Knot Theory," pp. 229–232, especially p. 232; Shankar, "Simple Derivation of the Baxter-Model Free Energy"; and Pokrovsky and Bashilov, "Star-Triangle Relations," in which the S-matrix equivalents ($\pi u/I'$ is the rapidity) are nicely presented.

55. It would be worthwhile, or so I would hope, to extract the physics buried in Baxter's mode of solving these functional equations. The crucial fact is surely that, in finding the zeros, he expresses Λ as a product of theta functions, modular functions that provide the $k^{-N/2}$ factor in the duality transformation. I shall not go into the solution of these functional equations except to say that they make lovely use of the features of Jacobian elliptic functions, their double periodicity, and their expression in terms of theta functions. That the Rogers-Ramanujan identities play a central role for the hard-hexagon model only emphasizes the importance of q-series and theta functions in this realm.

56. Baxter and Enting, "399th Solution of the Ising Model." They refer to work of Fisher that partly anticipates their results. It is not clear if duality plays a role in this solution. See also Baxter, "Exactly Solved Models," for other solutions in this vein; Baxter, *Exactly Solved Models*, pp. 281–283.

57. Baxter and Enting, "399th Solution of the Ising Model," pp. 2466–2467, equation 7. Note that, for symmetric lattices at the same k, where the K_i or the L_i are equal and are connected by the star-triangle relation, then $K < L$, since the triangular constants lead to next-nearest-neighbor correlations, while the hexagonal constants are for nearest neighbors.

58. Kramers and Wannier, "Statistics of the Two-Dimensional Ferromagnet," p. 260.

59. Baxter, "Solving Models in Statistical Mechanics." Baxter, *Exactly Solved Models*, generalizes the topological transformation (pp. 281–282) and the functional equation method (p. 298).

60. Samuel, "Use of Anticommuting Variable Integrals," p. 2806.

61. Ibid., p. 2916.

62. See Stephen and Mittag, "New Representation"; Knops and Hilhorst, "Exact Differential Renormalization." Note that Onsager's *ia* (his "imaginary parameter") corresponds to my or Baxter's *iu*, the argument. More precisely, below the critical point, am(ia) = $2iK$, where am is the amplitude function. (Onsager, "Crystal Statistics," pp. 145–147).

63. Maxwell, "Address to the Mathematical and Physical Sections," pp. 216–217.

64. Crowe, *History of Vector Analysis*.

65. Maxwell, "On the Mathematical Classification of Physical Quantities," p. 258.

66. Baxter, *Exactly Solved Models*, p. 254, in discussing Suzuki's proposal for weak universality, in which $1/\xi$ replaces t as the variable of interest.

67. Onsager, "Ising Model in Two Dimensions," p. 6.

68. For example, we might suspect that the fact that elliptic functions are the only periodic functions, and that they have only two periods, suggests that at least they might play a different role in the three-dimensional Ising model. They are almost certainly needed for star-triangle relations and duality, which appear in all dimensions. In any case, these kinds of suspicions are notoriously unreliable.

69. Baxter, "Solvable Eight-Vertex Model," p. 316.

70. Mathematically, there are no n-periodic functions for $n > 2$; for $n = 2$, they are the elliptic functions (where trigonometric functions and exponentials are special cases of elliptic functions). Elliptic functions are, in this sense, the "only" periodic functions. Moreover, by a theorem due to Weierstrass, every analytic function of one complex variable that satisfies a polynomial functional equation (that is, $f(x + y)$, $f(x)$, $f(y)$ identically satisfy a polynomial equation) must be an elliptic function. So, for example, Baxter's functional equation for the eigenvalues of the transfer matrix is just such an equation: $\Lambda(u) \times \Lambda(u + I')$ = ... It is perhaps no wonder that elliptic functions play a central role in the solutions to the Ising model. That wonder is not only mathematical, it is physical as well.

71. Here I follow Whittaker, *Analytical Dynamics*, pp. 71–74; Greenhill, *Applications of Elliptic Functions*.

72. Whittaker, *Analytical Dynamics*, sec. 44; Greenhill, *Applications of Elliptic Functions*, secs. 1–8.

73. Here I follow Greenhill, *Applications of Elliptic Functions*, p. 77.

74. Lawden, *Elliptic Functions and Applications*, chap. 1. By the way, for a capacitor, its capacity is approximately ln q. On elliptic measures, see p. 104 of Lawden for spherical triangles (rather than the case here, hyperbolic ones). See

Onsager, "Crystal Statistics," pp. 145–147, for some germinal ideas on the (imaginary) argument, u (what he calls a).

75. See Longuet-Higgins and Fisher, "Lars Onsager," p. 456.

76. Baxter, *Exactly Solved Models*, p. 123. $\wp$ uniformizes elliptic curves.

77. Onsager, "Crystal Statistics," p. 144.

78. Whittaker and Watson, *Modern Analysis*, p. 454.

79. See Krieger, *Doing Physics*, p. 117, on diagrammatics. Schweber (*QED and the Men Who Made It*) has written poignantly of Feynman's use of visual pictures in his formulations.

80. Onsager, "Crystal Statistics," pp. 138–139, where Onsager is actually discussing an integral; he is not directly referring to the geometric fact here, but he does make a good deal of it in analyzing the solution.

81. Stephen and Mittag, "New Representation," p. 1951.

82. Onsager, "Crystal Statistics," pp. 135–136. Onsager uses both his triangle and a diagram that displays the values of the amplitude, δ^*, as a function of ω and temperature (fig. 5 of his paper, my figure 7.12).

83. There is another payoff. The relevant functions prove to be meromorphic (their only singularities are poles), and so one might express the relevant elliptic functions as ratios of entire functions, analytic everywhere, namely, the theta functions, just what Baxter needs to do his work.

84. In effect, there is only one coupling constant at the critical point since there is no finite-length scale operative there (the correlation length is infinite), and ordinary circular or trigonometric functions (with their single period) do the needed work.

85. Baxter, *Exactly Solved Models*, p. 123. Note that k is built into the definition of u, and $I' = I'(k)$, and so the sum can actually work. If the us are computed from the hexagonal constants, $\{L_i\}$, then $\Sigma u_i = 2I'$. In this case, for a rectangular lattice derived from a hexagon, $L_3 = \infty$ and $u_3 = I'$, so $u_1 + u_2 = I'$ as before.

86. Baxter, *Exactly Solved Models*, p. 123; see also p. 210.

87. Stephen and Mittag ("New Representation," p. 1944) claim that the natural geometry is Euclidean, for which see their own triangle.

88. Baxter, *Exactly Solved Models*, equation 7.13.4, p. 124, puts it as $\Sigma u_i = I'$ (the quarter-period magnitude). Geometrically, the star-triangle relation can be seen to be a statement about the conservation of the elliptic measures of a hyperbolic triangle when it transformed to its polar opposite.

89. Onsager, "Ising Model in Two Dimensions," p. 6.

90. Baxter, *Exactly Solved Models*, p. 292. For an even deeper discussion of these geometric facts, connecting them to elliptic functions, see Abraham and Martin-Löf, "Transfer Matrix," pp. 253–254, especially figures 2 and 3. The transfer matrices are seen to operate in a two-dimensional hyperbolic space (projected, a la Klein, into a circle) and Onsager's triangle appears "naturally" in this context.

91. Krieger, *Doing Physics*, pp. 116–117.

92. Onsager, "Crystal Statistics," p. 146.

93. Onsager, "Crystal Statistics," p. 140. I have changed the notation slightly to be consistent with my text.

94. Baxter, *Exactly Solved Models*, p. 91.

95. But see also the papers by Kramers and Wannier, "Statistics of the Two-Dimensional Ferromagnet," and by Montroll, "Statistical Mechanics of Nearest Neighbor Systems," in which some of these models are worked out as well, in quite ingenious and illuminating approximations.

96. Date et al., "Solvable Lattice Models," pp. 295, 301.

97. Work on the Bethe *Ansatz* and ice models is not mentioned in my discussion, yet it is part of the earlier period. For more on what was going on in the 1970s, see Baxter, "Solvable Models in Statistical Mechanics."

Developments in particle physics and gauge field theory made the Ising model an important test case, and so there also developed other ways of viewing the Ising model. For example, Constant $\times k$ and u are in effect action-angle variables for the Ising model, as they are for planetary motion or for the pendulum. In the latter two cases, they are measures of the angular momentum, and the elliptic functions are in effect the canonical transformations to action-angle variables. Moreover, it would seem that the N_q and ϵ_q for the Ising model are the other action-angle variables, where the N_q are charges and the ϵ_q are angles (as the $\gamma(\omega)$ surely are for the Ising model in Kaufman's formulation). See Grady, "Infinite Set of Conserved Charges in the Ising Model." The nth charge is the nth partial derivative with respect to u of the logarithm of the transfer matrix.

98. Some of the most suggestive of the analogies, which I have mentioned elsewhere in the text: the transfer matrix of the statistical mechanical system can be seen to be a quantum mechanical evolution operator, the logarithm of its maximal eigenvalue being proportional to the ground-state energy of the associated quantum mechanical system; spin-spin correlation functions are vacuum expectation values measuring field fluctuations; the setting in of long-range order is a symmetry-breaking event, and implies the appearance of new particles; if we go to imaginary time for the heat equation, then the Schrödinger equation and the heat/diffusion equation are "the same"; the partition function can be shown to be "the same" as the Feynman sum-of-histories formula for the amplitude in quantum mechanics.

Deguchi, Wadati, and Akutsu ("Knot Theory," pp. 229–232) nicely show the connections in terms of the star-triangle relation and S-matrix notions. The analogy is exact.

99. Winner, *Autonomous Technology*.

CHAPTER EIGHT

1. This conception of a gas of paths or polymers, interacting when they cross, is due to Symanzik. See Brydges, Fröhlich, and Spencer, "Random Walk Representation of Classical Spin Systems."

2. For example, "a non-trivial solvable model reveals an essence of the physical phenomena under consideration. Third, solvability often brings about not only a new physics but also a new mathematics" (Deguchi, Wadati, and Akutsu, "Knot Theory," p. 194).

3. From a private communication by Eric Livingston, July 1994.

4. See, for example, Parisi, *Statistical Field Theory*; Zinn-Justin, *Quantum Field Theory and Critical Phenomena*; Simon, *Statistical Mechanics of Lattice Gases*.

5. I have not discussed features that might have once seemed significant, but that turned out to be not so crucial.

6. A cursory reading of Rota, *Gian-Carlo Rota on Combinatorics*, makes clear how rich are the mathematical structures developed to describe these processes, and that they are pregnant with possibilities for physics. See, for example, Simon, *Statistical Mechanics of Lattice Gases*, pp. 464–470.

7. See Itzykson and Drouffe, *Statistical Field Theory*, part 1, for an interpretation in terms of homology.

8. Andrews, *q-Series*.

9. McKean, "Kramers-Wannier Duality," p. 775. A more developed and technical formulation is to be found in Gruber, Hintermann, and Merlini, *Group Analysis*. The e^{ikx} factor in the Fourier transform is now $\exp(i \times \text{dual-polygon} \times \text{polygon})$. Concerning volumes in phase space, the $k^{-N/2}$ that appears in duality relations is formally just the Jacobian of the Fourier transform that leads to the Poisson sum that is the duality relation. See note 1, p. 311.

It should be recalled that the theta functions appear in solutions to the heat equation, and their appearance there is as Fourier transforms, for which see Lawden, *Elliptic Functions and Applications*, chap. 1.

10. See also Feynman and Hibbs, *Quantum Mechanics and Path Integrals*, chap. 10, where the analogy is beautifully presented.

11. Georgii, *Gibbs Measures and Phase Transitions*, pp. 1–6. I have eliminated all italics in this quote.

12. A state may consist of multiple phases or be a pure phase (which is usually taken to be translationally invariant), but what do we mean by this notion, in technical detail? Or ice and water at the freezing point is an "extreme" state (at the corner of a simplex) and not a statistical mixture of thermodynamic properties. That the ice and water have spatial boundaries is not a given. See Georgii, *Gibbs Measures and Phase Transitions*, pp. 119–120; Simon, *Statistical Mechanics of Lattice Gases*, chap. 3, especially pp. 282–283; and references therein.

13. See Wightman, introduction to Israel, *Convexity in the Theory of Lattice Gases*; Simon, *Statistical Mechanics of Lattice Gases*. The characterization of the DLR equations is from Parisi, *Statistical Field Theory*, p. 27. Perhaps the crucial figure here is David Ruelle.

We have to make precise just what we might mean by basic notions such as the thermodynamic limit, so that they accord with the actual physics. Wightman (pp. 1x–1xi) puts it nicely:

> These definitions seem reasonable, although they leave open the question how the equilibrium state is to be defined. However, they are, in fact, quite subtle as one sees if one attempts to describe the behavior of a sequence of states of larger and larger finite systems such that, in the thermodynamic limit, they converge to such a mixed state of the actually infinite system. If, for concreteness, one thinks of a mixture of liquid and gas pure phases, then the large system has to consist mainly of a large droplet of liquid and the rest gas in proportion α and $(1 - \alpha)$. However, one also expects a finite probability for a more complicated distribution of droplets and it is the asymptotic probability distribution of the droplets as the system becomes larger and larger that determines the thermodynamic limit is a mixture.

14. Baxter, "Solvable Models in Statistical Mechanics," p. 14.

15. Recall that Samuel remarked that everything you wanted to know about the Ising model is expressible as a Pfaffian ("Use of Anticommuting Variable Integrals," p. 2816).

16. D'Ariano, Montorsi, and Rasetti, *Integrable Systems in Statistical Mechanics*, p. 2.

17. Au-Yang and Perk, "Chiral Potts Model Revisited," p. 18. Baxter provides two diagrams, which connect the various models and the various modes of solution, insisting on the unity of the constitutions of matter ("Solving Models in Statistical Mechanics," pp. 97, 101).

18. I have taken this from Yang and Ge, *Braid Group*.

19. Those technical details have yet to be sifted for their essential meaning. See, for example, Itzykson and Drouffe, *Statistical Field Theory*, part 2, chap. 9, "Conformal Invariance," or for the Ising model, *Statistical Field Theory*, part 1, chap. 2, "Grassmannian Integrals and the Two-Dimensional Ising Model." What becomes apparent is how powerful techniques and modes of presentation efface the history of a problem, while making what was once hard to see now manifest and inarguable. As I have indicated in the text, I have deliberately decided not to present these constitutions here because they are, for the moment, too automatic.

20. Ferdinand and Fisher, "Bounded and Inhomogeneous Ising Models," p. 837. They call q a modulus, but I believe they mean a nome. In any case, such a q would provide for "finite size scaling." Note that if $q = \exp(i\pi\tau)$, as is conventional, then the Jacobian imaginary transformation in effect inverts τ, exchanging m and n (note that $k \to \sqrt{(1 - k^2)}$). Fisher also makes lovely use of the conformal mapping in his review article, "Nature of Critical Points."

21. See, for example, Itzykson and Drouffe, *Statistical Field Theory*, part 2, chap. 9, which has a lengthy discussion of conformal field theory.

22. Date et al., "Solvable Lattice Models," p. 305, referring to Andrews, Baxter, and Forrester, "Eight-Vertcx SOS Model"; Belavin, Polyakov, and

Zamolodchikov, "Infinite Conformal Symmetry"; Kac and Peterson, "Infinite-Dimensional Lie Algebras."

23. More complex lattice systems require further variables, but they either are like k, signifying commuting transfer matrices, or are like u, indexing those matrices. Again, note that after the triangle-to-star transformation, the sum of the us (defined now by $\{L_i\}$) is $2I'$, the difference from I' readily handled by the elliptic functions. Note that the formula for defining k from $\{K_i\}$ is not the same as that defining k from $\{L_i\}$. As indicated earlier, the us (or u/I') are rapidities in the S-matrix analogy. See, for example, Pokrovsky and Bashilov, "Star-Triangle Relations." See chapter 5, note 34.

24. Stephen and Mittag use commutativity to find Pfaffian forms, while Baxter's techniques are algebraic and involve solving functional equations.

25. Baxter, *Exactly Solved Models*, p. 124, 84–86. Note that $I' = I'(k)$.

26. It is straightforward, although not news, to show that this relationship means that a rectangular lattice may be built up of a horizontal link and then a vertical one, or of a vertical link and then a horizontal one, as long as the coupling constants are chosen correctly. Namely, if K_1 is horizontal and K_2 is vertical, then let L_2 (defined by the star-triangle relation) be the horizontal coupling and L_1 be the vertical coupling. Then $K_1 = L_2$ and $K_2 = L_1$.

27. Some more complex linkages are readily included within this framework.

28. Here I follow Deguchi, Wadati, and Akutsu, "Knot Theory." By the way, figure 8.2, drawn from Deguchi et al., is reminiscent of Green and Hurst's *Order-Disorder*, p. 175, fig. 4.5, about counting bonds (separately from sites).

29. We are in an "infinite momentum (or rapidity) frame," effectively one-dimensional.

30. See chapter 4, note 44; chapter 5, note 34. See Yang and Yang, "One-Dimensional Chain of Anisotropic Spin-Spin Interactions"; Yang, "S-Matrix for the One-Dimensional N-Body Problem"; McGuire, "Study of Exactly Soluble One-Dimensional N-Body Problems." The source is Bethe, "Zur Theorie der Metalle."

31. See, for example, Baxter, *Exactly Solved Models*, chap. 10.

32. See Andrews, *q-Series*; Ehrenpreis and Gunning, *Theta Functions*.

33. Montroll, "Statistical Mechanics of Nearest Neighbor Systems"; Kramers and Wannier, "Statistics of the Two-Dimensional Ferromagnet."

34. The Ashkin and Lamb paper appeared in the *Physical Review*, while the Lassettre and Howe paper appeared in the *Journal of Chemical Physics*. This observation is not probative, but it is suggestive.

35. Zinn-Justin, *Quantum Field Theory and Critical Phenomena*, p. 19. Moreover, if the ground state is unique and isolated, repeated application of $\exp - \beta H_{qm}$ projects out the ground state, for β large, much as repeated application of $V(H_{sm})$ projects out the maximal eigenvector.

36. Newell and Montroll, "Theory of the Ising Model," p. 377. Of course, classical perturbation theory is the foundation for quantum mechanical perturbation theory, but I think it is fair to say that in this context the classical origins are forgotten.

37. Domb, "Self-Avoiding Walks and Scaling." Feynman and Hibbs, *Quantum Mechanics and Path Integrals*, p. 276, has a particularly nice discussion of the analogy. "What we are doing is providing a physical analogue for the mathematical expression." $\beta\hbar$ is taken to be a time (it has the dimensions of time). And then,

> Consider all the possible paths, or "motions," by which the system can travel between the initial and final configurations in the "time" $\beta\hbar$. The density matrix is a sum of contributions from each motion, the contribution from a particular motion being the "time" integral of the "energy" divided by $\hbar$ for the path in question.
>
> The partition function is derived by considering only those cases in which the final configuration is the same as the initial configuration, and we sum over all possible initial configurations.

38. Simon, *Statistical Mechanics of Lattice Gases*, p. 443, attributes the zero-component idea to deGennes.

39. Temperley, "Two-Dimensional Ising Models," pp. 232–233.

40. Note that, if we add a magnetic field, then many of the symmetries of these eigenvectors have to be abandoned.

41. See Lieb and Mattis, *Mathematical Physics in One Dimension*, pp. 339–342, for a discussion of Tomonaga's original work on interacting fermions. All of this is reminiscent of the Dirac particle-hole picture of pairs, or of inner shell vacancies in atoms that produce X-rays when an electron falls into them. See Kragh, *Dirac*.

The mixing angle ϕ_q (of order with disorder) is given in Schultz, Mattis, and Lieb, "Two-Dimensional Ising Model," p. 863. It is small for high temperatures, and zero for $q = 0, \pi$. It has a low-q peak ($\phi_q < \pi/4$) for lower temperatures. "At" the critical temperature, it peaks at $q = 0$ (and $\phi_q = \pi/4$ there) and is a fairly straight line to zero at $q = \pi$. Then the peak moves discontinuously to $\pi/2$ at $q = 0$, and as the temperature gets lower the curve becomes straighter. Mixture is greatest for $\phi_q = \pi/4$. In sum, as the temperature gets lower, more and more order is mixed in. Its behavior is much as for Onsager's disorder amplitude ("Crystal Statistics," p. 135), δ^*: $\phi_q(K) \approx \delta_q^*(K^*)/2$. See my figure 7.12. ($\phi_q \approx \phi_{SM} \approx \phi_W/2$, where *SM* is for Stephen and Mittag, "New Representation," p. 1946, and *W* is for Wannier, *Statistical Physics*, p. 362. Note that Onsager's γ [or ϵ_q] is closely related to the disorder amplitude, δ^*, as can be seen from his triangle, figure 6.1 here.) And, as Onsager (p. 137) points out: "As regards the variation of δ^* with the temperature for a fixed $\omega[q]$, we note that as the former increases from 0 to ∞, δ^* increases from 0 to $\pi - \omega$. For small values of ω, nearly all of this increase takes place over a small interval of temperatures which includes the critical point."

42. See Simon, *Statistical Mechanics of Lattice Gases*.

43. For a nice presentation about resistance and accommodation to nature, see Pickering, *Mangle of Practice*.

44. Parisi, *Field Theory, Disorder, and Simulations*. On cellular automata and the Ising model, see the papers in Wolfram, *Theory and Application of Cellular Automata*. In particular, our earlier explanation of the relationship of a two-dimensional lattice with a random walk to a one-dimensional row particle is realized in a cellular automaton.

45. Krieger, *Marginalism and Discontinuity*; Krieger, *Doing Physics*.

46. I do not believe this is a statement about human cognitive limits.

47. Buchwald, *From Maxwell to Microphysics*.

48. On re-proving, see Rota, "Pernicious Influence."

49. This formulation concerning mathematical logic seems to be sourced in G. Kreisel's work ("Survey of Proof Theory," "Mathematical Significance of Consistency Proofs"). Quoted from Barwise, *Handbook of Mathematical Logic*, p. 820.

50. Rota, "Kant's Synthesis A Priori and Husserl's Phenomenology of Fulfillment."

NOTES ADDED IN PROOF

1. In reading G. W. Mackey, "Harmonic Analysis as the Exploitation of Symmetry—An Historical Survey" (*Rice University Studies* 64 [Spring–Summer 1978]: 73–228), harmonic analysis would appear to be the crucial theme in the Ising model solution: harmonic analysis over groups, and group characters, and duality (cf. Rota, *Gian-Carlo Rota on Combinatorics*, on Möbius inversion); convolution, converting differential equations into algebraic ones; generating functions (Laplace); the Poisson summation formula; automorphic group (transformations that take a regular solid into itself), the modular group (automorphic group with rational coefficients) and elliptic functions, complex multiplication and addition theorems (for which see S. G. Vlăduţ, *Kronecker's Jugendtraum and Modular Functions* [New York: Gordon and Breach, 1991]); uniformization and monodromy (or single-valuedness) of elliptic curves ($y^2 = 4x^3 - g_2 x - g_3$) by Weierstrass elliptic functions (Whittaker and Watson, *Modern Analysis*, chap. 20), $\wp(z; g_2, g_3)$, which after many further developments eventually leads to the proof of Fermat's last theorem (A. Wiles, "Modular Elliptic Curves and Fermat's Last Theorem" [*Annals of Mathematics* 141 (1995): 443–551]). If one were acquainted with 1940s-era harmonic analysis, group theory, and elliptic functions, as sketched by Mackey, then Onsager's 1944 moves might seem quite natural and the subsequent fifty years of work would also appear natural.

2. There is a lovely parallel between the problem of proving the stability of the solar system (a problem in harmonic analysis) and the stability of matter: "This is, if truth be told, more of a mathematical question than a physical one. Even if one were to discover a general and rigorous proof, one could not conclude that the solar system is eternal. It may, in fact, be subject to forces other than those of Newton [such as seas and tides]." Henri Poincaré, quote in H. Poincaré, *New Methods of Celestial Mechanics*, ed. D. Goroff (New York: American Institute of Physics, 1993), p. 180.

Goroff (p. 172), relates a beautiful analogy concerning uniform convergence (and the problem of when to turn off the external magnetic field in a ferromagnet), due to Arnol'd. Given a bucket filled with water and having a small hole in the bottom, eventually all the water escapes. But if we treated this problem perturbatively, and first went to infinite time and then to zero hole size (for "far-sighted geometers"), or first went to zero hole size and then infinite time (for "finite-lived observers"), we would get different results. There is a "lack of uniformity of one limit with respect to the other."

3. In 1902 David Hilbert set forth his Sixth Problem, "the problem of developing mathematically the limiting processes, [in Boltzmann] merely indicated, which lead from the atomistic view to the laws of motion of continua. Conversely one might try to derive the laws of motion of rigid bodies by a limiting process from a system of axioms depending upon the idea of continuously varying conditions of a material filling all space continuously, these conditions being defined by parameters." D. Hilbert, "Mathematical Problems" (*Bulletin of the American Mathematical Society* 8 (1902): 437–479), at pp. 450–451. In this mathematical context, we might note that in modern number theory (Dirichlet L-functions), the combinatorial (and random walk) account of the Ising model, and the duality/star-triangle/k account, are more generally related. The partition function "encodes" combinatorial information much as does the Riemann zeta function; the partition function also encodes modular symmetry much as do theta functions. And the Riemann zeta function can be shown to be a (Mellin) transform of the theta function. (A. Knapp, Elliptic Curves [Princeton: Princeton University Press, 1992].)

REFERENCES

Abraham, D. B., and A. Martin-Löf. "The Transfer Matrix for a Pure Phase in the Two-Dimensional Ising Model." *Communications in Mathematical Physics* 32 (1973): 245–268.

Anderson, P. W. *Basic Notions of Condensed Matter Physics*. Menlo Park, CA: Benjamin/Cummings, 1984.

Andrews, G. *q-Series*. Providence: American Mathematical Society, 1986.

Andrews, G., R. J. Baxter, and P. J. Forrester. "Eight-Vertex SOS Model and Generalized Rogers-Ramanujan-Type Identities." *Journal of Statistical Physics* 35 (1984): 193–266.

Arthur, W. B. "Competing Technologies, Increasing Returns, and Lock-in by Historical Events." *Economic Journal* 99 (1989): 116–131.

Ashkin, J., and W. E. Lamb, Jr. "The Propagation of Order in Crystal Lattices." *Physical Review* 64 (1943): 159–178.

Au-Yang, H., and J. H. H. Perk. "The Chiral Potts Model Revisited." *Journal of Statistical Physics* 78 (1995): 17–78.

———, "Onsager's Star-Triangle Equation: Master Key to Integrability." In *Integrable Systems in Quantum Field Theory and Statistical Mechanics*, ed. M. Jimbo, T. Miwa, and A. Tsuchiya. Advanced Studies in Pure Mathematics 19. Tokyo: Kinokuniya, 1989, pp. 57–94.

Barouch, E. "Ising Model in Presence of Magnetic Field." *Physica* D1 (1980): 333–337.

Barrow, J. D., and F. Tipler. *The Anthropic Cosmological Principle*. Oxford: Clarendon, 1986.

Barwise, J., ed. *Handbook of Mathematical Logic*. Amsterdam: North Holland, 1977.

Baxter, R. J. *Exactly Solved Models in Statistical Mechanics*. London: Academic Press, 1982.

———, "Exactly Solved Models in Statistical Mechanics." In *Integrable Systems in Statistical Mechanics*, ed. G. D'Ariano, A. Montorsi, and M. Rasetti. Singapore: World Scientific, 1985, pp. 5–63.

———, "Solvable Eight-Vertex Model on an Arbitrary Planar Lattice." *Philosophical Transactions of the Royal Society* 289A (1978): 315–344.

———, "Solvable Models in Statistical Mechanics, from Onsager Onward." *Journal of Statistical Physics* 78 (1995): 7–16.

———, "Solving Models in Statistical Mechanics." In *Integrable Systems in Quantum Field Theory and Statistical Mechanics*, ed. M. Jimbo, T. Miwa, and A. Tsuchiya. Advanced Studies in Pure Mathematics 19. Tokyo: Kinokuniya, 1989, pp. 95–116.

313

Baxter, R. J., and I. Enting. "399th Solution of the Ising Model." *Journal of Physics* A11 (1978): 2463–2473.

Belavin, A. A., A. M. Polyakov, and A. B. Zamolodchikov. "Infinite Conformal Symmetry in Two-Dimensional Quantum Field Theory." *Nuclear Physics* B241 (1984): 333–380.

Benettin, B., G. Gallavotti, G. Jona-Lasinio, and A. L. Stella. "On the Onsager-Yang Value of the Spontaneous Magnetization." *Communications in Mathematical Physics* 30 (1973): 45–54.

Bethe, H. "Zur Theorie der Metalle: 1, Eigenwerte und Eigenfunktionen der linearen Atomkette." *Zeitschrift für Physik* 71 (1931): 205–226.

Brush, S. "History of the Lenz-Ising Model." *Reviews of Modern Physics* 39 (1967): 883–893.

———, *Statistical Physics and the Atomic Theory of Matter*. Princeton: Princeton University Press, 1983.

Brydges, D., J. Fröhlich, and T. Spencer. "The Random Walk Representation of Classical Spin Systems and Correlation Inequalities." *Communications in Mathematical Physics* 83 (1982): 123–150.

Buchwald, J. *From Maxwell to Microphysics*. Chicago: University of Chicago Press, 1985.

Caianiello, E. R. *Combinatorics and Renormalization in Quantum Field Theory*. Reading, MA: Benjamin, 1973.

Cao, T. Y., and S. S. Schweber. "The Conceptual Foundations and the Philosophical Aspects of Renormalization Theory." *Synthese* 97 (1993): 33–108.

Cartwright, N. *How the Laws of Physics Lie*. Oxford: Clarendon, 1983.

———, *Nature's Capacities and Their Measurement*. Oxford: Clarendon, 1989.

Ceperley, D. M. "Path Integrals in the Theory of Condensed Helium." *Reviews of Modern Physics* 67 (1995): 279–355.

Chandler, D. *Introduction to Modern Statistical Mechanics*. New York: Oxford University Press, 1987.

Crowe, M. J. *History of Vector Analysis*. Notre Dame, IN: University of Notre Dame Press, 1967.

Cushing, J. T. *Theory Construction and Selection in Modern Physics: The S Matrix*. Cambridge: Cambridge University Press, 1990.

D'Ariano, G., A. Montorsi, and M. Rasetti, eds. *Integrable Systems in Statistical Mechanics*. Singapore: World Scientific, 1985.

Darrigol, O. *From c-Numbers to q-Numbers*. Berkeley: University of California Press, 1992.

Darwin, C. G., and R. H. Fowler. "On the Partition of Energy." Parts 1 and 2. *Philosophical Magazine* 44 (1922): 450–479, 823–842.

Date, E., M. Jimbo, T. Miwa, and M. Okado. "Solvable Lattice Models." In *Theta Functions: Bowdoin 1987*, ed. L. Ehrenpreis and R. C. Gunning. Proceedings of Symposia in Pure Mathematics, vol. 49, part 1. Providence: American Mathematical Society, 1989, pp. 295–332.

Deguchi, T., M. Wadati, and Y. Akutsu. "Knot Theory Based on Solvable Lattice Models at Criticality." In *Integrable Systems in Quantum Field Theory and Statistical Mechanics*, ed. M. Jimbo, T. Miwa, and A. Tsuchiya. Advanced Studies in Pure Mathematics 19. Tokyo: Kinokuniya, 1989, pp. 193–285.

Dolan, L., and M. Grady. "Conserved Charges from Self-Duality." *Physical Review* D25 (1982): 1587–1604.

Domb, C. "Self-Avoiding Walks and Scaling." In *Critical Phenomena in Alloys, Magnets, and Superconductors*, ed. R. E. Mills, E. Ascher, and R. I. Jaffee. New York: McGraw-Hill, 1971, pp. 89–103.

Dresden, M. *H. A. Kramers: Between Tradition and Revolution*. New York: Springer, 1987.

———, "Kramers's Contributions to Statistical Mechanics." *Physics Today* 41 (September 1988): 26–33.

Dyson, F. J. "Existence of a Phase Transition in a One-Dimensional Ising Ferromagnet." *Communications in Mathematical Physics* 12 (1969): 91–107.

———, "Ground-State Energy of a Finite System of Charged Particles." *Journal of Mathematical Physics* 8 (1967): 1538–1545.

———, "Missed Opportunities." *Bulletin of the American Mathematical Society* 78 (1972): 635–652.

———, *The Origins of Life*. New York: Cambridge University Press, 1985.

———, "Time without End: Physics and Biology in an Open Universe." *Reviews of Modern Physics* 51 (1979): 447–460.

Dyson, F. J., and A. Lenard. "Stability of Matter." Part 1. *Journal of Mathematical Physics* 8 (1967): 423–434.

Ehrenfest, P. *Collected Scientific Papers*. Amsterdam: North Holland, 1959.

Ehrenpreis, L., and R. C. Gunning, eds. *Theta Functions: Bowdoin 1987.* Proceedings of Symposia in Pure Mathematics, vol. 49, part 1. Providence: American Mathematical Society, 1989.

Feller, W. *An Introduction to Probability Theory and Its Applications*. Vol. 1, 3d ed. Vol. 2, 2d ed. New York: Wiley, 1968, 1971.

Ferdinand, A. E., and M. E. Fisher. "Bounded and Inhomogeneous Ising Models: Part 1, Specific-Heat Anomaly of a Finite Lattice." *Physical Review* 185 (1969): 832–846.

Feynman, R. P. *Lectures on Gravitation*. Pasadena: California Institute of Technology, 1971.

———, *QED*. Princeton: Princeton University Press, 1985.

———, *Statistical Mechanics*. Reading, MA: Benjamin, 1972.

Feynman, R. P., and A. Hibbs. *Quantum Mechanics and Path Integrals*. New York: McGraw-Hill, 1965.

Feynman, R. P., R. B. Leighton, and M. Sands. *The Feynman Lectures on Physics*. Reading, MA: Addison-Wesley, 1963–1965.

Field, H. *Realism, Mathematics, and Modality*. Oxford: Blackwell, 1989.

———, *Science without Numbers*. Princeton: Princeton University Press, 1980.

Fisher, M. E. "The Free Energy of a Macroscopic System." *Archive for Rational Mechanics and Analysis* 17 (1964): 377–410.

———, "The Nature of Critical Points." In *Lectures in Theoretical Physics*, ed. W. E. Brittin. Boulder: University of Colorado Press, 1965, pp. 1–159.

——, "Statistical Mechanics of Dimers on a Plane Lattice." *Physical Review* 124 (1961): 1664–1672.

——, "Theory of Condensation and the Critical Point." *Physics* 3 (1967): 255–283.

——, "The Theory of Equilibrium Critical Phenomena." *Reports on Progress in Physics* 30 (1967): 615–730.

Fisher, M. E., and D. Ruelle. "The Stability of Many-Particle Systems," *Journal of Mathematical Physics* 7 (1966): 260–270.

Fowler, R. H. "A Simple Extension of Fourier's Integral Theorem and Some Physical Applications, in Particular to the Theory of Quanta." *Proceedings of the Royal Society of London*, ser. A99 (1921): 462–471.

——, *Statistical Mechanics*. 1936. Reprint. Cambridge: Cambridge University Press, 1955.

Frank, R. G., Jr. "Instruments, Nerve Action, and the All-or-None Principle." *Osiris* 9 (1994): 208–235.

Franklin, A. *The Neglect of Experiment*. New York: Cambridge University Press, 1986.

Galison, P. *How Experiments End*. Chicago: University of Chicago Press, 1987.

Gaudin, M. *La fonction d'onde de Bethe*. Paris: Masson, 1983.

Georgii, H.-O. *Gibbs Measures and Phase Transitions*. Berlin: De Gruyter, 1988.

Glimm, J., and A. Jaffe. *Quantum Physics: A Functional Integral Point of View*. New York: Springer, 1981.

Goldstein, H. *Classical Mechanics*. Reading, MA: Addison-Wesley, 1959.

Grady, M. "Infinite Set of Conserved Charges in the Ising Model." *Physical Review* D25 (1982): 1103–1113.

Green, H. S., and C. A. Hurst. *Order-Disorder Phenomena*. London: Interscience, 1964.

Greenhill, A. G. *The Applications of Elliptic Functions*. 1892. Reprint, New York: Dover, 1959.

Griffiths, R. B. "Microcanonical Ensemble in Quantum Statistical Mechanics." *Journal of Mathematical Physics* 6 (1965): 1447–1454.

——, "Peierls Proof of Spontaneous Magnetization in a Two-Dimensional Ising Ferromagnet." *Physical Review* 136 (1964): A437–439.

——, "A Proof That the Free Energy of a Spin System Is Extensive." *Journal of Mathematical Physics* 5 (1964): 1215–1222.

——, "Spontaneous Magnetization in Idealized Ferromagnets." *Physical Review* 152 (1966): 240–246.

Gruber, C., A. Hintermann, and D. Merlini. *Group Analysis of Classical Lattice Systems*. Lecture Notes in Physics 60. Berlin: Springer, 1977.

Hacking, I. *Representing and Intervening: Introductory Topics in the Philosophy of Natural Science*. New York: Cambridge University Press, 1983.

Hankins, T. *William Rowan Hamilton*. Baltimore: Johns Hopkins University Press, 1980.

Haymet, A. "Freezing." *Science* 236 (1987): 1076–1080.

Hersh, R. "Some Proposals for Reviving the Philosophy of Mathematics." *Advances in Mathematics* 31 (1979): 31–50.

Hessen, B. "The Social and Economic Roots of Newton's 'Principia.'" In *Science at the Cross Roads*, ed. N. Bukharin. 1931. Reprint, London: Cass, 1981.

Hildenbrand, W. "Cores." In *The New Palgrave*. Vol. 1. New York: Stockton, 1987, pp. 666–670.

Hilhorst, H. J., M. Schick, and J. M. J. van Leeuwen. "Differential Form of Real-Space Renormalization: Exact Results for Two-Dimensional Ising Models." *Physical Review Letters* 40 (1978): 1605–1608.

———, "Exact Renormalization Group Equations for the Two-Dimensional Ising Model." *Physical Review* B19 (1979): 2749–2763.

Hoddeson, L., E. Braun, J. Teichmann, and S. Weart. *Out of the Crystal Maze: Chapters from the History of Solid State Physics*. New York: Oxford University Press, 1992.

Houtappel, R. M. F. "Order-Disorder in Hexagonal Lattices." *Physica* 16 (1950): 425–455.

Huang, K. *Statistical Mechanics*. New York: Wiley, 1963.

Hurst, C. A. "Applicability of the Pfaffian Method to Combinatorial Problems on a Lattice." *Journal of Mathematical Physics* 5 (1964): 90–100.

———, "New Approach to the Ising Problem." *Journal of Mathematical Physics* 7 (1966): 305–310.

Hurst, C. A., and H. S. Green. "New Solution of the Ising Problem from a Rectangular Lattice." *Journal of Chemical Physics* 33 (1960): 1059–1062.

Iagolnitzer, D., and B. Souillard. "Lee-Yang Theory and Normal Fluctuations." *Physical Review* B19 (1979): 1515–1518.

Itzykson, C. "Ising Fermions." Part 1. *Nuclear Physics* B210 (1982): 448–476.

Itzykson, C., and J.-M. Drouffe. *Statistical Field Theory*. Vol. 1. *From Brownian Motion to Renormalization and Lattice Gauge Theory*. Vol. 2. *Strong Coupling, Monte Carlo Methods, Conformal Field Theory, and Random Systems* Cambridge: Cambridge University Press, 1989.

Jaffe, A., and F. Quinn. "Theoretical Mathematics: Toward a Cultural Synthesis of Mathematics and Theoretical Physics." *Bulletin of the American Mathematical Society* 29 (1993): 1–13.

Kac, M., and J. C. Ward. "A Combinatorial Solution of the Two-Dimensional Ising Model." *Physical Review* 88 (1952): 1332–1337.

Kac, V. G., and D. G. Peterson. "Infinite-Dimensional Lie Algebras, Theta Functions, and Modular Forms." *Advances in Mathematics* 53 (1984): 125–264.

Kadanoff, L. P., W. Götze, D. Hamblin, R. Hecht, E. A. S. Lewis, V. V. Palaciauskas, M. Rayl, and J. Swift. "Static Phenomena near Critical Points: Theory and Experiment." *Review of Modern Physics* 39 (1967): 395–431.

Kadanoff, L. P., and A. Houghton. "Numerical Evaluations of the Critical Properties of the Two-Dimensional Ising Model." *Physical Review* B11 (1975): 377–386.

Kastelyn, P. W. "Dimer Statistics and Phase Transitions." *Journal of Mathematical Physics* 4 (1963): 287–293.

Kaufman, B. "Crystal Statistics: Part 2, Partition Function Evaluated by Spinor Analysis." *Physical Review* 76 (1949): 1232–1243.

Kaufman, B., and L. Onsager. "Crystal Statistics: Part 3, Short-Range Order in a Binary Ising Lattice." *Physical Review* 76 (1949): 1244–1252.

Kirkwood, J. G. "Order and Disorder in Binary Solid Solutions." *Journal of Chemical Physics* 6 (1938): 70–75.

Knops, J. J. F., and H. J. Hilhorst, "Exact Differential Renormalization of the Ising Model at Criticality." *Physical Review* B19 (1979): 3689–3699.

Kragh, H. *Dirac*. Cambridge: Cambridge University Press, 1990.

Kramers, H. A. *Quantum Mechanics*. Amsterdam: North Holland, 1957.

Kramers, H. A., and G. H. Wannier. "Statistics of the Two-Dimensional Ferromagnet." Parts 1 and 2. *Physical Review* 60 (1941): 252–276.

Kreisel, G. "Mathematical Significance of Consistency Proofs." *Journal of Symbolic Logic* 23 (1958): 155–182.

———, "A Survey of Proof Theory." *Journal of Symbolic Logic* 33 (1968): 321–388.

Krieger, M. H. "Could the Probability of Doom Be Zero or One?" *Journal of Philosophy* 92 (1995): 382–387.

———, *Doing Physics: How Physicists Take Hold of the World*. Bloomington: Indiana University Press, 1992.

———, *Marginalism and Discontinuity: Tools for the Crafts of Knowledge and Decision*. New York: Russell Sage Foundation, 1989.

———, "Theorems as Meaningful Cultural Artifacts: Making the World Additive." *Synthese* 88 (1991): 135–154.

Kuhn, T. S. *Black-Body Theory and the Quantum Discontinuity. 1894–1912*. New York: Oxford University Press, 1978.

———, *The Structure of Scientific Revolutions*. 2d ed. Chicago: University of Chicago Press, 1970.

Landau, L., and E. Lifshitz. *Statistical Physics*. Part 1. Oxford: Pergamon, 1980.

Landsberg, P. T. "Is Equilibrium Always an Entropy Maximum?" *Journal of Statistical Physics* 35 (1984): 159–169.

Lassettre, E. N., and J. P. Howe, "Thermodynamic Properties of Binary Solid Solutions on the Basis of the Nearest Neighbor Approximation." *Journal of Chemical Physics* 9 (1941): 747–754.

Lawden, D. F. *Elliptic Functions and Applications*. New York: Springer, 1989.

Lebowitz, J. L. "Statistical Mechanics: A Review of Selected Rigorous Results." *Annual Reviews of Physical Chemistry* 19 (1968): 389–418.

Lebowitz, J. L., and E. H. Lieb. "Existence of Thermodynamics for Real Matter with Coulomb Forces." *Physical Review Letters* 22 (1969): 631–634.

Lebowitz, J. L., and A. Martin-Löf. "On the Uniqueness of the Equilibrium State for Ising Spin Systems." *Communications in Mathematical Physics* 25 (1972): 276–282.

Lee, T. D. *Selected Papers*. Ed. G. Feinberg. Boston: Birkhauser, 1986.

Lee, T. D., and C. N. Yang, "Statistical Theory of Equations of State and Phase Transitions: Part 2, Lattice Gas and Ising Model." *Physical Review* 87 (1952): 410–419.

Lenard, A., and F. J. Dyson. "Stability of Matter." Part 2. *Journal of Mathematical Physics* 9 (1968): 698–711.

Lévy-Leblond, J.-M. "Nonsaturation of Gravitational Forces." *Journal of Mathematical Physics* 10 (1969): 806–812.

Lieb, E. H. "The Stability of Matter." *Reviews of Modern Physics* 48 (1976): 553–569.

——, "The Stability of Matter: From Atoms to Stars." *Bulletin of the American Mathematical Society* 22 (1990): 1–49.

——, *The Stability of Matter: From Atoms to Stars: Selecta of Elliott H. Lieb.* Ed. W. Thirring. Berlin: Springer, 1991.

Lieb, E. H., and J. L. Lebowitz. "The Constitution of Matter: Existence of Thermodynamics for Systems Composed of Electrons and Nuclei." *Advances in Mathematics* 9 (1972): 316–398.

——, "Lectures on the Thermodynamic Limit for Coulomb Systems." In *Statistical Mechanics and Mathematical Problems.* Lecture Notes in Physics 20. Berlin: Springer, 1973, pp. 136–161.

Lieb, E. H., and D. C. Mattis. *Mathematical Physics in One Dimension.* New York: Academic Press, 1966.

——, "Theory of Ferromagnetism and the Ordering of Electronic Energy Levels." *Physical Review* 125 (1962): 164–172.

Lieb, E. H., T. D. Schultz, and D. C. Mattis. "Two Soluble Models of an Antiferromagnetic Chain." *Annals of Physics* 16 (1961): 407–466.

Lieb, E. H., and W. Thirring. "Bound for Kinetic Energy of Fermions Which Proves the Stability of Matter." *Physical Review Letters* 35 (1975): 687–689.

Liston, M. "Taking Mathematical Fictions Seriously." *Synthese* 95 (1993): 433–458.

Livingston, E. *The Ethnomethodological Foundations of Mathematics.* London: Routledge and Kegan Paul, 1986.

Longuet-Higgins, H. C., and M. E. Fisher. "Lars Onsager." *Biographical Memoirs of the Royal Society* (1978): 443–471.

McCoy, B. M., and T. T. Wu. *The Two-Dimensional Ising Model.* Cambridge: Harvard University Press, 1973.

McGuire, J. B. "Study of Exactly Soluble One-Dimensional N-Body Problems." *Journal of Mathematical Physics* 5 (1964): 622–636.

McKean, H. P., Jr. "Kramers-Wannier Duality for the Two-Dimensional Ising Model as an Instance of Poisson's Summation Formula." *Journal of Mathematical Physics* 5 (1964): 775–776.

Maddy, P. *Realism in Mathematics.* Oxford: Clarendon, 1992.

Maillard, J.-M. "Introduction." *International Journal of Modern Physics* B7, nos. 20, 21 (1993): v–vi.

Mattis, D. C. *The Theory of Magnetism.* 2 vols. Berlin: Springer, 1985–1988.

Mattis, D. C., and E. H. Lieb. "Exact Solution of a Many-Fermion System and Its Associated Boson Field." *Journal of Mathematical Physics* 6 (1965): 304–312.

Maxwell, J. C. "Address to the Mathematical and Physical Sections of the British Association," 1890. In *Collected Scientific Papers*. New York: Dover, 1960, pp. 215–229.

———, "On the Mathematical Classification on Physical Quantities." In *Collected Scientific Papers*. New York: Dover, 1960, pp. 257–266.

Merlini, D., and C. Gruber. "Spin-1/2 Lattice System: Group Structure and Duality Relation." *Journal of Mathematical Physics* 13 (1972): 1814–1823.

Montroll, E. W. "Statistical Mechanics of Nearest Neighbor Systems." *Journal of Chemical Physics* 9 (1941): 706–721.

Montroll, E. W., R. B. Potts, and J. C. Ward. "Correlations and Spontaneous Magnetization of the Two-Dimensional Ising Model." *Journal of Mathematical Physics* 4 (1963): 308–322.

Nambu, Y. "A Note on the Eigenvalue Problem in Crystal Statistics." *Progress of Theoretical Physics* 5 (1950): 1–13.

Newell, G. F., and E. W. Montroll. "On the Theory of the Ising Model of Ferromagnetism." *Reviews of Modern Physics* 25 (1953): 353–389.

Onsager, L. "Crystal Statistics: Part 1, A Two-Dimensional Model with an Order-Disorder Transition." *Physical Review* 65 (1944): 117–149.

———, "Electrostatic Interaction of Molecules." *Journal of Physical Chemistry* 43 (1939): 189–196.

———, "The Ising Model in Two Dimensions." In *Critical Phenomena in Alloys, Magnets, and Superconductors*, ed. R. E. Mills, E. Ascher, and R. I. Jaffee. New York: McGraw-Hill, 1971, pp. 3–12.

Pais, A. "On a Plane Ising Lattice with First and Second Interactions." *Proceedings of the National Academy of Science* 49 (1963): 34–38.

Parisi, G. *Field Theory, Disorder, and Simulations*. Singapore: World Scientific, 1992.

———, *Statistical Field Theory*. Menlo Park, CA: Benjamin/Cummings, 1988.

Peierls, R. *More Surprises in Theoretical Physics*. Princeton: Princeton University Press, 1991.

———, "On Ising's Model of Ferromagnetism." *Proceedings of the Cambridge Philosophical Society* 32 (1936): 477–481.

Pfeuty, P. "Ising Chain in Transverse Field." *Annals of Physics* 57 (1970): 79–90.

Pickering, A. *Constructing Quarks*. Chicago: University of Chicago Press, 1984.

———, *The Mangle of Practice: Time, Agency, and Science*. Chicago: University of Chicago Press, 1995.

Plechko, V. N. "Simple Solution of Two-Dimensional Ising Model on a Torus in Terms of Grassmann Integrals." *Theoretical and Mathematical Physics* 64 (1985): 748–756.

Pokrovsky, S. V., and Y. A. Bashilov. "Star-Triangle Relations in the Exactly Solvable Statistical Models." *Communications in Mathematical Physics* 84 (1982): 103–132.

Reed, M., and B. Simon. *Methods of Modern Mathematical Physics*. Part 1. Functional Analysis. New York: Academic Press, 1980.

Roberts, J. "Large Economies." In *The New Palgrave*. Vol. 3. New York: Stockton, 1987, pp. 13–14.

Rota, G.-C. *Gian-Carlo Rota on Combinatorics*. Ed. J. P. S. Kung. Boston: Birkhauser, 1995.

———, "Kant's Synthesis A Priori and Husserl's Phenomenology of Fulfillment." In *Indiscrete Thoughts*, ed. A. Kostant. Boston: Birkhauser, 1996.

———, "Mathematical Beauty." In *Indiscrete Thoughts*, ed. A. Kostant. Boston: Birkhauser, 1996.

———, "The Pernicious Influence of Mathematics on Philosophy." *Synthese* 88 (1991): 165–178.

Rotman, B. *Ad Infinitum*. Stanford: Stanford University Press, 1993.

Ruelle, D. "States of Classical Statistical Mechanics." *Journal of Mathematical Physics* 8 (1967): 1657–1668.

———, *Statistical Mechanics: Rigorous Results*. New York: Benjamin, 1969.

———, *Thermodynamic Formalism*. Reading, MA: Addison-Wesley, 1978.

Saarloos, W., and D. A. Kurtze. "Location of Zeros in the Complex Temperature Plane: Absence of Lee-Yang Theorem." *Journal of Physics* A17 (1984): 1301–1311.

Samuel, S. "Grand Partition Function in Field Theory with Applications to Sine-Gordon Theory." *Physical Review*, D18 (1978): 1916–1932.

———, "The Use of Anticommuting Variable Integrals in Statistical Mechanics." Parts 1–3. *Journal of Mathematical Physics* 21 (1980): 2806–2833.

Savit, R. "Duality in Field Theory and Statistical Physics." *Reviews of Modern Physics* 52 (1980): 453–487.

Schrödinger, E. *Statistical Thermodynamics*. Cambridge: Cambridge University Press, 1944.

Schultz, T. D., D. C. Mattis, and E. H. Lieb. "Two-Dimensional Ising Model as a Soluble Problem of Many Fermions." *Reviews of Modern Physics* 36 (1964): 856–871.

Schweber, S. S. "Physics, Community, and the Crisis in Physical Theory." *Physics Today* 46 (November 1993): 34–40.

———, *QED and the Men Who Made It*. Princeton: Princeton University Press, 1994.

Schwinger, J. "On the Euclidean Structure of Relativistic Field Theory." *Proceedings of the National Academy of Science* 44 (1958): 956–965.

Schwinger, J., and D. S. Saxon. *Discontinuities in Waveguides*. New York: Gordon and Breach, 1968.

Shankar, R. "Renormalization Group Approach to Interacting Fermions." *Reviews of Modern Physics* 66 (1994): 129–192.

———, "Simple Derivation of the Baxter-Model Free Energy." *Physical Review Letters* 47 (1981): 1177–1180.

Siegel, D. M. *Innovation in Maxwell's Electromagnetic Theory*. New York: Cambridge University Press, 1991.

Simon, B. *The Statistical Mechanics of Lattice Gases*. Vol. 1. Princeton: Princeton University Press, 1993.

Spruch, L. "Pedagogic Notes on Thomas-Fermi Theory." *Reviews of Modern Physics* 63 (1991): 151–209.

Srednicki, M. "Hidden Fermions in $Z(2)$ Theories." *Physical Review* D21 (1980): 2878–2884.

Steiner, M. "The Applicability of Mathematics to Natural Science." *Journal of Philosophy* 86 (1989): 449–480.

Stephen, M. J., and L. Mittag. "New Representation of the Solution to the Ising Model." *Journal of Mathematical Physics* 13 (1972): 1944–1951.

Stephenson, J., and R. Couzens. "Partition Function Zeros for the Two-Dimensional Ising Model." *Physica* A129 (1984): 201–210.

Sternberg, S. *Group Theory and Physics*. New York: Cambridge University Press, 1994.

Symanzik, K. "Euclidean Quantum Field Theory: Part 1, Equations for a Scalar Model." *Journal of Mathematical Physics* 7 (1966): 510–525.

Temperley, H. N. V. "Two-Dimensional Ising Models." In *Phase Transitions and Critical Phenomena*, vol. 1, *Exact Results*, ed. C. Domb and M. S. Green. London: Academic Press, 1972, pp. 227–267.

Temperley, H. N. V., and M. E. Fisher. "Dimer Problem in Statistical Mechanics: An Exact Result." *Philosophical Magazine* 6 (1961): 1061–1063.

Thacker, H. B. "Exact Integrability in Quantum Field Theory and Statistical Systems." *Reviews of Modern Physics* 53 (1981): 253–285.

Thompson, C. J. *Mathematical Statistical Mechanics*. Princeton: Princeton University Press, 1972.

Tragesser, R. *Husserl and Realism in Logic and Mathematics*. Cambridge: Cambridge University Press, 1984.

———, *Phenomenology and Logic*. Ithaca, NY: Cornell University Press, 1977.

Uhlenbeck, G. E., and G. W. Ford. *Lectures in Statistical Mechanics*. Providence: American Mathematical Society, 1963.

Van der Waerden, B. "Die lange Reichweite der regelmassigen Atomanordnung in Mischkristallen." *Zeitschrift fur Physik* 118 (1941/1942): 473–488.

Van Hove, L. "Quelques propriétés générales de l'intégrale de configuration d'un système de particules avec interaction." *Physica* 15 (1949): 951–961.

———, "Sur l'intégrale de configuration pour les systèmes de particules à une dimension." *Physica* 16 (1950): 137–143.

Wadati, M., T. Deguchi, and Y. Akutsu. "Exactly Solvable Models and Knot Theory." *Physics Reports* 180 (1989): 248–332.

Wannier, G. H. "Antiferromagnetism: The Triangular Ising Net." *Physical Review* 79 (1950): 357–364.

———, *Statistical Physics*. New York: Wiley, 1966.

———, "The Statistical Problem in Cooperative Phenomena." *Reviews of Modern Physics* 17 (1945): 50–60.

Wegner, F. J. "Duality in Generalized Ising Models and Phase Transitions without Local Order Parameter." *Journal of Mathematical Physics* 12 (1971): 2259–2272.

Weinberg, S. *Dreams of a Final Theory*. New York: Pantheon, 1992.

———, "Why the Renormalization Group Is a Good Thing." In *Asymptotic Realms of Physics*, ed. A. H. Guth, K. Huang, and R. L. Jaffe. Cambridge: MIT Press, 1983, pp. 1–19.

Weng, C-Y., R. B. Griffiths, and M. E. Fisher. "Critical Temperature of Anisotropic Ising Lattices: Part 1, Lower Bounds." *Physical Review* 162 (1967): 475–479.

Wheeler, J. A. *A Journey into Gravity and Spacetime*. New York: Scientific American Library, 1990.

Wheeler, J. A., and R. P. Feynman. "Interaction with the Absorber as the Mechanism of Radiation." *Reviews of Modern Physics* 17 (1945): 157–181.

Whittaker, E. T. *A Treatise on the Analytical Dynamics of Particles and Rigid Bodies*. 4th ed. 1937. Reprint. Cambridge: Cambridge University Press, 1988.

Whittaker, E. T., and G. N. Watson. *A Course of Modern Analysis*. 4th ed. 1927. Reprint. Cambridge: Cambridge University Press, 1990.

Widom, B. "Equation of State in the Neighborhood of the Critical Point." *Journal of Chemical Physics* 41 (1965): 3898–3905.

Wightman, A. S. Introduction to R. B. Israel, *Convexity in the Theory of Lattice Gases*. Princeton: Princeton University Press, 1979, pp. ix–lxxxv.

Wilson, K. G. "Problems in Physics with Many Scales of Length." *Scientific American* (August 1979): 158–179.

———, "The Renormalization Group and Critical Phenomena." *Reviews of Modern Physics* 55 (1983): 583–600.

———, "The Renormalization Group: Critical Phenomena and the Kondo Problem." *Reviews of Modern Physics* 47 (1975): 773–840.

Wilson, K. G., and J. Kogut. "The Renormalization Group and the ϵ Expansion." *Physics Reports* 12C (1974): 76–200.

Winner, L. *Autonomous Technology*. Cambridge: MIT Press, 1977.

Wolfram, S. *Theory and Application of Cellular Automata*. Singapore: World Scientific, 1986.

Yang, C. N. "One-Dimensional Delta Function Interaction." In *Critical Phenomena in Alloys, Magnets, and Superconductors*, ed. R. E. Mills, E. Ascher, and R. I. Jaffee. New York: McGraw-Hill, 1971, pp. 13–21.

———, *Selected Papers, 1945–1990, with Commentary*. San Francisco: W. H. Freeman, 1983.

———, "S-Matrix for the One-Dimensional N-Body Problem with Repulsive or Delta-Function Interaction." *Physical Review* 168 (1968): 1920–1923.

———, "The Spontaneous Magnetization of a Two-Dimensional Ising Model." *Physical Review* 85 (1952): 808–816.

Yang, C. N., and M. L. Ge. *Braid Group, Knot Theory, and Statistical Mechanics*. Singapore: World Scientific, 1989.

Yang, C. N., and T. D. Lee. "Statistical Theory of Equations of State and Phase Transitions: Part 1, Theory of Condensation." *Physical Review* 87 (1952): 404–409.

Yang, C. N., and C. P. Yang. "One-Dimensional Chain of Anisotropic Spin-Spin Interactions." *Physical Review* 150 (1966): 321–327.

Zeitlin, F. Introduction to J.-P. Vernant, *Mortals and Immortals*, ed. F. Zeitlin. Princeton: Princeton University Press, 1992, pp. 3–24.

Zinn-Justin, J. *Quantum Field Theory and Critical Phenomena*. Oxford: Clarendon, 1993.

INDEX

Definitions are indicated by page numbers in italics. Since symbol usage reflects the conventions of the various authors, I have indicated the different meanings as they appear.